DISCARD

ACS SYMPOSIUM SERIES **969**

Modern NMR Spectroscopy in Education

David Rovnyak, Editor
Bucknell University

Robert Stockland Jr., Editor
Bucknell University

**Sponsored by the
Division of Chemical Education, Inc.**

American Chemical Society, Washington, DC

ISBN: 978–0–8412–3995–1

Foreword

The ACS Symposium Series was first published in 1974 to provide a mechanism for publishing symposia quickly in book form. The purpose of the series is to publish timely, comprehensive books developed from ACS sponsored symposia based on current scientific research. Occasionally, books are developed from symposia sponsored by other organizations when the topic is of keen interest to the chemistry audience.

Before agreeing to publish a book, the proposed table of contents is reviewed for appropriate and comprehensive coverage and for interest to the audience. Some papers may be excluded to better focus the book; others may be added to provide comprehensiveness. When appropriate, overview or introductory chapters are added. Drafts of chapters are peer-reviewed prior to final acceptance or rejection, and manuscripts are prepared in camera-ready format.

As a rule, only original research papers and original review papers are included in the volumes. Verbatim reproductions of previously published papers are not accepted.

ACS Books Department

Contents

Teaching NMR with Technology

Modern NMR in Laboratory Development: Physical Chemistry

Modern NMR in Laboratory Development: Advanced Organic Chemistry

Modern NMR in Laboratory Development: Biochemistry and Biophysics

Modern NMR in Laboratory Development: Inorganic Chemistry

Permanent Magnet Fourier Transform–NMR

Indexes

Preface

Today NMR is a transformative, research-enabling tool in the chemical sciences, due in part to extraordinary advances over the past 10 to 20 years. This book hopes to answer a call to action to train the next generation of scientists to be able to realize the full potential of modern NMR.

In his introduction which follows, Professor Wagner describes first-hand the extraordinary potential for NMR-enabled research to answer significant and pressing scientific questions in the public interest. He further illustrates the present state of exciting, ongoing advancements of NMR technology. We urge you to consider the impact of Professor Wagner's chapter in illustrating the need for educators to more broadly reflect modern NMR applications and methods in their curricula. The desire to address this need is one of the common threads linking together the various chapters of this book. A second major theme emerges among these chapters as well. It is the use of innovative NMR laboratory experiments to achieve improved student learning in all areas of chemistry. In other words, our authors were motivated by far more than just performing NMR for its own sake, but by the need to achieve new milestones in student learning in chemistry.

Success Leads to Challenges and Opportunities

Educators incorporating further NMR training into curricula face challenges which are linked to the successes and technological advances in the ongoing development of NMR spectroscopy.

First, cost has limited the availability of research grade instrumentation to educators. No principally undergraduate institution (PUI) operates a cryogenic probe at this time, and many offer FT-NMR at proton frequencies

< 90 MHz. Furthermore, NMR instruments with four independent RF channels and specialized probes, crucially needed for biomolecular NMR studies, are rarely found at PUIs. In this book biomolecular NMR is made significantly more accessible, for example, by Harris and Driscoll who describe a web resource that allows anyone to perform a full NMR analysis of ubiquitin. Fortunately one of the most important advances has been widely adopted: magnetic field gradients. Gradients are increasingly available in NMR instruments at PUIs. Their utility in achieving automated, computer-optimized shimming, in studying molecular dynamics, and in enabling powerful new experiments is difficult to understate. Maki and Loening introduce gradient diffusion methods as well as several applications. Odeh and Li give a further example of diffusion NMR in coursework.

Second, the multidisciplinary nature of modern applications of NMR spectroscopy may appear to be a barrier to fitting NMR into core curricula. Faculty across the nation are presently discussing how to integrate NMR across the chemistry curriculum. This discussion leads to many significant issues, some of which are beyond the scope of this book: major curricular changes may ultimately be needed to adapt to the broader trend of complex technological and methodological advances permeating chemistry. However, a central role of this book is to examine and contextualize NMR in the curriculum. Fisher and Fish share their case study of NMR across the curriculum. Mueller and coworkers discuss combined theoretical and experimental priorities for integrating NMR throughout a curriculum. Steinmetz and O'Leary show how they carry NMR into advanced and seminar courses, while Goldman and Dominey demonstrate that NMR can be extremely successful in exciting and teaching students in non-majors courses. The chapter by Hanson compellingly shows how the use of automated sample-handling can be invaluable in achieving pedagogical goals.

Of special significance are the discussions and case studies given by Esselman and Mencer on incorporating NMR into high school chemistry instruction. It is not yet widely known that NMR can be found in many states' educational standards. We believe this chapter should stimulate a wider and needed effort to enhance the role of NMR in high school science education.

Finally, modern NMR methods may appear too complex for achieving meaningful, hands-on student experimentation. The advancement of NMR spectroscopy has, quite candidly, been based upon extensive use of non-trivial theoretical tools. Many chapters indicate that pedagogically significant advances in NMR theory can be incorporated into curricula.

Mueller and coworkers promote the need to train students in the use of the product operator method. We also applaud Zee and Howard who introduce tensors into a laboratory experiment. Moreover, Gaede shows that exchange calculations are accessible in an undergraduate setting. And Dominey, Abrams, Kanters and Goldman introduce powerful computer visualization tools to advance students' theoretical comprehension.

Experimentation for Advanced Laboratories

Whereas NMR is often widely and effectively featured in first- and second-year coursework, there has been a paucity of options for performing NMR experimentation in advanced coursework. In the symposia and discussions that led to this book, many faculty stressed that a significant issue in incorporating NMR across a curriculum is the need for more laboratory development in the latter years of an undergraduate degree. This book aims to describe a rich set of new experiments, tested and evaluated, that address this need.

In physical chemistry, Grushow and Sheats describe a suite of experiments that drive student learning in numerous areas. DeVore and coworkers give an innovative gas phase experiment with elegant thermodynamics. Maki and Loening describe the accessibility of diffusion measurements for the study of molecular transport and give a sound introduction to the topic.

Chapters by Harris, Trumbo, Zee and Howard, and Rovnyak and coworkers address the need for more options for NMR in biochemistry and biophysics. Significantly, these chapters have a theme of increasing student learning at the interface of biochemistry and structural biology.

Options for advanced organic instruction are given in chapters by Gaede, Odeh and Li, and O'Leary and Steinmetz. Where O'Leary and Steinmetz push the boundary for organic structure solving with students, Gaede and Odeh and Li join other authors in this book in developing experiments that allow students to witness and measure molecular dynamics.

Metal promoted transformations comprise a wide variety of key processes in chemical and biological systems. The use of NMR spectroscopy to probe the coordination environment and stereochemistry of the metal complexes facilitates a greater understanding of the critical reaction parameters. Despite the importance of this technique, inorganic NMR spectroscopy often receives only a cursory glance in educational settings. To address this deficiency, the chapters by Nataro, Minelli,

Hawrelak, and Uffelman outline the synthesis and characterization of metal containing complexes that offer a range of unique spectral properties.

The material presented in these chapters can be incorporated into both lecture and laboratory courses in order to broaden the coverage and depth of the NMR discussion.

Permanent Magnet FT-NMR

Permanent magnet NMR spectrometers equipped with a pulsed, Fourier-Transform upgrade (Anasazi Instruments, Inc.) are widely used in chemistry education and are an area of a good deal of significant and recent pedagogical research, carefully reviewed by Abe and Contratto, who also show a new imaging experiment. Chapters by Niece and Moyna, Collins and Amel, and Goldman and Dominey all show the versatility and power of this instrumentation in achieving pedagogical goals.

David S. Rovnyak
Robert A. Stockland, Jr.
Department of Chemistry
Bucknell University
Lewisburg, PA 17837

About the Book

The 2005 National Meeting of the American Chemical Society in Washington, D.C. featured two symposia on nuclear magnetic resonance (NMR) in the Division of Chemical Education. One was "Evolution of Modern Nuclear Magnetic Resonance in Undergraduate Learning," organized by both of us. This session came about from Dr. Matt Fisher's suggestion and we are indebted to him for creating that opportunity. We were delighted when the interest in this session grew to a day-long symposium of uniformly high-quality talks. The other was "Innovative Experiments and Approaches for Permanent Magnet NMR Spectrometers in High School and Undergraduate Courses," organized by Dr. Steven P. Lee, which showcased the growth and successes of permanent magnet FT-NMR in education. Both sessions came about by independent means, and we began to feel that it was more than a coincidence. The combined 18 talks reflected a strong national interest in promoting innovative NMR instruction in chemistry curricula. Our perceptions were further reinforced as we learned of a number of potential presenters eager to share their work, but unable to attend the symposium due to prior commitments.

There are many current efforts that could not be included here. We had intended at one point to review all prior work in NMR education in an appendix, and came to realize that this was beyond our time and means. There are many NMR-oriented papers already in the educational literature, but one of our concerns is that there have been only a limited number of NMR experiments published for more advanced coursework. The chapters here provide a wealth of ideas for incorporating NMR into the third and fourth years of a chemistry degree.

We wish to express our deepest thanks to the contributing authors for their hard work, which can be seen in the quality and care shown in these chapters. We are grateful there is such a wonderful network of colleagues as passionate about NMR in education as we are. We also thank the countless peer reviewers for their careful efforts, sometimes working nearly under duress to meet our deadlines.

We thank particularly Steve Lee for his contribution to this book. Steve worked extensively to bring on board papers from his symposium and elsewhere, and his efforts considerably strengthened the breadth and content of the chapters included here.

The patience, hard work, advocacy and encouragement of Dara Moore from ACS Books is very gratefully acknowledged. Broadly, it has been a uniformly pleasant experience to develop this book with the ACS Books division.

David S. Rovnyak   *Robert A. Stockland, Jr.*
Department of Chemistry
Bucknell University, Lewisburg, PA 17837

****

While our symposium was on August 30, 2005, just a few days prior, on August 23, my son Henry was born. When I have become overwhelmed by the magnitude of this project I have often rediscovered my energy and focus through him. I am very grateful for the crucial support of my wife Jennifer; I do not know how this would have been done without her encouragement, enthusiasm, and perspective.

David S. Rovnyak
December, 2006

During the construction of this book, my wife Barb was an invaluable resource. Her unique viewpoint was critical to the completion of the text, and she has my deepest gratitude.

Robert A. Stockland, Jr.
December, 2006

Modern NMR Spectroscopy
in Education

Chapter 1

Modern NMR in Undergraduate Education: Introduction

Gerhard Wagner

Elkan R. Blout Professor, Department of Biological Chemistry
and Molecular Pharmacology, Harvard Medical School,
240 Longwood Avenue, Boston, MA 02115

The rapid increase of knowledge has made most scientists specialists of a small area of expertise, and a comprehensive knowledge of all sciences is impossible today. In contrast, some of the greatest thinkers of past centuries, such as Newton, Leibnitz or Goethe had a near universal knowledge and engaged in a wide spectrum of intellectual activities, often ranging from mathematics, physics, chemistry, astronomy, biology or geology to philosophy and politics. Indeed, Gottfried Wilhelm Leibnitz (1646-1716), the inventor of differentials, was called a "Walking Encyclopedia" by his employer George Luis of Hanover who later became King George I of England. In an earlier unsuccessful political mission as an envoy of the Elector of Mainz he tried to convince the king Louis XIV of France to give up attacking the Alsace and pursue the conquest of Egypt instead. Scientists are hardly politically active any more to the better or worse, and fortunately we have now easy access to encyclopedic knowledge through the internet. However, most scientists miss the joy to be active in a wide spectrum of sciences. Rare exceptions are those of us who are engaged in nuclear magnetic resonance spectroscopy, which draws its power from a wide range of fields in science and mathematics.

It is one of the beauties of NMR that it provides an intellectual playing field entertaining physicists, chemists, mathematicians, electrical and mechanical engineers, biochemists, biologists and physicians. A single individual may be active in one particular of these areas or cover a wide spectrum. Using NMR skillfully requires basic physical understanding. Those who like it can dig into quantum mechanics, density matrix theory, spin dynamics, average Hamiltonian theory, or advanced data processing. On the other hand, a biologist may prefer to solve a beautiful protein structure without needing to understand the physical basis of the method in much detail, and the results obtained can be of high significance for elucidating biological processes or designing drugs for fighting human disease.

I did not know much about the potential of NMR and the dramatic development of this technology was hardly foreseeable when I became interested in this method. I was a physics undergraduate student at the Technical University of Munich when I became attracted by resonance phenomena. However, when I considered pursuing NMR spectroscopy for a Ph.D. project I was asked: "Why do you want to do NMR? Everything has been done in NMR already."

By then indeed, NMR was established as an analytical tool to characterize and confirm the chemical structures of small molecules. Much was known about the features that define an NMR spectrum, such as chemical shifts and scalar couplings [1], spin decoupling, the nuclear Overhauser effect, and relaxation phenomena in general [2]. Also most basic features of solid-state NMR spectra were known [3]. Thus, my supervisors discouraged me to enter this field. However, I had seen some ^{1}H NMR spectra of the protein hemoglobin and was startled by the complexity and plethora of signals found in the spectra. I thought that there must be a lot of wonderful information hidden in these signals. I imagined it could be exciting to go treasure hunting, and I decided to pursue a career in NMR spectroscopy.

The development of NMR in the 1970s and 1980s was to a large extent driven by spectroscopy with proteins and to some degree with nucleic acids. Protein NMR spectra were complex and challenging, and demanded new technologies. It was obvious that there was a lot of information in protein NMR spectra, in particular when magnets of higher fields became available, resulting in a better dispersion of the resonances. However, it was entirely unclear what the numerous spectral features would eventually reveal. There was hope that structural information could be obtained from analyzing chemical shifts since the basic principles responsible for the variation of resonance positions were known, such as ring-current shifts, effects of electronegativity and hydrogen bonding. However, this turned out to be inadequate for deriving solid structural information. Scalar coupling constants were recognized early on as a valuable source of structural information and much effort went into deriving quantitative relations with dihedral angles.[4] Indeed such information was found sufficient for determining structures of small cyclic peptides [5] but was inadequate for obtaining structural information for proteins.

The single most important obstacle preventing access to the treasures of protein NMR spectra was the lack of technologies to assign the NMR resonances to individual nuclei of a protein. The most powerful assignment technology available early on was spin decoupling, which could identify pairs or spins connected by scalar coupling. However, locating these spin pairs in a protein was impossible and additional methods had to be explored. Initially, assignments were only attempted for proteins with known crystal structures. Chemical modifications were introduced to create binding sites for paramagnetic lanthanides, and the effects on protein resonances were used to obtain early assignments of a few resonances [6].

The breakthrough with protein assignments came with the proper use of the nuclear Overhauser effect (NOE) in combination with spin decoupling. It had been known for a long time that saturation of individual spins caused changes of resonance intensities of nearby spins. Originally, in these experiments a particular resonance was irradiated for several seconds, and many resonance were found to experience intensity changes due to direct NOEs and multi-step spin diffusion. Thus, scientists were afraid that NOEs would only be marginally useful for measuring intra-protein distances accurately enough to have an impact for assignments and structure elucidation [7, 8]. However, this fear was proven unsubstantiated when it was shown that NOE experiments with short irradiation times could be quite selective and yield quantitative distance information [9]. Using this approach combined with spin decoupling allowed the first sequence-specific assignment of a large part of a small protein in simple 1D NMR experiments [10]. However the real break-through came only with the development of 2D and multi-dimensional NMR, which was developed by physical chemists [11, 12].

An important aspect of the development of NMR was the possibility to play with spins. The increasing sophistication of NMR spectrometers allowed physically oriented scientists to manipulate ensembles of nuclear spins and follow the response on the computer screen. Thus, NMR was essentially "hands-on quantum mechanics". As a consequence of such efforts, two-dimensional NMR was invented, which was extended to multi-dimensional NMR experiments. Initially, 2D NMR methods were all homonuclear and allowed solving protein structures of up to 10 or 15 kDa. Crucial for a further development was the ability to label proteins and nucleic acids with the stable isotopes ^{15}N and ^{13}C. This and more elaborate labeling methods, together with increasingly sophisticated experiments made possible determination of protein and nucleic acid structures in the molecular weight range up to 50 kDa and beyond.

Despite many spectacular success stories of NMR spectroscopy one has to think where this technique can provide unique and most valuable contributions. NMR can solve protein structures but what is the impact when we have X-ray crystallography as a structural method that is probably faster and less limited by molecular size? Clearly there are many proteins that do not crystallize and often their structures can be solved with NMR. Getting a structure is clearly faster than not getting it at all.

However, both NMR and crystallography are quite complementary. As an example of mutual benefit, we were recently approached by a local crystallographer asking whether we could have a look at a protein that is found at the tip of a rotavirus spike. They had tried to crystallize the protein for a year without success. Placed in the magnet it exhibited excellent NMR spectra, and the structure could be solved. Inspecting the molecular model revealed that there were flexible tails that might have prevented crystallization. More importantly, the structure had similarity to a fold found in sialic acid-binding proteins. Realizing this, a simple sialic acid was added, and the flexible tails were trimmed. This version of the protein immediately crystallized, and the co-structure could quickly be solved with X-ray methods.[13]

Probably the most valuable and unique power of NMR lies in detecting and characterizing protein interactions. In contrast to crystallography, where crystal contacts may be erroneously interpreted as binding interfaces, NMR only sees true and physiological contacts.[14] Even very weak interactions with equilibrium dissociation constants up to 10 mM can be seen as small chemical shift changes. NMR is the most reliable and only technique to directly identify and characterize such weak interactions. This is usually pursued by observing changes in 2D ^{1}H-^{15}N or ^{1}H-^{13}C correlated spectra upon titration with a putative ligand. NMR is a "litmus test" for proving protein interactions that may have been suggested based on other biochemical techniques, and often such claimed interactions turn out to be not existent.

Often protein interactions are weak and transient to allow for rearrangement of components during the assembly and alteration of mega-complexes that may accomplish different functions during a well-defined time course of events. Examples are the assembly of the pre-initiation complex of eukaryotic translation initiation where numerous factors join and leave the 40S subunit to ready the ribosome for protein synthesis. Many of these factors or their domains are small enough to be analyzable by NMR and multiple interfaces can be identified with a variety of NMR mapping experiments.[15] This can reveal topologies of factor association in dynamic assemblies of mega complexes.

Another powerful aspect of NMR is its ability to characterize weak interactions with small chemicals. Small-molecule libraries have become increasingly available to academic laboratories and can be exploited for probing cellular processes. Inhibitors of protein-protein interactions can be found with simple high-throughput screening methods, such as using fluorescence-polarization.[16, 17] However, the inhibitors that can be found are typically rather weak, in the low μM range. Thus, the complexes with target proteins may not be suitable for crystal-structure analysis but can be characterized with NMR methods.

Designing potent inhibitors of protein interactions could become a powerful approach for developing novel molecules of therapeutic value. The information about protein complexes with small molecules obtainable with NMR is suitable for a rational improvement of the affinities of these inhibitors. This is a field for fruitful collaborations between structural biologists, chemists and medicinal chemists and has the potential of contributing new means for fighting human disease.

A rather new and promising application of NMR is in the field of metabolomics, which tries to characterize the state of whole cells, organs or even whole living beings based on the entirety of all metabolites, the metabolome.[18] While this is most efficiently pursued with mass spectroscopy, NMR is complementary and can provide quantitative measurements of metabolite levels from 1D and 2D spectra, and it can be employed to identify the chemical structures of so far unknown metabolites. Usually two classes of samples are compared, such as from sick and healthy individuals, or from animals with and without a drug treatment, or from cells with and without a mutant protein. Considering the large number of metabolites found in body fluids or cell extracts, data are usually analyzed with multi-variate statistical tools, such as principle component analysis (PCA) or partial least squares discriminant analysis (PLSDA). This reveals the molecules for which concentrations are most different between the two groups of samples. Mass spectroscopy and NMR can then identify these molecules, which might be biomarkers of a disease. This approach has the potential of elucidating disease pathways and even identifying new drug targets. To obtain valid results it is important to use ultimate care in sample preparation and statistical analysis.

NMR has made dramatic and unexpected advances in a large range of areas. For quite a while its uses in chemistry and structural biology were separate. But recently the two fields are crossing over with the rising interest in chemical biology and metabolomics. Technical advances have been astounding, with the development of ultra high field magnets that are now available at 22.3 (950 MHz) and will soon be at 23.488 Ts (1 GHz) or above. However, most projects can be carried out at lower fields, such as at 400, 500 and 600 MHz spectrometers. Cryogenically-cooled probes have boosted the sensitivity of spectrometers up to five fold, and the sophistication of pulse sequences, data acquisition and processing is booming. NMR is a field with a wide spectrum of applications and certainly will continue to come up with unexpected and fascinating new achievements.

References

1. Gutowsky, H.S., McCall, D.W., and Slichter, C.P., Nuclear Magnetic Resonance Multiplets in Liquids. The Journal of Chemical Physics, 1953. 21(2): p. 279-292.
2. Solomon, I., Relaxation processes in a system of two spins. Phys. Rev., 1955. 99: p. 559-565.
3. Abragam, A., The Principles of Nuclear Magnetism. The International series of Monographs on Physics, ed. W.C. Marshall and D.H. Wilkinson. 1961, London: Oxford University Press.
4. Bystrov, V.F., Spin-spin coupling and the conformational states of peptide systems. Progr. NMR Spectrosc., 1976. 10: p. 41-81.

6

5. Meraldi, J.P., Schwyzer, R., Tun-Kyi, A., and Wuthrich, K., Conformational studies of cyclic pentapeptides by proton magnetic resonance spectroscopy. Helv Chim Acta, 1972. 55(6): p. 1962-1973.

6. Marinetti, T.D., Snyder, G.H., and Sykes, B.D., Nitrotyrosine chelation of nuclear magnetic resonance shift probes in proteins: application to bovine pancreatic trypsin inhibitor. Biochemistry, 1977. 16(4): p. 647-653.

7. Kalk, A. and Berendsen, H.J.C., Proton magnetic relaxation and spin diffusion in proteins. J. Magn. Resonance, 1976. 24: p. 346-366.

8. Sykes, B.D., Hull, W.E., and Snyder, G.H., Experimental evidence for the role of cross-relaxation in proton nuclear magnetic resonance spin lattice relaxation time measurements in proteins. Biophys J, 1978. 21(2): p. 137-146.

9. Wagner, G. and Wüthrich, K., Truncated driven nuclear overhauser effect (TOE). A new technique for studies of selective ^{1}H-^{1}H Overhauser effects in the presence of spin diffusion. J. Magn. Reson., 1979. 33: p. 675-680.

10. Dubs, A., Wagner, G., and Wuthrich, K., Individual assignments of amide proton resonances in the proton NMR spectrum of the basic pancreatic trypsin inhibitor. Biochim Biophys Acta, 1979. 577(1): p. 177-194.

11. Jeener, J. in Ampére Interational Summer School. 1971. Basko Polje, Yugoslavia.

12. Aue, W.P., Bartholdi, E., and Ernst, R.R., Two-dimensional Spectroscopy. Application to NMR. J. Chem. Phys., 1976. 64: p. 2229-2246.

13. Dormitzer, P.R., Sun, Z.Y., Wagner, G., and Harrison, S.C., The rhesus rotavirus VP4 sialic acid binding domain has a galectin fold with a novel carbohydrate binding site. Embo J, 2002. 21(5): p. 885-897.

14. Ferentz, A., Opperman, T., Walker, G., and Wagner, G., Dimerization of the UmuD' protein in solution and its implications for regulation of SOS mutagenesis. Nature Structural Biology, 1997. 4(12): p. 979-983.

15. Marintchev, A. and Wagner, G., Translation initiation: structures, mechanisms and evolution. Q Rev Biophys, 2004. 37(3-4): p. 197-284.

16. Lugovskoy, A.A., Degterev, A.I., Fahmy, A.F., Zhou, P., Gross, J.D., Yuan, J., and Wagner, G., A novel approach for characterizing protein ligand complexes: molecular basis for specificity of small-molecule Bcl-2 inhibitors. J Am Chem Soc, 2002. 124(7): p. 1234-1240.

17. Moerke, N.J., Aktas, H., Chen, H., Cantel, S., Reibarkh, M., Fahmy, A., Gross, J.D., Degterev, A., Yuan, J., Chorev, M., Halperin, J.A., and Wagner, G., Small Molecule Inhibition of the Interaction Between the Translation Initiation Factors eIF4E and eIF4G. Cell, 2007: p. in press.

18. Nicholson, J.K., Connelly, J., Lindon, J.C., and Holmes, E., Metabonomics: a platform for studying drug toxicity and gene function. Nat Rev Drug Discov, 2002. 1(2): p. 153-161.

NMR across the Curriculum

Chapter 2

Enhancing Undergraduate Pedagogy with NMR across the Curriculum

Matthew A. Fisher and Daryle H. Fish

Department of Chemistry, Saint Vincent College, 300 Fraser Purchase Road, Latrobe, PA 15650

The use of NMR in the undergraduate curriculum provides opportunities to create significant learning experiences for students from their freshman year through the time of graduation. We describe how Saint Vincent College has incorporated NMR based experiments and activities in general chemistry, organic chemistry, and biochemistry in a manner that provides students with a range of NMR experiences and fosters connections between chemical concepts across a course and between courses.

Introduction

Since its initial development in the 1940's, NMR has become one of the most important methods for determining the structure of molecules. As described in the recent National Research Council report *Beyond the Molecular Frontier: Challenges for Chemistry and Chemical Engineering* (*1*), NMR has revolutionized how chemists determine the structure of molecules in solution, in solids, and even in the human body. Four Nobel prizes have been awarded for work related to NMR - Felix Bloch (1952 - Physics), E. M. Purcell (1952 - Physics), Richard Ernst (1991 - Chemistry), Kurt Wüthrich (2002 - Chemistry), Paul Lauterbur (2003 - Medicine/Physiology), and Sir Peter Mansfield (2003 - Medicine/Physiology). Even the guidelines for undergraduate professional education developed by the American Chemical Society's Committee on

Professional Training single out NMR when they state "Nuclear magnetic resonance spectroscopy has become an indispensable experimental method for chemistry. Approved chemistry programs must have an operational NMR spectrometer." (2)

The central importance of NMR is reflected in the number of articles published in the *Journal of Chemical Education* that use this technique. Since mid-1995, 444 articles have been published in the *Journal* that have NMR as a keyword. A similar search of the Project Chemlab database, which includes laboratories published in the *Journal*, gave 315 results; 165 were laboratories for organic chemistry while 48 were for inorganic chemistry and 42 were for physical chemistry.

While the application of NMR in individual courses is well established in the undergraduate chemistry curriculum, we believe that NMR has the potential to serve as a unifying and integrative thread as well. Our experience at Saint Vincent College has helped us see that thoughtful incorporation of NMR throughout the undergraduate curriculum can be a major tool in the creation of significant learning experiences for undergraduate students.

What Characterizes Best Practice and Significant Learning in Undergraduate Education?

Much attention has been directed over the last 15 years as to what characterizes both significant learning experiences and "best practice" in undergraduate science education. Project Kaleidoscope, a major leader in this work, has described(3) "what works" in undergraduate science education as being characterized in part by:

- learning that is experiential and places great emphasis on investigation throughout the curriculum, starting with the very first course.
- learning that is personally meaningful for students and faculty, that makes connections to other fields, and that can be linked to practical applications

The much cited report *How People Learn* produced by the National Research Council in 1999(4) identified the following as essential to effective learning:

- an environment centered on learners and that is designed to "help students make connections between their previous knowledge and their current academic tasks."
- teaching knowledge in multiple contexts rather than in just a single context. The use of multiple contexts for learning helps students identify the relevant features of concepts in a more flexible manner.

And the final report issued from the 2003 "Exploring the Molecular Vision" conference sponsored by the ACS Society Committee on Education stated as part of the conference's conclusions regarding preparation for the profession that "problem solving is one of the most empowering learning experiences for science students" and "a deep knowledge of chemistry and the ability to pursue cutting-edge chemical research in teams have become even more central to innovation at the forefront of science and technology."(5).

The incorporation of NMR throughout the undergraduate curriculum offers several ways to achieve these characteristics. NMR is a technique that is both "hands on" and at the same time potentially involves a substantial body of theory to interpret different types of experimental results. The multiple levels of information contained in an NMR spectrum and the variety of ways that NMR can be used provide a data-rich environment for students to work in. Using NMR to make connections between different chemistry courses, various sub-disciplines of chemistry, or chemistry and other disciplines is relatively straightforward. Analysis of spectral data is clearly a problem solving activity and one that can be constructed at varying levels of difficulty. And using NMR in multiple courses provides students with an opportunity to see the same fundamental chemical concepts in different contexts.

In *Designing Significant Learning Experiences* (6), Dee Fink puts forth a taxonomy of significant learning that includes six different types of learning. Among the kinds of learning that make up the taxonomy are foundational knowledge (facts, terms, concepts, principles), application (problem solving and decision making) and integration (making connections among ideas, subjects, etc.). Using NMR across the undergraduate curriculum clearly helps students develop foundational knowledge and engage in a variety of applications. At the same time, NMR also offers possibilities in regards to how students integrate the ideas from various courses. Biology majors who first encounter NMR in organic chemistry and then encounter it again in biochemistry are likely to see the connections between the two disciplines differently than students who only encounter NMR in organic chemistry. For all of the reasons given here, we believe that it is in every chemistry department's best interest to explore collaboratively how NMR can be incorporated throughout the undergraduate curriculum.

NMR in the Saint Vincent College Chemistry Curriculum

Saint Vincent College is a Catholic liberal arts college enrolling approximately 1500 undergraduates. In 1998, the Chemistry department at Saint Vincent College began a comprehensive program for ongoing assessment of student learning and continuous program improvement that includes alumni surveys, a comprehensive laboratory practicum, and assessment of senior theses and presentations. The assessment indicated that our graduating seniors had inadequate preparation in the area of spectral interpretation, especially NMR spectroscopy. In addition, the assessment results showed that students had a limited exposure to two-dimensional and multi-pulse NMR techniques. In the spring of 2003 we received a CCLI-Adaptation and Implementation grant from NSF to integrate FT-NMR throughout our curriculum by purchasing the Anasazi EFT-NMR upgrade for our 60 Mhz fixed magnet.

There were several goals that we had at the onset of this project. We believed it was important that students in all science, technology, engineering, and math majors would have a basic understanding of the uses of one-dimensional NMR spectroscopy. In addition, we wanted all biology and chemistry majors to be able to operate an FT-NMR and interpret proton, carbon, DEPT, COSY, and HETCOR spectra. Finally, we wanted our chemistry majors to be able to choose the appropriate NMR experiments to address specific experimental needs. To accomplish these goals we have, over the past two years, incorporated a series of experiments into general chemistry, organic chemistry, and the first semester of biochemistry.

General Chemistry

Students in general chemistry are introduced to NMR through an activity that focuses on the concepts of NMR activity and nuclear structure, bond polarity, and chemical shift. Earlier work published in the *Journal of Chemical Education* by Davis and Moore described how Mercer University had incorporated FT-NMR into general chemistry through an experiment that studied electronegativity through the additive effects of halogens on the ^1H chemical shift of methane (7). While we opted to use a similar approach, we chose to connect these concepts to the determination of protein structure through closer examination of chemical shifts in amino acids. In addition, we chose to use the POGIL approach for this activity. POGIL (process oriented guided inquiry learning) is a pedagogical approach that seeks to "simultaneously teach content and key process skills such as the ability to think analytically and work effectively as part of a collaborative team." (8,9) The activity we developed starts with a review of bonding, dipole moments, and Lewis dot structures. Students are then asked to determine the number of protons and neutrons of 3 NMR active and 3 NMR inactive nuclei. From this, students are asked to devise a rule for determining whether or not a particular nuclei is NMR active. In the next phase of the activity, students are asked to compare the chemical shift of methylene protons next to a halogen for a series of alkyl halides that differ in the number and chemical identity of the halogens. From this data students are asked to construct a plot of ^1H chemical shift as a function of the sum of electronegativities for the atoms attached to a single carbon. Figure 1 shows a typical plot constructed by students as part of this activity.

Finally, students are asked to apply their model to an amino acid by determining the chemical shift of a proton attached to the first carbon in a side chain and evaluating the polarity of the side chain as a whole. Amino acids uses for this activity include alanine, valine, leucine, serine, cysteine, threonine, and phenylalanine.

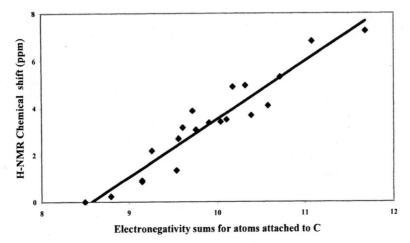

Figure 1. Effect of total electronegativity of atoms attached to carbon on ^1H chemical shift.

Organic Chemistry

Students in organic chemistry are introduced to ^1H, ^{13}C, and DEPT NMR spectroscopy in the first semester laboratory course. These techniques are used throughout the first semester as the primary means for structurally characterizing compounds. In the second semester lab, students are introduced to 2D NMR through the use of COSY to identify an unknown. Branz and colleagues published an experiment in the *Journal of Chemical Education (10)* where they described what they called a "double unknown" experiment where students use a unknown alcohol and an unknown carboxylic acid to synthesize an ester whose structure is unknown. COSY spectroscopy was then used to determine the structure of the product. While we found this experiment very attractive for several reasons, we saw one difficulty. Because ester synthesis is not covered in the typical sophomore organic course until late in the spring semester, using the experiment as published would mean that students taking organic chemistry would not encounter 2D NMR until late in the year. To allow us to introduce 2D NMR earlier in the spring semester, we modified this experiment so that students synthesize an unknown ether using reagents whose structure is not known to the students. Otherwise, we implemented this experiment as Branz and colleagues describe it.

To provide additional opportunities to work with spectral data and how structure and spectroscopy are related, students in the spring semester are given problem sets that require them to develop synthetic routes for compounds such as Albuterol, Atenalol, Ibuprofen, and Prozac. As part of these problem sets, students are given NMR spectral data that they must work with in order to answer some specific questions. The details of what NMR data is given and what students are asked to do with it varies from problem set to problem set. In one case, students are asked to predict the chemical shift, splitting pattern, and integration in the ^1H NMR for ibuprofen alcohol, an immediate precursor to ibuprofen. Students then are asked to assign all the peaks in the ^1H and ^{13}C NMR for ibuprofen itself. In another problem set, students are given the structures of two components found in cough medicine along with the COSY spectrum for one of the drugs. Students are asked to draw the correct structure of the molecule and assign all of the protons other than -OH and -NH.

Upper Level Courses

The fall semester biochemistry course at Saint Vincent focuses heavily on experimental techniques in protein chemistry. To provide students with "hands-on" experience in the process of determining the three dimensional structure of a protein, we have incorporated into that course the experiment published by Rehart and Gerig in the *Journal of Chemical Education (11)*. In this experiment, data from 1D-NMR, COSY, TOCSY, and NOESY are used to develop a three dimensional model for the structure of the octapeptide angiotensin in DMSO. Through a web site, the authors provide spectra that they collected on a 300 MHz instrument. We ask our students to engage in the "paper and pencil" process of assigning peaks in the 1D-NMR, COSY, and TOCSY spectra. Then students use those peak assignments to examine NOESY spectra for information that will provide distance constraints for the model. While Rehart and Gerig do give information on how students can collect the data themselves on a high field instrument, that aspect isn't possible for us using the Anazasi setup. To utilize "dead time" while other experiments (such as chromatographic separations) are running, we spread this activity out over a period of roughly four weeks.

In the near future we plan to incorporate another NMR experiment in the fall biochemistry lab course. This experiment, originally published by Giles et al. in the *Journal of Chemical Education (12)* uses ^{13}C NMR as a means to track metabolic flux through glycolysis in yeast cells grown under different osmolarities. As the external osmolarity of the growth medium is altered, the cells respond by shifting the balance among metabolic pathways such that the ratio of ^{13}C labeled ethanol to ^{13}C labeled glycerol changes. We see this experiment as a nice complement to the protein structure activity we already use, and a pedagogically sound way to introduce students to the use of NMR as a probe for the dynamics of metabolic pathways.

Advanced Physical Methods is a project-based laboratory course formed by integrating elements of what were originally separate courses - Physical Chemistry II Laboratory and Instrumental Laboratory - into a single course. The NMR experiment we plan to incorporate into this course will be a kinetics experiment, either esterification of trifluoroacetic acid (*13*) or hydrolysis of an orthoester (*14*), with NMR used as the primary means of measuring reaction rate.

Table I provides a summary of the various NMR activities that are being incorporated into the four year curriculum at Saint Vincent College.

Table I. NMR activities in the Saint Vincent chemistry curriculum

Course	Activity	Assessment
general chemistry	correlating chemical shift with electronegativity	student worksheet
organic chemistry	structure determination using 1D (^{1}H, ^{13}C) and 2D COSY spectra	lab reports problem sets
biochemistry	structure determination using TOCSY and NOESY, *metabolic analysis using in vivo NMR*	lab reports
other upper level courses	*reaction kinetics monitored by NMR*	lab reports
end of undergraduate education	ability to interpret spectra and determine structure	lab skills assessment exam

NOTE: Activities in italics are in process of being incorporated into curriculum at time this chapter was written.

What Have We Learned?

Three years after receiving the grant from NSF, we have learned several things that we believe will be of use to other departments seeking to integrate NMR throughout the undergraduate curriculum:

Don't expect to do everything at once. This statement applies equally to the implementation of new experiments and how students are introduced to the concept of NMR. In terms of introducing new experiments, we have successfully implemented more than half of the experiments that we eventually expect will be part of our curriculum. Still to be incorporated into our Advanced Physical Methods course is a kinetics experiment using NMR, and a metabolic experiment in the biochemistry lab that uses ^{13}C NMR to track flux through glycolysis in yeast cells.

In the same manner, we have found that NMR concepts are probably best introduced in small blocks, with opportunities for practice worked in wherever possible. Our General Chemistry activity focuses very heavily on chemical shift and ignores other important topics like splitting or the explanation of NMR in terms of bulk magnetization vectors. The angiotensin activity in biochemistry lab assumes some prior exposure to basic 2D NMR techniques.

Not all NMR related activities require having students physically collect the data; sometime pencil and paper is equally effective. If we insisted that General Chemistry students actually run their own NMR, we would be concerned about the potential confusion and clouding of learning that might develop. If we insisted that students in the biochemistry lab had to collect their own spectral data to determine the conformation of angiotensin, we would never be able to incorporate this experiment into our curriculum. "Paper and pencil" activites can be just as thought-provoking and demanding of critical thinking and problem solving skills as an activity where students collect the data themselves. The key is how the assignment is designed so that it both builds on prior learning and challenges students to use concepts in new ways or new contexts.

An assessment process can provide valuable information and feedback. Our department assessment plan initially helped us see more clearly the weaknesses that our graduating seniors had in regards to NMR and helped us make a stronger case for the resources necessary to implement NMR throughout the curriculum. Three years into this project, we have seen significant increases in the percentage of graduating seniors who can assign the correct structure for a given set of NMR data. We also observed that students who encountered our revised curriculum starting in their sophomore year were more successful as seniors in correctly determining a structure from NMR data than they were as second semester sophomores. This result suggests that student mastery of NMR concepts and skills required both time and encountering NMR in multiple settings. Our hope is that assessment will provide insight into what students actually learn in this area so that we can, in an informed manner, make any needed modifications. Our goal has always been to integrate NMR into our curriculum so that it fosters significant learning, so a carefully thought out assessment strategy provides a way to understand "what works" and what needs revision.

Assessment of Student Learning

The question of assessment is one that many chemistry faculty struggle with, and so it is worth looking more closely at this issue in the context of NMR and pedagogy. Student learning related to NMR has the potential to be particularly challenging, as it involves a complex mixture of laboratory skills, conceptual understanding, and problem solving abilities. But that also means that student learning related to NMR offers some unique assessment opportunities.

The National Research Council report *Evaluating and Improving Undergraduate Teaching in Science, Technology, Engineering, and Mathematics* (*15*) strongly encourages faculty to use outcome assessment as the basis for determining what students have learned from a unit, a course, or a series of courses. Outcome assessment starts with the articulation of student learning objectives worded in terms of what students will be able to do after learning the material in question. After the goals have been framed in terms of student learning, then appropriate assessment measures are utilized.

Walvoord has pointed out that the grading of student work can be a valuable assessment tool if criteria used to evaluate the work are made explicit (*16, 17*) The most systematic way to accomplish this is the approach known as "primary trait analysis", which is described in more detail in several sources (*15, 17*). In using primary trait analysis, an instructor first identifies the traits that he or she feels are most important to evaluate, then develops a scoring system (typically two to five points) for each trait. The result is a rubric that provides explicit criteria for evaluating student work and that can readily be communicated to other parties such as students, accrediting bodies, or organizations such as the National Science Foundation. We are beginning to think about how we might develop some resources for primary trait analysis of student work involving NMR that could be used by various members of the department.

While primary trait analysis is the approach to grading student work that most effectively connects with assessment, it is not the only assessment strategy that should be considered. Much recent research in cognitive science has demonstrated the importance of formative assessments in helping students learn more effectively. Formative assessment, which is provided more frequently and immediately than the traditional summative assessments such as exams, can be a powerful tool for both helping to reinforce student understanding of NMR concepts and addressing student misconceptions in a timely fashion. Our efforts at providing formative assessment have focused on both the use of assignments that involve using NMR concepts and data - such as the OTC drug synthesis problems described above – as well as frequent use of NMR in lab reports.

Grant Wiggins (*18*) has called for the use of authentic tasks in the process of assessment because they serve to provide direction, coherence, and motivation for the work involved in learning. NMR is particularly amenable to being assessed in the context of authentic tasks. Consider a project-based laboratory where in the later stages students must decide which NMR experiments to run in order to unambiguously determine the structure of a compound. This is a task that is both authentic (something chemists routinely do in their work) and provides instructors with the opportunity to gather information as to the reasons why students chose particular techniques and how they interpreted the data. We have found this approach to assessment particularly valuable in our work at Saint Vincent. Experiments such as the "double unknown" synthesis of an ether and the determination of the 3D structure of angiotensin involve using NMR concepts and data in the same manner as research scientists. In addition, our department's assessment of the laboratory skill level of graduating seniors requires students to interpret NMR spectra that are given to them by the faculty, rather than simply answering questions in an exam format. This also more closely mirrors how chemists routinely use NMR concepts and data in their own work.

In addition to direct measures of student learning such as the ones described in the preceding paragraphs, indirect measures can also be useful in the context of NMR. Asking students how much they thought they learned, can be useful in gathering information about what students have learned over a longer period of time - a series of courses or an entire four year curriculum. A particularly useful indirect measure is the Knowledge Survey developed by Nuhfer and Knipp (*19*), which could be used with seniors to gather some evidence as to what they feel they have learned about NMR over the course of their entire undergraduate experience. We are exploring the possibility of incorporating a Knowledge Survey that would include questions specific to NMR as part of our department's assessment of graduating seniors.

Conclusions

The past ten years have seen a number of reports, including the National Research Council's *Transforming Undergraduate Education in Science, Mathematics, Engineering, and Technology* (20) and *BIO 2010: Transforming Undergraduate Education for Future Research Biologists* (*21*), that have called for reforms in undergraduate science education that focus on inquiry-based activities and interdisciplinary connections. The incorporation of NMR throughout the undergraduate curriculum offers unique opportunities to address these challenges in ways that challenge students, engage them in the process of doing science, and making connections between various scientific disciplines. In addition, NMR-based activities lend themselves to using a variety of assessment approaches to gather information about student learning. Our ongoing experience at Saint Vincent College has shown us that incorporating NMR across the undergraduate curriculum can lead to significant improvements that ultimately help our students develop a better understanding of chemistry and its role in scientific research.

Acknowledgements

The work described in this article was supported in part by NSF-CCLI-A&I Grant #0310756 and a grant from the Spectroscopy Society of Pittsburgh.

References

1. National Research Council Board on Chemical Sciences and Technology, *Beyond the Molecular Frontier: Challenges for Chemistry and Chemical Engineering;* National Academy Press: Washington, DC, 2003.
2. American Chemical Society Committee on Professional Training "Undergraduate Professional Education in Chemistry: Guidelines and Evaluation Procedures" http://www.chemistry.org/portal/resources/ACS/ ACSContent/education/cpt/guidelines_spring2003.pdf (accessed 21 February 2006)
3. Project Kaleidoscope. *What Works: Building Natural Science Communities;* Washington, DC, 1991
4. National Research Council Committee on Developments in the Science of Learning. *How People Learn;* National Academy Press: Washington, DC, 1999
5. American Chemical Society Committee on Education. *Exploring Molecular Vision Final Conference Report.* Exploring Molecular Vision Home Page. http://www.chemistry.org/portal/a/c/s/1/acsdisplay.html? DOC=education%5Csoced%5Cmolecularvision.html (accessed June 2006)
6. Fink, L.D. *Creating Significant Learning Experiences;* Jossey-Bass: San Francisco, CA, 2003
7. Davis, D.S.; Moore, D.E. *J. Chem. Educ.* **1999**, *76*, 1617-1618
8. Spencer, J.N. *J. Chem. Educ.* **1999**, *76*, 566-569.
9. Farrell, J.J.; Moog, R.S.; Spencer, J.N. *J. Chem. Educ.* **1999**, *76*, 570-574.
10. Branz, S.E.; Miele, R.G.; Okuda, R.K.; Straus, D.A. *J. Chem. Educ.* **1995**, *72*, 659-661
11. Rehart, A.M.; Gerig, J.T. *J. Chem. Ed.*, **2000**, *77*, 892-894.
12. Giles, B.J.; Matsche, Z.; Egeland, R.D.; Reed, R.A.; Morioka, S.S.; Taber, R.L. *J. Chem. Educ.* **1999**, *76,* 1564-1566
13. Minter, D.E.; Villarreal, M.C. *J. Chem. Ed.*, **1985**, *62*, 911-912
14. Potts, R.A.; Schaller, R.A. *J. Chem. Ed.*, **1993**, *70*, 421
15. National Research Council Center for Education. Evaluating and Improving Undergraduate Teaching in Science, Technology, Engineering, and Mathematics; National Academy Press, Washington, DC, 2003.
16. Walvoord, B.E. *Assessment Clear and Simple;* Jossey-Bass: San Francisco, CA, 2004
17. Walvoord, B.E.; Anderson, V.J. *Effective Grading: A Tool for Learning and Assessment;* Jossey-Bass: San Francisco, CA, 1998
18. Wiggins, G. *Educative Assessment: Designing Assessments to Inform and Improve Student Performance;* Jossey-Bass: San Francisco, CA 1998, pp 21-24.
19. Nuhfer, E.B.; Knipp, D., *To Improve the Academy* **2003**, *21*, pp 59-78.

20. National Research Council Committee on Undergraduate Science Education. *Transforming Undergraduate Education in Science, Mathematics, Engineering, and Technology;* National Academy Press: Washington, DC, 1999
21. National Research Council Board on Life Sciences. *BIO 2010: Transforming Undergraduate Education for Future Research Biologists;* National Academy Press: Washington, DC, 2003

Chapter 3

Toward the Integration of Liquid- and Solid-State NMR across the Undergraduate Curriculum

N. M. Washton, K. C. Earnheart, D. G. Sykes,
M. Ucak-Astarlioglu, and K. T. Mueller

Department of Chemistry, Penn State University, 104 Chemistry Building,
University Park, PA 16802

This work addresses issues of fundamental reform in the undergraduate chemistry curriculum via advanced integration of liquid- and solid-state nuclear magnetic resonance (NMR) experiments and theory. Our integrated curricula comprises a suite of NMR laboratory exercises (initially targeted within the physical and analytical chemistry laboratories) coupled with advanced NMR theory. This curriculum provides multiple layers of instructional merit from basic structural characterization to important physical chemistry concepts. For programs lacking solid-state NMR capabilities, we have developed schematics and construction materials for a low-cost, broadband NMR probe that is compatible with existing liquid-state spectrometers. This probe has been used to detect a wide range of NMR-sensitive nuclei in liquid samples, and in the future will be utilized to study deuterium NMR lineshapes in solids.

Introduction

Nuclear magnetic resonance (NMR) spectroscopy has undergone a dramatic transformation in the last 20 years from a specialized tool utilized only in advanced research laboratories to a standard analytical tool now being routinely applied to new materials, forensic samples, pharmaceuticals, foodstuffs and environmental samples. There are many sophisticated NMR techniques that have become indispensable to researchers in chemistry and allied fields, and therefore it is a critical element of undergraduate chemistry education to incorporate some of these techniques into the laboratory curriculum. Given its importance, it is not surprising that there has been an explosion in the number of undergraduate programs requesting funds to purchase new or upgrade existing NMR instruments, but most of the implementation has been course or discipline specific. Addition of both advanced and non-traditional NMR methods in undergraduate chemistry courses requires a comprehensive implementation plan across the curriculum. To date, the coherent curriculum structure necessary for transfer of NMR knowledge has been absent. Our work, the beginning stages of which are described here, focuses on the development of a cohesive curricular plan across the undergraduate curriculum for the implementation of NMR theory and experiments

NMR provides a wide range of information about chemical systems [1], and this is reflected by numerous published experiments for undergraduate laboratory courses. In the majority of these, NMR has been used to aid in structure determination of organic molecules, but less common are experiments that extend problem solving abilities by taking advantage of the inherent predictive capabilities of NMR. Recent examples of such value-added experiments include conformational analysis of brominated cyclohexanone products [2] and a discovery approach to the concept of shift additivity for a series of aromatic compounds [3]. One especially notable example for novice level students is a guided inquiry introduction to NMR in which students compare a series of related compounds that increase in number of peaks (acetone, acetic acid, etc.) and then use this information to help them select the product, and by-products, of an aspirin synthesis [4]. In addition, there has been a recent increase in the number of published organic laboratory projects that relate to other fields such as inorganic chemistry and biochemistry. In particular, the synthesis and subsequent characterization of organometallic compounds encourages students to explore multinuclear NMR [5]. Similarly, at the interface with biochemical research, Peterman et al. have reported a laboratory-based enzyme study using ^{19}F NMR [6], providing another example of a situation where using NMR-active species other than the conventional ^{1}H or ^{13}C is advantageous.

For chemistry students, physical organic chemistry is an advanced area of study where NMR can extend experimental approaches to the understanding of molecular behavior and reaction mechanisms. Adapted for the undergraduate laboratory by Chechik [7], the reaction rates of both the deuteration and bromination of hexanone with NMR are determined, and using these data the bond being broken in the rate-limiting step is identified. NMR spectroscopy also provides detailed information on bonding, as demonstated by Mosher [8]. In this work, data from NMR experiments demostrate the different orbital mixing available to carbon atoms in an organic molecule, as the J-coupling between C and H is related to percent s character in model systems. Once the "standard" J values for sp, sp^2 and sp^3 hybridization are determined, they can be compared by students to values obtained for other non-standard C-H bonds.

NMR is an ideal instrument to examine dynamic molecular processes, and many experiments in this area that are accessible to undergraduates fall at the interface of organic and physical chemistry. For example, determining the rotational energy barrier for an amide bond using analysis of NMR lineshapes as a function of temperature is a classic experiment for physical chemistry [9]. However, such examples are rare in the undergraduate curriculum and are still being improved upon. Morris and Erickson [10] have recently reported a modification that adds saturation transfer techniques (and necessary spin-lattice relaxation rate determination) to improve the results that students achieve for the enthalpy and entropy of activation measured in such an experiment. In a more simplified study, Weil [11] has found that bond rotation of a hydrogen-bonded picryl system can also demonstrate the conformational exchange of protons through lineshape analysis at various temperatures coupled with spin-spin relaxation rate ($1/T_2$) measurements. Electron exchange can be monitored in an undergraduate laboratory setting by NMR as well, as described by Jameson [12] where the fast-exchange process between a diamagnetic species (ferrocene), which oxidizes to form a paramagnetic species (ferrocenium ion), can be monitored by relating shifts and line-widths of peaks from each species to determine the mole fraction of each.

State-of-the-art NMR research also explores systems extending far beyond the standard liquid-state samples found in most undergraduate laboratories. A few reports that reflect this change are appearing in the literature, and include examples of solid-state NMR and magnetic resonance imaging (MRI) experiments. Solid-state NMR is an invaluable analysis technique for materials science, and a recent experiment reported by Anderson et al [13] introduces students to ^{31}P NMR studies of sodium phosphate glasses. In these experiments, the students analyze percent composition of phosphate species in the condensed phase by matching their ^{31}P NMR spectra to computer-simulated spectra of various compositions. A much less common technique in the undergraduate curriculum is MRI, and Quist has developed an imaging experiment that can be performed on a 100 MHz NMR instrument [14]. The phantom sample for this study consists of cylinders of water, constructed by filling cylindrical holes in a poly(tetrafluoroethylene) plug that is then placed in a normal NMR tube. After applying gradients, created using the x-, y-, and z-shims on the spectrometer, an image is reconstructed through back-projection of the spectra.

Focussing in particular on the physical chemistry laboratory curriculum, a survey of the *Journal of Chemical Education* and *The Chemical Educator* reveals that hundreds of physical chemistry laboratory exercises have been published within the last few years. Most of these activities are designed to enhance the relevance of physical chemistry by introducing instructional activities in thermodynamics and laser spectroscopy that reflect contemporary practice in physical chemistry. However, a survey of the same journals reveals few published laboratory exercises using NMR to study crucial topics in physical chemistry, such as chemical kinetics [15, 16] and thermodynamics [17, 18]. As a consequence, the educational potential of one of the most elegant and sophisticated tools for probing chemical systems has never been fully achieved, and instruments found in many undergraduate university laboratories could be at risk of being nothing more than expensive data loggers. *The main reasons for this weak emphasis on NMR instruction across the curriculum, even in institutions with large NMR research facilities, has been the lack of a cohesive curricular plan and the lack of NMR expertise among the faculty responsible for the development of instructional material.*

The goal of our research and teaching team has been to close the widening gap of unrelated organic, analytical, and physical chemistry laboratory experiments by developing a set of tools and NMR laboratory exercises that could be implemented within the physical chemistry laboratory (leading also to vertical integration within a subsequent instrumental analysis course). The curriculum we developed guides students through a series of linked hands-on exercises that promote a basic understanding of NMR theory and relaxation processes in chemical systems. Students use Mathematica^{TM}, a powerful mathematical computing platform, for reduction of spectral data to obtain relaxation and correlation times. Instructional activities focus on liquid-state NMR and use the vector model to describe single- and multiple-pulse sequences (including ^{1}H-decoupled, DEPT, INEPT, COSY, and INADEQUATE NMR) [1]. In our curriculum at Penn State University, the physical chemistry experience provides an excellent foundation upon which an instrumental analysis course expands to include the product operator formalism, solid-state NMR, and more in-depth analyses and understanding of data acquisition and processing. Recognizing that not all instructional programs have solid-state capabilities, we have also developed a tunable, low-cost NMR probe that is compatible with existing liquid-state spectrometers. We have demonstrated that ^{2}H signals, along with sugnals from a wide range of other nuclei, can be obtained using these simple, static probes, and have begun the development of a set of experiments using a number of inexpensive, ^{2}H-labeled solids. The spectra of these solids are sensitive to molecular motion and serve as excellent instructional aids for advanced topics such as motion in the solid state (ring flips, rotations of methyl groups, etc.).

In this project, our team is working toward the ultimate integration of NMR spectroscopy throughout the undergraduate science curriculum, focusing here on physical and related analytical chemistry laboratory courses. A primary goal of this work has been to teach students to think critically about instrumental-based chemical analysis in general, and better prepare students for graduate-level research and industry careers. A broad-ranging objective of this initiative is the integration of NMR spectroscopy throughout the undergraduate science curriculum while concurrently increasing synergy between courses that contain overlapping or prerequisite material.

A Brief Description of the Curricular Development

During this project, we developed and tested curricular materials focused on NMR methodologies (tutorials and theory) and laboratory implementation (experiments). Much of the work completed has been developed and tested in the Chemical Spectroscopy (Chem. 426) and Experimental Physical Chemistry (Chem. 457) courses at Penn State University, where the Experimental Physical Chemistry course is a prerequisite for the more analytically-oriented spectroscopy course.

Specific, new curricular materials introduced into these courses address the following topics:

Physical Chemistry – Topics covered in this course include an introduction to Mathematica^{TM}, Boltzmann distributions, the vector approach to NMR, a brief introduction to product operators, relaxation in NMR (T_1 and T_2), and the study of kinetics with NMR. The curricular materials produced include Mathematica^{TM} tutorials where problem sets with solutions are provided, as well as basic laboratory exercises with step-by-step instructions

Chemical Spectroscopy – Topics covered in this course include an introduction to NMR theory, the full vector description of NMR spin systems, a product operator description with reference to the corresponding vector model, solid-state NMR, and instrumental design considerations and operation. This material also includes problem sets with solutions and laboratory exercises with step-by-step instructions.

The notes for these courses are modular in concept, meaning that it is not necessary to present a vector description of NMR prior to a discussion of, for example, product operators (although the notes, as a whole, are meant to provide an integrated and comprehensive treatment of NMR). These notes may be accessed by interested parties, and are available on-line at

http://research.chem.psu.edu/ktmgroup/chemed/

or by request from the authors.

We have also designed, with the assistance of undergraduate students carrying out special projects within the Chemical Spectroscopy course, an NMR probe kit that we intend to make available for purchase. In the design stage, we assigned four sets of two students (a total of eight students over three successive spring semesters) a semester-long research project to develop an inexpensive

NMR probe. In addition, students who had shown particular interest in applications of NMR in physical chemistry were given the opportunity to conduct special projects, in groups of two, on relaxation in NMR (T_1/T_2 experiments), the study of chemical kinetics, or the measurement of diffusion constants in solution.

Examples of Curricular Developments

Physical Chemistry Laboratory

Placing abstract concepts within an experimental framework, this course aims to make physical chemistry more self-explanatory through the coupling of theory and experimentation for science and engineering students. This course mainly utilizes concepts in kinetics, thermodynamics, surface chemistry, and spectroscopy. Having an NMR module launched in this physical chemistry laboratory course not only makes the spectroscopic theory more transparent, but also continues the integration of NMR with the rest of the chemistry curriculum beyond the normal termination of these concepts after their use in advanced organic laboratory courses.

NMR incorporation within a physical chemistry framework takes place in both the lecture and laboratory parts of this ourse. Lecture implementation begins with the physical chemical applications of NMR, and continues with a theoretical component by introducing the vector model (semi-classical picture), the rotating frame, the effect of RF pulses, the Bloch equations, and a brief introduction to product operators (quantum mechanical picture), In addition, the students practice Mathematica^{TM} tutorials on linear and nonlinear regression, relaxation time measurement, and Fourier transforms for analysis of data collected from the NMR experiments. The laboratory implementation utilizes the analysis of the spin echo pulse sequence, measurement of nuclear spin relaxation times, and the interpretation of experimental findings using Mathematica^{TM}.

At the novice level, qualitative descriptions of NMR are most commonly accomplished using vector notation (Figure 1) and the right hand rule. However, only the simplest NMR experiments may be accurately described in this manner. Given the rigor of the full (but proper) density matrix descriptors, one cannot move directly from the simple physical model of vectors to the density matrix formalism. Product operators form a necessary and illustrative bridge between the two descriptions, allowing students to retain their physical intuition of vector diagrams while moving on to a formalism that accurately describes the behavior of complex spin systems. Background material covering sine, cosine, and exponential decay functions, as well as the imaginary plane were introduced prior to the concept of spin evolution as described by product operators (Figure 2).

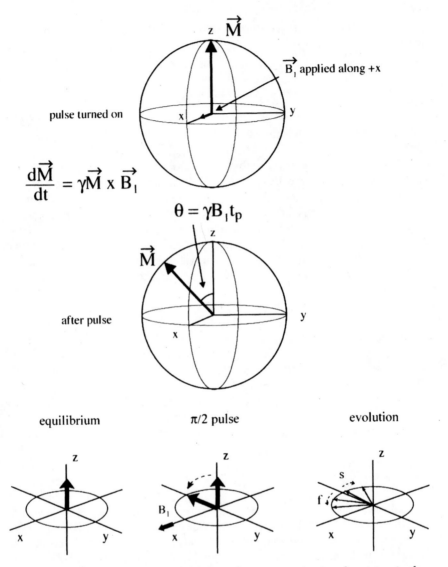

Figure 1. Traditional vector model describing magnetization dynamics in the rotating frame.

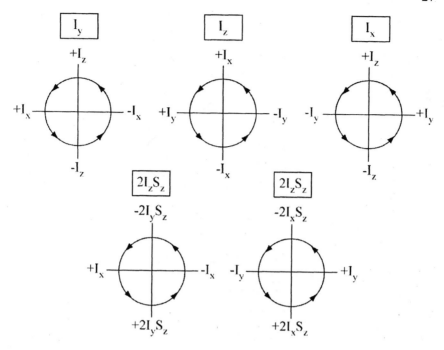

Figure 2. Spin evolution described by product operator formalism.

Students were given product operator tables and simple circle diagrams of spin evolution under the influences of pulses, chemical shift, and J coupling. Figure 3 represents a pulse sequence and product operator assignment given to students for working through a simplified version of a COSY experiment using the product operator formalism, and neglecting chemical shift interactions to focus on scalar coupling effects. In a classroom lecture, the students were first stepped through a simpler, heteronuclear INEPT experiment, where magnetization transfer and anti-phase magnetization were introduced. A Mathematica[TM] tutorial also led the students through the application of Fourier transforms to signals that provide simple single resonance lines as well as the "anti-phase doublets" obtained in more complicated experiments. Although a published product operator formalism in Mathematica (POMA) [19] would have accomplished this task with ease, we chose to require students to work through the simplified homonuclear COSY experiment (again, disregarding chemical shift evolution) without the use of software tools. The problem statement and solution are reproduced in Figures 3 and 4, respectively. Pattern recognition in spin evolution as described by product operators is relatively straightforward, and yet we found this task to be very difficult for students. Transferal of skills from mathematics courses was predominantly absent, resulting in an obstacle for

28

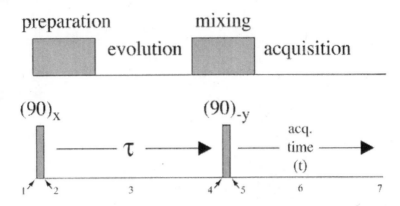

Figure 3. Schematic of a two-dimensional NMR experiment and the simplified COSY-type problem statement (the solution appears in Figure 4).

Assignment: Given the pulse sequence above for two coupled ^{1}H spins (call them 1 and 2), determine the product operator representation of the spin systems at all points in time labeled on the above pulse sequence. Assume that the first delay, τ, is chosen to be $1/(2J)$, and also calculate the acquired signal (time interval 6) at a specific time 7 where $t = \tau = 1/(2J)$.

the coalescence of new information with already learned skills. This problem is being addressed as we revise the curriculum for subsequent semesters.

Undergraduate students commonly assume that data reduction is trivial given the widespread use of pre-packaged software running analytical instruments. However, use of such software may be limited to only the most common data acquisitions, necessitating separate data analyses for more complex experiments. Although many scientific graphing and data analysis software packages are available, we determined that Mathematica^{TM} presented the best combination of analysis capabilities and platform transparency. Students are accustomed to graphical user interfaces, whereby clicking on icons generates graphs, analysis, fitting, etc., without their ever understanding the underlying mathematics behind the tasks. This leads to a "black box" approach to data analysis, which we attempted to circumvent by using Mathematica^{TM}. Students were provided relaxation data sets in ascii format to determine the T_1 and T_2 relaxation times for protons in methylene chloride. This required a non-linear fitting routine utilizing user defined functions with graphical output of raw data and best-fit curves. Figure 5 demonstrates both the conceptual picture introduced to the students, as well as typical code generated within Marhematica^{TM} for a T_1 analysis. More complete Mathematica^{TM}-based workbooks and problem sets used within the physical chemistry laboratory may be accessed at the Mueller Group chemical education site.

1. Before the first pulse, the two spin system is represented by the spin operators corresponding to z magnetization for spins 1 and 2:

$Iz_1 + Iz_2$

2. Application of a $\pi/2$ pulse along the x direction in the rotating frame produces $-y$ magnetization for each spin:

$$Iz_1 + Iz_2 \xrightarrow{\left(\frac{\pi}{2}\right)_x} -Iy_1 - Iy_2$$

3. Evolution then proceeds only under the J-coupling Hamiltonian, where with an evolution angle from the "circle diagrams" given by πJt:

$$-Iy_1 - Iy_2 \xrightarrow{\pi Jt}$$

$$|-Iy_1 \cos \pi Jt + 2Ix_1Iz_2 \sin \pi Jt| + |-Iy_2 \cos \pi Jt + 2Ix_2Iz_1 \sin \pi Jt|$$

4. After a delay time of $t = \tau = 1/(2J)$ the sine terms equal unity and the cosine terms vanish, so that:

$$|-Iy_1 \cos \pi Jt + 2Ix_1Iz_2 \sin \pi Jt| + |-Iy_2 \cos \pi Jt + 2Ix_2Iz_1 \sin \pi Jt|$$

$$\xrightarrow{t=1/(2J)} 2Ix_1Iz_2 + 2Ix_2Iz_1$$

5. Application of a $\pi/2$ pulse along the $-y$ axis yields:

$$2Ix_1Iz_2 + 2Ix_2Iz_1 \xrightarrow{\left(\frac{\pi}{2}\right)_{-y}} -2Iz_1Ix_2 - 2Iz_2Ix_1$$

6. We then acquire the signal over the course of time, and further evolution under the J-coupling Hamiltonian gives:

$$-2Iz_1Ix_2 - 2Iz_2Ix_1 \xrightarrow{\pi Jt}$$

$$-Iz_1Ix_2 \cos \pi Jt + Iy_2 \sin \pi Jt$$

$$-2Iz_2Ix_1 \cos \pi Jt + Iy_1 \sin \pi Jt$$

7. Substitution of $t = \tau = 1/(2J)$ yields,

$$Iy_2 + Iy_1$$

demonstrating the re-emergence of observable magnetization, and (by keeping spin terms in the same order) the fact that the observable magnetization on each spin arose from transfer from the other spin via the J coupling.

Figure 4. Annotated solution to the coupled spin problem.

Analytical Chemistry Laboratory

A recent poll indicates that the percentage of instructional time spent on electronics in instrumental analysis courses has increased by 10% over the past ten years [20]. This trend reflects the increase in sophistication of the analytical tools at our disposal, which requires a higher level of understanding to manipulate and troubleshoot instrumentation. In order to increase student interest in electronics and promote an appreciation for the practical applications of electronics in chemistry, electronics-based research projects have been introduced into the analytical courses (e.g., Chemical Spectroscopy and Separation courses). The projects require the students to build, from scratch, calibrated and quantitative instruments. One such project has been the development of a simple, low-cost multinuclear probe that provides senior-level undergraduate students a hands-on opportunity to assemble a solid-state static NMR probe and perform experiments with the probe they have built. The probe can be tuned to any NMR-active nucleus by adjusting a pair of capacitors (tune and match) and/or changing a "snap-in" inductance coil (Figure 6). The probe is a basic single-resonance static probe consisting of the probe enclosure and supports, semi-rigid coaxial cable, and a simple tank (LC) circuit. Since a narrow-bore (54 mm) NMR spectrometer is used, the probe has a 47-mm inner diameter with an aluminum sheath that is 38.7 cm long. The aluminum tube is designed such that the top portion of the tube may be removed to change samples and/or coils. The tube is attached to an aluminum base that can be mounted to the NMR shim stack with brass screws. The support shelf and capacitor shelf are made from copper-clad PC board; these shelves have holes drilled through them for the support rods, tuning rods and coaxial cables.

The coil was constructed out of 18-gauge silver wire with an inner diameter slightly larger than a 5-mm NMR sample tube. The coil was easily tuned to the deuterium resonance in a 9.4 T magnetic field ($v_L = 61.4$ MHz) using an Agilent network analyzer, although a simpler amateur radio SWR analyzer (such as the MFJ-239B available from MFJ Enterprises, Inc.) could be purchased and used successfully for this step. The probe has been designed, at this time, to be compatible with a Bruker Avance 400 MHz spectrometer.

The "snap-in" inductor coil enables the probe to be tuned to v_L in the range 39-102 MHz. To-date students have collected ^{35}Cl spectra of NaCl and CHCl₃, ^{2}H spectra of D₂O and toluene-d₈, ^{127}I spectra of NaI and HI, and ^{45}Sc spectra of ScCl₃. Figure 7 shows the ^{13}C spectra of ^{13}C-enriched CH₃I (A) and unenriched CH₃I (B). In addition to the proton-coupled splitting, additional fine-splitting due to ^{127}I is observed in the spectra of the isotopically-enriched sample (Figure 7A). Probe schematics and additional examples of student work in this area are also available at the previously listed web site.

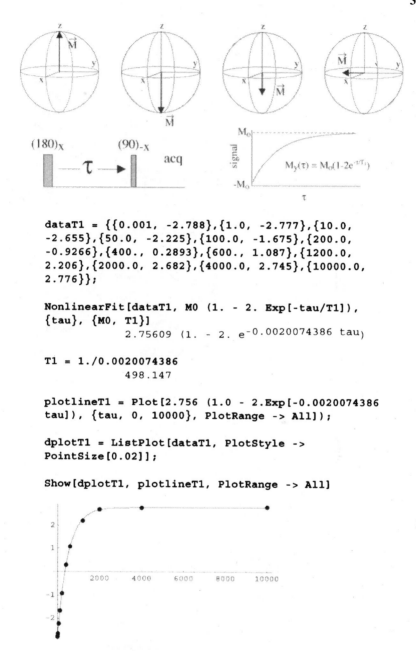

```
dataT1 = {{0.001, -2.788},{1.0, -2.777},{10.0,
-2.655},{50.0, -2.225},{100.0, -1.675},{200.0,
-0.9266},{400., 0.2893},{600., 1.087},{1200.0,
2.206},{2000.0, 2.682},{4000.0, 2.745},{10000.0,
2.776}};
```

```
NonlinearFit[dataT1, M0 (1. - 2. Exp[-tau/T1]),
{tau}, {M0, T1}]
```
$$2.75609 \ (1. \ - \ 2. \ e^{-0.0020074386 \ tau})$$

```
T1 = 1./0.0020074386
```
$$498.147$$

```
plotlineT1 = Plot[2.756 (1.0 - 2.Exp[-0.0020074386
tau]), {tau, 0, 10000}, PlotRange -> All]);
```

```
dplotT1 = ListPlot[dataT1, PlotStyle ->
PointSize[0.02]];
```

```
Show[dplotT1, plotlineT1, PlotRange -> All]
```

Figure 5. Top: Vector and schematic representation of a T_1 inversion recovery pulse sequence. Bottom: Mathematica input for T_1 relaxation data analysis.

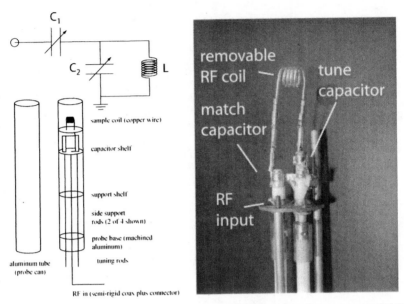

Figure 6. Left: Schematic diagram of a single channel multi-resonance NMR probe. Right: Photograph of a probe built by undergraduate students.

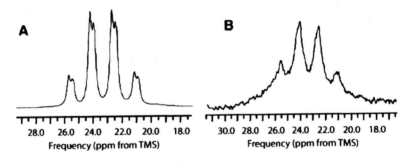

Figure 7. A: isotopically-enriched ^{13}C spectrum of CH_3I. B: non-isotopically-enriched CH_3I (spectra acquired by Rebekah McLaughlin, and Keith Reber).

Assessments

In order to relate depth of understanding of NMR to changes in problem-solving abilities, we have developed a suite of assessments to measure both constructs. These will be tested using both online and laboratory testing venues. Assessing depth of understanding for students using our materials is the first step. Traditional testing of content knowledge will be done through coursework in the form of written exams and assignments. In addition, students will be tested on their abilities to run experiments on the spectrometer and to interpret many types of NMR data. Results on these tests may vary based on achievement level, major, and career plans, and so this information will also be collected and compared. Problem-solving skills must be tested with sufficiently complex items in order to be valid to this study. Students will be asked to propose experiments that could be used to solve chemistry-related problems and detail expected results. Translating NMR data into meaningful results (related to experimental hypotheses) is key to solving problems and this will also be evaluated. Additionally, a standardized assignment will be designed to measure retention over time, while a transfer task will be used to measure knowledge transfer from course to course.

Learner characteristics can be an important intermediary between understanding and problem-solving. Increased depth of understanding can lead to changes in student characteristics such as self-efficacy, task confidence, and attitudes (towards the technique, course, chemistry, and science). These, in turn, can effect motivation and goals toward problem solving. Items on task confidence and attitude have been developed by modifying corresponding sections of the Chemistry Attitudes and Experiences Questionnaire (CAEQ) [21]. Students' self-efficacy for specific NMR tasks will be measured based on recommendations for self-efficacy measures found in the literature [22]. Attitudes toward the course (effectiveness, cohesiveness, etc.) will also be evaluated, since course-specific factors (e.g. lecturers, teaching assistants, etc.) could vary and influence responses.

Conclusions

We have presented the beginning steps of our program aiming to increase the curricular connections between chemistry laboratory courses on the organic, analytical, and physical areas. Lesson materials and laboratory exercises have been generated, and will be assessed in the future for correlation to student learning characteristics and problem solving abilities using NMR. One area of concern is the transference of knowledge from previous courses, predominantly those in the physics and mathematics fields, and some materials are being targeted to bridge the gap in language and understanding between these subjects and physical chemistry. A popular outcome from this work has been the

34

development of a low-cost, static NMR probe that can be built by students in an analytical or physical chemistry laboratory class. This exercise promotes hands-on understanding of electronics, probe engineering, and the interfacing of new equipment with an existing NMR spectrometer.

This project is novel in its cross-curricular approach and should be accessible to many undergraduate universities. Although in the past only large research universities had access to NMR facilities, the number of smaller universities that own or have access to these instruments has been on the rise over the past 10 years. A search of NSF grants awarded to projects related to NMR gives 345 hits [23] and the majority of these funds were used to purchase or upgrade NMR spectrometers for the university. Packaging and distribution of materials to other universities is integral to this project and course notes and other materials are available online, as noted above, through Penn State websites. In addition to course materials, student built probes are being used as prototypes for kits that we intend to market, making it easier for universities across the country to try out this element of our program. Though there are many reports of NMR experiments that have been added to laboratory courses, these rarely contain any evaluation or assessment. The assessment component of this project is an important feature that could attract additional advocates, and these results will be disseminated in the future.

Acknowledgements

This work was performed with funding from the National Science Foundation Course, Curriculum, and Laboratory Improvement program, through grant DUE-0341487. The 400 MHz NMR spectrometer within the Department of Chemistry undergraduate laboratories was purchased in part with funds from the National Science Foundation, grant DUE-9981068. The authors also thank Dr. Jackie Bortiatynski, Dr. Alan Benesi, and Dr. Katie Masters, Rebekah McLaughlin, and Keith Reber for their input and work on this project.

Literature Cited

1. Ernst, R. R.; Bodenhausen, G.; Wokaun, A. *Principles of Nuclear Magnetic Resonance in One and Two Dimensions*; International Series of Monographs on Chemistry; Clarendon Press; Oxford, 1990.
2. Ball, D. B.; Miller, R. M. *J. Chem. Ed.* **2004**, *81*, 121-125.
3. Bosch, E. *J. Chem. Ed.* **2000**, *77*, 890-892.
4. Parmentier, L. E.; Lisensky, G. C.; Spencer, B. *J. Chem. Ed.* **1998**, *75*, 470-471.
5. Brittingham, K. A.; Schreiner, S.; Gallaher, T. N. *J. Chem. Ed.* **1995**, *72*, 941-944.

6. Peterman, K. E.; Lentz, K.; Duncan, J. *J. Chem. Ed.* **1998**, *75*, 1283-1284.
7. Atkinson, D.; Chechik, V. *J. Chem. Ed.* **2004**, *81*, 1030-1033.
8. Mosher, M. D.; Ojha, S. *J. Chem. Ed.* **1998**, *75*, 888-890.
9. Gasparro, F. P.; Kolodny, N. H. *J. Chem. Ed.* **1977**, *54*, 258-261.
10. Morris, K. F.; Erickson, L. E. *J. Chem. Ed.* **1996**, *73*, 471-473.
11. Brown, K. C.; Tyson, R. L.; Weil, J. A. *J. Chem. Ed.* **1998**, *75*, 1632-1635.
12. Jameson, D. L.; Anand, R. *J. Chem. Ed.* **2000**, *77*, 88-89.
13. Anderson, S. E.; Saiki, D.; Eckert, H.; Meise-Gresch, K. *J. Chem. Ed.* **2004**, *81*, 1034-1036.
14. Quist, P. O. *J. Chem. Ed.* **1996**, *73*, 751-752.
15. Lorigan, G. A.; Minto, R. E.; Zhang, W. *J. Chem. Ed.* **2001**, *78*, 956.
16. Jarek, R. L.; Flesher, R. J.; Shin, S. K. *J. Chem. Ed.* **2002**, *79*, 306-307.
17. Ball, D. B.; Miller, R. M. *J. Chem. Ed.* **2002**, *79*, 665-666.
18. Vaughn, J. B. *J. Chem. Ed.* **2002**, *79*, 306-307.
19. Guntert, P.; Schaefer, N.; Otting, G.; Wutrich, K. *J. Magn. Reson., Ser. A* **1993**, *101*, 103-105.
20. Girard, J. E.; Diamant, C. T. *J. Chem. Ed.* **2000**, *77*, 646-648.
21. Dalgety, J.; Coll, R.; Jones, A. *J. Res. Sci. Teach.* **2003**, *40*, 649-668.
22. Pajares, F. *Rev. Ed. Research* **1996**, *66*, 543-479.
23. *Project Information Resource System (PIRS), NSF Division of Undergraduate Education (keyword: NMR, discipline: chemistry)*, URL https://www.ehr.nsf.gov/pirs_prs_web/search/.

Chapter 4

Elective and Capstone Undergraduate Experiences in NMR Spectroscopy

A Curriculum That Prepares Students for Independent Research in Magnetic Resonance

Daniel J. O'Leary and Wayne E. Steinmetz

Department of Chemistry, Pomona College, 645 North College Avenue, Claremont, CA 91711

This chapter describes two undergraduate courses that provide students with hands-on experience with routine and advanced experiments in nuclear magnetic resonance.

Introduction

Pomona College is a member of The Claremont Colleges, a consortium which includes five undergraduate colleges (Pomona, Pitzer, Claremont McKenna, Scripps, and Harvey Mudd) and three chemistry departments (Pomona, Harvey Mudd, and the Joint Science Department). The chemistry curriculum at Pomona College is taught in a traditional manner, with students taking two semesters of general chemistry in the first year, followed by a year of organic chemistry. Required coursework includes a year of physical chemistry, a year-long capstone laboratory experience, three semesters of calculus, a year of physics, and a senior thesis project. Students are also required to take three elective courses, which can be taken at Pomona, the Joint Science Department, or Harvey Mudd College. NMR spectroscopy is introduced to all students during their second-year organic chemistry course. This chapter describes the 'NMR curriculum' for Pomona students who have moved beyond these introductory experiences. For these students, NMR can be encountered via an elective course or within the context of a required capstone laboratory experience. Accordingly, the chapter is organized into two sections describing each element of the curriculum.

Chemistry 172: An Elective NMR Spectroscopy Course

Within the consortium, elective courses are taught either as traditional 'full courses' that meet three hours a week (not including laboratory time) or as 'half-courses', which meet once a week for ninety minutes. Since 1996, the NMR spectroscopy half-course has been taught at Pomona College every other year. This course is taught from the perspective of an organic chemist. During alternate years, Professor Mary Hatcher-Skeers teaches an NMR half-course at the Joint Science Department; her course emphasizes the mathematical and physical aspects of NMR. The only prerequisite for the Pomona course is first-semester organic chemistry. Originally designed for juniors and seniors, several years ago the course was opened to second-semester sophomores to encourage them to consider majoring in chemistry or molecular biology by providing early access to upper-division courses. The course enrollment has been reasonably strong, with an average enrollment of 13 students per year over six iterations of the course. Detailed information may be found on the course Web page (*1*).

The Pomona course aims to provide students with a grounding in liquid-state NMR, with an emphasis on modern ^{1}H and ^{13}C 1D and 2D methods. The goal is to get students excited about the power of NMR within the realm of probing molecular structure. This is accomplished, within the constraints of a half-course, by: (i) in-class discussion of selected readings from an accessible text, (ii) designing non-repetitive problem sets that test students' understanding of theory and spectral interpretation by their analysis of "real" digital data, (iii) employing a challenging 'capstone' spectral assignment problem, and (iv) using a student-designed hands-on research project.

In a typical semester, ten class meetings are reserved for discussions from the assigned text, which for the life of the course has been Friebolin's *Basic One- and Two-Dimensional NMR Spectroscopy* (*2*). Weekly problem sets are written to test students on their understanding of spectral interpretation and aspects of NMR theory. To engender a more lively in-class interaction, the problem sets are due two days after the associated topics are discussed in class. Wherever possible, students are supplied with digital data which they process with the shareware NMR processing package Mestre-C (*3*). A course schedule is shown in Table 1.

In the first class meeting, time is taken to review NMR concepts covered in the first-semester organic chemistry course: the physical basis of NMR, origin of diamagnetic shielding, spin-spin coupling and the n+1 rule, topological descriptors, basic ^{1}H and ^{13}C spectroscopy, and interpretation of 2D homonuclear and heteronuclear data. A more detailed discussion of the physical basis of NMR and the pulsed FT method is covered during the second meeting: the concept of spin angular momentum, the nuclear magnetic moment, spin-1/2 energy level quantization, and the basic pulsed FT experiment.

The third class is used to discuss ^1H and ^{13}C chemical shifts. Terms such as *deshielded, low field*, and *high frequency* are defined. Shielding anisotropy and other contributions to the chemical shift are presented. Topological descriptors are revisited and the origin of anisochrony in diastereotopic systems is discussed. One of the problem set questions for this section reads: *"The data in folder dj63 were recorded for a compound that participates in the Krebs Cycle. Identify the compound and assign the peaks in the proton and carbon spectrum."* The proton spectrum, recorded in DMSO-d_6, consists of an AB quartet and two broad low field resonances. Only one compound (citric acid) is consistent with the spectrum, but students do not find the question particularly easy, especially this early in the course.

Table 1. Course Schedule for Chemistry 172 "NMR Spectroscopy" (2004)

Topic/Activity	*Reading/Assignment*
Introduction: Review of NMR from Organic Chemistry	
Fundamentals	Chapter 1
	Problem Set 1 due
The Chemical Shift	Chapter 2
	Problem Set 2 due
The Coupling Constant & Spin System Classification	Chapters 3, 4
	Problem Set 3 due
Visit of Prof. Kurt Wüthrich	**Lecture write-up due**
Spectral Assignment: 1D Spectra	Chapters 5, 6
	Problem Set 4 due
Spectral Assignment: Multiple Pulse 1D Methods	Chapter 8
	Problem Set 5 due
	Project Proposal due
Spectral Assignment: Multiple Pulse 2D Methods	Chapter 9
Spectral Assignment: Multiple Pulse 2D Methods	Chapter 9
NMR Relaxation and the Nuclear Overhauser Effect	Chapters 7, 10
	Problem Set 6 due
Journal Article Presentations (two weeks)	
In-Class Examination	
Project Presentations (two weeks)	

NOTE: Bold text is used to identify assignment deadlines and in-class activities.

A presentation of coupling constants in the fourth class focuses on the utility of scalar couplings in structural studies: the estimation of C-H orbital hybridization via $^1J_{CH}$ and the Karplus ^3J relationship for dihedral angle determination. The importance of orbitals in relaying spin information is emphasized. In the second part of this class, the "alphabetical" method of spin system assignment is discussed and the AB quartet is presented in some detail. The spectral simulation subroutine within Mestre-C is introduced and reinforced by the problem set. For example, a question on Problem Set 3 asks students to simulate the complex pattern observed for the diastereotopic C-3 protons in *meso*-2,4-pentanediol (Figure 1). Exercises such as these complement the more typical NMR structural elucidation problems that invariably begin with some hint of a structural formula.

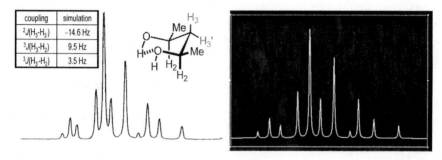

Figure 1. Student simulation (right) of the experimental ^1H NMR spectrum (left) for the diastereotopic C-3 protons in meso-2,4-pentanediol.

In the 2004 course, we were fortunate to have NMR pioneer Professor Kurt Wüthrich provide a series of departmental lectures which were woven into Chemistry 172. Students were asked to write a summary of two of these lectures. In prior years, this class period was used for a discussion of dynamic NMR or NMR imaging (scheduled later in the semester).

Four classes are dedicated to spectral assignment. Topics include homonuclear decoupling experiments and broad band proton decoupling in ^{13}C spectroscopy. The use of empirical correlations for predicting ^1H and ^{13}C chemical shifts is presented. Multiple-pulse 1D methods are discussed in the context of ^{13}C experiments such as the J-modulated spin-echo and INEPT experiments. Although these experiments are not as widely used as the DEPT variant, they are still a useful vehicle for familiarizing students with the vector model, J-modulated dephasing, and the phase behavior of radiofrequency pulses. A significant amount of class time is spent surveying the plethora of 2D methods and their use in structure elucidation. Effort is made to take the discussion of 2D NMR beyond that of simple pattern recognition by providing a qualititative description of how diagonal and off-diagonal peaks are manifest in 2D spectra. The genesis and benefits of indirect detection are also discussed. Density matrices and product operators are avoided.

A final class section is dedicated to the topic of NMR relaxation and the nuclear Overhauser effect. Relaxation mechanisms are discussed and their nontrivial manifestation in spectra is described (for example, why are amide N-H proton resonances so broad?). A significant amount of class time is used to develop a qualitative appreciation for the 1D NOE experiment, a discussion that includes the saturating effect of selective irradiation, redistribution of energy level populations via relaxation, and spectral subtraction.

A 'capstone' problem set is then used to integrate and apply the lecture topics on a molecule with complex spectral features. Morphine is one such example (Figure 2). Students are supplied with 1D ^{1}H and ^{13}C data, as well as 2D COSY, TOCSY, ROESY, HMQC, and HMBC data. They are encouraged to work in groups outside of class, and they emerge from this exercise with newfound confidence in their ability to interpret NMR data.

A student-designed research project is used to provide students with hands-on spectrometer experience. Depending upon the course enrollment, students work alone or in pairs. Midway through the term, students are asked to develop a brief project proposal. One or two meetings with the instructor are usually required to define a practical, challenging, and affordable molecule for study. Several examples are shown in Figure 2. Taxol and reserpine are examples of biologically active molecules with challenging NMR spectral assignments. The class project has also been integrated with faculty research interests at Pomona and beyond.

For example, one student broadened their senior thesis studies of the conformational behavior of erythomycin analogs by designing a project incorporating selective-excitation Overhauser measurements. Another student expressed an interest in learning about the interface of mathematics and NMR and wrote a software program to determine exchange rates in a multi-site hydroxyl-containing system. In collaboration with Professor Robert H. Grubbs of the California Institute of Technology, the same student later used this program in a study of dynamics in a ruthenium-olefin adduct (4). With careful planning, student-designed projects are sometimes matched with an external research group during the semester. In this way, students with an interest in functional peptides were able to study the conformational behavior of a peptide catalyst from the group of Professor Scott Miller at Yale University. Another group, interested in topologically interesting organic molecules, was able to secure a catenane sample from Professor Fraser Stoddart at the University of California, Los Angeles.

One memorable project involved a student pair who discovered a discrepancy in the 1972 literature assignments for the geminal methyl groups in the antibiotic ampicillin (Figure 2). Using a slate of modern NMR experiments,

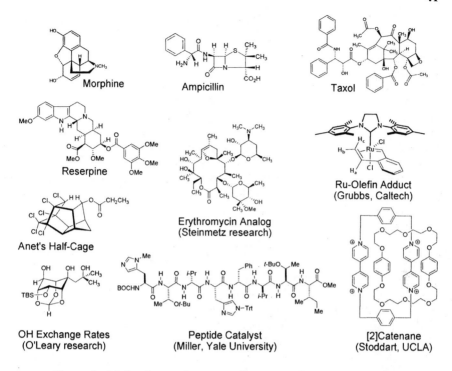

Morphine

Ampicillin

Taxol

Reserpine

Erythromycin Analog
(Steinmetz research)

Ru-Olefin Adduct
(Grubbs, Caltech)

Anet's Half-Cage

OH Exchange Rates
(O'Leary research)

Peptide Catalyst
(Miller, Yale University)

[2]Catenane
(Stoddart, UCLA)

Figure 2. Molecules used in research projects for the elective NMR spectroscopy course.

they were able to provide compelling evidence for a reassignment and subsequently published their findings (5).

Students interested in the nuclear Overhauser effect can elect to study an isodrin derivative used in Professor Frank Anet's 1965 landmark demonstration of the nuclear Overhauser effect in organic molecules. This compound (Figure 2) is readily prepared and is useful for demonstrating a large (*ca.* 46%) Overhauser effect (6,7). As will be described in the second part of this chapter, we occasionally use this molecule for a double resonance experiment in the capstone laboratory experience.

In addition to the project proposal and hands-on experience with data acquisition and analysis, each research group must also conduct a literature search and present a relevant paper to the class (Table 1). The timing of these presentations tends to coincide with when the groups are hard at work on the spectrometer. This overlap gets students excited about the paper they are presenting, as they usually feel "connected" to the paper. To complete their projects, students provide the class with a one-page abstract and a 15 minute presentation of their experimental findings.

In conclusion, the half-course designation of this course does present certain challenges with respect to breadth and depth. The limited number of class meetings requires special attention to the selection of topics and construction of problem sets. These activities are designed to provide students with a suitable background for taking NMR into their own hands by undertaking a mini-research project of their own design. Some of these projects are time-intensive but are easily accommodated with late-afternoon and overnight sequential runs. Because these projects are self-designed, students are highly motivated in their studies. Some students have been able to add new information to the chemical literature or to a research group's efforts. When undergraduates realize that they or their classmates have gained this ability, then their education suddenly has context, relevance, and a sense of excitement.

Chemistry 160: A Capstone Laboratory Course

Chemistry 160, Physical Measurements and Analysis, is a capstone laboratory course required of all chemistry majors at Pomona. The course focuses on experimental methods employed in the current practice of chemistry and the use of statistics in the design of experiments and the analysis of data. The multi-pronged research project nature of each experiment points out that real problems require a range of methods for their solution. Our approach allows us to survey a considerable expanse of the experimental landscape with only 12 experiments. Students work with individual faculty and have hands-on experience with research-level instrumentation. Many of the experiments are open ended so the course serves as a natural bridge to graduate work in chemistry or a position in industry. The course is divided into two half-courses; each has 6 experiments and a statistics module. The students write a paper in the form of an article in a refereed journal for four of the 12 experiments. Presently, an examination of linear regression and the experiments devoted to experimental physical chemistry are grouped into Chemistry 160b, Experimental Physical Chemistry. The current experiments include powder X-ray diffraction, the adsorption of gases on a zeolite, the electronic spectrum of gaseous iodine, stopped-flow kinetics, and NMR spectroscopy. The syllabus for the course and the experimental protocols can be found on the course Web page (*8*). The assignment of 2 out of the 6 experiments to NMR spectroscopy indicates the importance that the Pomona faculty attaches to the field.

The first NMR experiment is a determination of the barrier to internal rotation in *N,N*-dimethylacetamide from analyzing the temperature dependence of its 400 MHz proton NMR line shape (*9*). The experiment addresses a number of issues: introduction to the operation of a FT spectrometer, NMR as a novel probe of chemical dynamics, non-linear regression, and an Eyring analysis of the lifetimes obtained as a function of temperature. By the end of the experiment, the student is able to perform elementary tasks on the spectrometer such as shimming after a change in temperature. Normally the students measure spectra in the range 65-100 °C at 5 °C intervals and complete the experiment in one afternoon.

The student extracts rate constants for the internal rotation by fitting the line shape at each temperature to a set of parameters including the unimolecular rate constant for internal rotation. We employ the full, non-linear model originally derived by Gutowsky and Holm (10). The fit requires the application of non-linear regression and the experiment is the principal vehicle in Chemistry 160b for illustrating this topic. Students presently use the regression module of NCSS, a robust and comprehensive commercial package that runs on a PC (11).

Our graduates report that in time they forget the content and specific techniques covered in their courses but retain the approach to problem solving. In this sense, the non-linear regression is the most useful component of the experiment. The data analysis conveys several lessons. Non-linear regression is the preferred approach when the original relationship is non-linear and its application is straightforward with the assistance of modern software. The data analysis requires prior estimates of the answer. This is an important aspect of many problems in the physical sciences, including chemical equilibrium. The student ends up with both a value for the parameter and a measure of its uncertainty.

The student uses the activated complex (ACT) model, $k = \exp(-\Delta G^{\ddagger}/RT)$, to analyze the temperature dependence of the rate constants and obtains values of the enthalpy and entropy of activation. This is the only experiment in the curriculum in which the student performs a full Eyring analysis. In this case, a simple molecular orbital argument argues for the existence of a barrier. Examination of the translational, rotational, and vibration contributions to entropy leads to the conclusion that the entropy of activation should be small and slightly negative. This result is discussed with the student before the final analysis so she/he is not totally surprised by the large relative error in the entropy of activation.

The second NMR experiment addresses the theme of multi-pulse experiments and relaxation in spectroscopy. We point out that relaxation is a phenomenon that characterizes all types of spectroscopy but is particularly accessible to measurement in NMR where measurements on the second to millisecond time scale yields measures of dynamics on the nanosecond time scale. In one variant of the experiment, the student is provided with a 1 mol % solution of H_2O in DMSO-d_6 in a degassed and sealed NMR tube and measures T_1 and T_2 at one temperature. The system was carefully selected so that relaxation is dominated by the dipole-dipole mechanism. A tutorial on the instrument includes a review of the techniques learned in the previous experiment as well as the salient features of the new exercise. The student measures in order the 90° pulse time, T_1 via the method of inversion recovery and composite 180° pulse, and finally T_2 with a Carr-Purcell spin echo pulse sequence. We do not provide the delay times and the student determines a useful set of values in a set of survey runs. We point out that this is the case with all experiments in kinetics. The students extract from the data values of T_1, T_2, and τ_c, the rotational correlation time. The reading assigned for the experiment is the lucid discussion in Carrington and McLachlan, a text which is regrettably out of print (12). The calculation of τ_c requires a value for the proton-distance in water. To determine this distance, the student uses Spartan (13) to compute a structure with a Hartree-Fock Hamiltonian and a 3-21G* basis set.

The experiment has a second part in which the student measures the viscosity η of a 1% solution of H_2O in DMSO with an Ostwald viscometer. This result and estimate of V_m, the molecular volume of water derived from Spartan, yield an estimate of τ_c. The calculation is based on the celebrated relationship between V_m and τ_c originally derived by Peter Debye, $\tau_c = \eta V_m/k_B T$. The agreement of the values derived from NMR and viscosity data is quite good. We cover several bases with this alternate measurement. We incorporate the measurement of viscosity and density into the curriculum in a meaningful context. We provide an additional example of the benefits derived from the marriage of experimental chemistry and molecular modeling. This enterprise has become a prominent feature of Quantitative Structure-Activity Relationships (QSAR) research in the pharmaceutical industry (14).

An alternative experiment uses Anet's half-cage isodrin derivative (Figure 2) as a vehicle for studying selective decoupling experiments and the nuclear Overhauser effect (NOE) (15). A prominent feature of this compound is the ultra-short nonbonded distance between the transannular hydrogen nuclei. An examination of selective homonuclear decoupling serves to introduce selective irradiation. A second experiment then measures the transannular NOE via a traditional 1D difference experiment. The setting of irradiation and relaxation delays for the measurement requires knowledge of T_1, which the student measures first using an inversion-recovery pulse sequence. The student also learns the freeze-pump-thaw technique using a J-Young NMR tube.

The student thus learns in this experiment that a proper measurement of the NOE requires facility with a range of techniques. Vacuum-line technique, introduced in the adsorption experiment, is revisited. We do not provide the student with a completely preconfigured method so she/he is introduced to pulse programs and their associated parameters.

In its most recent grant cycle oriented to primarily undergraduate institutions, the Howard Hughes Medical Institute (HHMI) invited proposals that would address the importance of physics and mathematics in the training of biology students. Many of the students majoring in chemistry and biological chemistry at Pomona pursue a career in the medical sciences after graduation. We identified magnetic resonance imaging (MRI) as a natural vehicle for covering topics in chemical physics while keeping their interests. FT

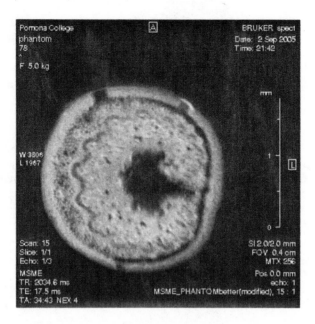

Figure 3. MRI image of the stem from Tupidanthus calyptratus acquired on a 400 MHz Bruker spectrometer with an Avance console and a 5 mm proton probe.

spectrometers are routinely equipped with hardware for producing gradients because of their utility in selecting coherence pathways. Gradients are also the essential feature of an NMR spectrometer that is configured as an imager. With financial assistance from HHMI, we purchased a 5 mm triple-axis gradient probe and a three-channel gradient amplifier and licenses for ParaVision, Bruker's professional imaging software.

In principle, our project to develop an MRI experiment is straightforward. However, the details of the execution are non-trivial. Those who wish to attempt imaging are encouraged to consult the authors and avoid some of the pitfalls of the installation. First, conventional probes in typical NMR spectrometers admit 5 mm NMR tubes which have an inner diameter of 4.2 mm. Our imaging experiments are limited to biological specimens such as plant stems and insect larvae that fit in the tube. Furthermore, we showed through mapping measurements that the active region of the probe covers ca. 10 mm in the axial direction. At first glance, this might appear to be a severe disadvantage. However, the small dimensions of the bore, 52 mm, translate to high homogeneity and therefore potentially high spatial resolution. In other words, a conventional NMR spectrometer modified for imaging functions as a MRI microscope. In favorable cases, e.g. plant materials where the specimen does not move, we have achieved a resolution of 0.01 mm (Figure 3). The construction of the NMR probe poses additional constraints. In hospital units, the gradient coils are water cooled and are driven by power supplies that can deliver up to 200 A! The gradient amplifier that comes with a spectrometer normally delivers up to 10 A and the coils are air cooled. Therefore, one has to pay careful attention to the duty cycle and power levels in order to preserve the integrity of the probe. These features require a mandatory relaxation delays of *ca.* 2 s; this translates to longer imaging times, e.g. 30 minutes instead of a few minutes.

The new MRI experiment will become the second NMR experiment and replace the experiments devoted to relaxation. However, relaxation is not ignored as the variation of T_2 in various organelles is an important tool for achieving contrast in MRI. Furthermore, a Carr-Purcell-Meiboom-Gill pulse sequence modified with gradients is the most frequently used imaging method. This method is known in the MRI community as MSME, multi-slice-multi-echo (16). In the new experiment, the student selects a plant specimen and employs the MSME method to generate a two-dimensional image of an axial slice. He/she then generates a second series of images where the echo time is systematically varied. The variation of the signal at selected locations or voxels with the echo time yields a value of T_2 for the water population in the selected voxel.

The new experiment demonstrates something known to all chemists. Many important tools in biology were developed by chemists and based on chemical principles. Embedded in the experiment is exposure to multi-dimensional NMR, relaxation theory, selective excitation, gradients, and multi-pulse experiments.

Conclusion

We anticipate a long-term emphasis on NMR in the chemistry curriculum at Pomona College. Our success results from a pool of excellent students, faculty trained in NMR, and an instrument suitable for a broad range of experiments, a Bruker 400 MHz spectrometer with an Avance DPX console. We have found that the introduction to NMR spectroscopy provided in Chemistry 172 and 160b has prepared students for post-baccalaureate activities such as graduate work in chemistry or research associate positions in industry. These courses have also significantly impacted a critical element of our curriculum, the senior thesis. Many students majoring in Chemistry and Molecular Biology have selected a senior thesis project making significant use of NMR. Most of these students have taken Chemistry 172 and 160b and are thus prepared to make meaningful progress at the start of their independent research projects.

References

1. www.chemistry.pomona.edu/Chemistry/172/172_home.html.
2. Friebolin, H. *Basic One- and Two-Dimensional NMR Spectroscopy*, 4th ed. Wiley-VCH: New York, NY, 2004.
3. www.mestrec.com.
4. Anderson, D. R.; Hickstein, D. D.; O'Leary, D. J.; Grubbs, R. H. *J. Am. Chem. Soc.* **2006**, *128*, 8386-8387.
5. Tung, J. C.; Gonzales, A.; Sadowsky, J.; O'Leary, D. J. *Magnetic Resonance in Chemistry*, **2000**, *38*, 126-128.
6. Chen, A. E.; O'Leary, D. J.; Miura, S. S.; Anet, F. A. L. *Concepts in Magn. Reson.*, **2000**, *12*, 1-5.
7. Samples of the half-cage isodrin derivative are available upon request from Dan O'Leary (doleary@pomona.edu).
8. pages.pomona.edu/~wsteinmetz/Chem160/index.htm.
9. Conti, F.; von Phillipsborn, W. *Helv. Chim. Acta* **1967**, *50*, 603-607. Ross, B. D.; True, N. S.; Matson, G. B. *J. Phys. Chem.* **1984**, *88*, 2675-2678.
10. Gutowsky, H. S.; Holm, C. H. *J. Chem. Phys.* **1956**, *25*, 1228-1234.
11. Hintze, J. L. *Number Cruncher Statistical System 2000*; NCSS: Kaysville, UT, 1998.
12. Carrington, A.; McLachlan, A. D. *Introduction to Magnetic Resonance*; Harper & Row: New York, NY, 1967.
13. www.wavefunction.com.
14. Hansch, C.; Leo, A. *Exploring QSAR*; American Chemical Society: Washington, DC, 1995.
15. Berger, S.; Braun, S. *200 and More NMR Experiments*; Wiley-VCH: Weimar, 2004.
16. Liang, Z.-P.; Lauterbur, P. C. *Principles of Magnetic Resonance Imaging*; IEEE Press: New York, NY, 2000.

Chapter 5

Teaching NMR Spectroscopy in a General Education Course for Nonmajors

Emma W. Goldman and Raymond N. Dominey

Department of Chemistry, University of Richmond, 28 Westhampton Way, Richmond, VA 23173

"Chemistry 111: Chemical Structure Determination." is a general education course for students not intending to major in the sciences. The course was developed in response to new general education requirements - specifically, to fullfill a need for one semester courses which introduce students to how scientists approach problems in their various fields, but which require minimal background and yet avoid the typical surveys of general chemistry. This course has been very popular and student satisfaction has been high. Students learn to use Anasazi modified 60MHz and 90MHz NMR spectrometers along with UV, IR, mass spectrometers, and chromatographic techniques to identify unknown organic compounds in real samples. Problem solving in both lecture and laboratory and teamwork are key components. Applications to real world chemistry are emphasized. A description of the course and topics covered will be presented, as well as an overview of the laboratory exercises, including examples of the group projects.

A number of studies about chemical education stress teaching chemistry by letting the students do chemistry. What better way to introduce students to how chemists solve problems than by letting them learn about, and do, organic structure determination. Part of the appeal of the many crime dramas currently on TV is the intrigue and intellectual satisfaction associated with any problem, or "puzzle", solving endeavor. Organic structure determination gives students the opportunity to learn some basic concepts and then apply them to solve "puzzles" -- in essence it gives students the opportunity to experience, and develop confidence in, their *own* ability to solve problems. The intellectual satisfaction achieved by such experiences is directly analogous to that which is associated with any *"eureka"*, or *"discovery"*, moment. While the currently popular Forensic Science TV Mysteries draw on this effect by providing a surrogate through which the audience can experience this via the scripted intellectual efforts of others, a course designed such as this Molecular Structure Determination (Chemistry Puzzle) Course provides an opportunity for the student to experience this via the efforts of their own reasoning and problem solving skills; thus providing a much higher level of intellectual satisfaction, and also having the added benefit of building confidence in that ability.

Even though this is a topic that generally is introduced first in chemistry majors' level organic chemistry, we have found that it is intellectually accessible for the general student population, even those without prior coursework in the sciences, and those not majoring in the sciences. By focusing on the problem (puzzle) solving aspects of organic structure determination, we have found that we can tightly focus the required background to a manageable handful of topics. In particular, we have found that students really only need a background in basic bonding and structure concepts, along with an appreciation for the fact that there is a one-to-one correlation between the language and symbols used by scientists to discuss molecules and the 3-dimensional structures of those molecules.

When the University of Richmond changed its general education requirement in the sciences to one semester of a lab-based course, the explicit goal was to allow students to learn how scientists approach problems in their respective fields. For chemistry, it seemed clear that many of the problem-solving skills intrinsic to all areas of chemistry are nicely captured by the intellectual activities associated with using spectrometric data for organic structure determination. In particular, NMR is one of the most versatile and powerful instruments for determining chemical structure. It is a technique used by chemists in a wide variety of fields from inorganic and materials, to organic, physical, and biochemistry, and even in the medical field. Thus, the spectrometric determination of organic structures seemed like an ideal topic by which to accomplish our goals. We, therefore set out to design a course that would <u>not</u> be a repeat of high school science courses, and one that would *"stand on its own"*, since we also knew from experience that we could not count on them remembering much from their high school science courses. A few other specific goals are also noteworthy and worth mentioning before we proceed with the description of the course. We wanted students to come away from the course with the ability to question what they read, and understand that there are limitations to any data that one can obtain, and thus that there is room for interpretation. And finally, through this course we wanted students to discover for themselves, and therefore genuinely recognize and appreciate, that the *sciences* indeed *do* make sense and have the potential of providing a palpable intellectual satisfaction that is capable of propelling some people into a career *doing* science.

Outline of Course

The course is divided into three basic parts (see Table I below)(1): Part I, Learning the Language of Atoms and Molecules; Part II, Learning About the Spectrometric Methods and the Approach to Solving Spectral Problems; and Part III, Applications of the Tools in the Real World.

Part I: Learning the Language of Atoms and Molecules

The first part of the course is an introduction to the basic principles and concepts concerning atoms, molecules, bonding, and intermolecular forces of attraction. This then extends into an introduction to the ideas central to molecular structure of simple organic compounds, the principle modes of bonding available to H, C, O, and N atoms, and simple nomenclature of organic compounds including that of isomers and the basic functional groups. Simple nomenclature is emphasized to follow a standard pattern (Prefixes-Base Name-Ending) with the corresponding explanation of each. The ideas associated with bonding and structure are extended further with the concept of polar molecules which provides a natural venue for introducing the idea of intermolecular forces and their importance in determining and predicting physical properties and solubilities of molecules. This first section concludes with a discussion of chromatography in a manner which builds on the principles of intermolecular forces and physical properties already discussed.

Part II: Learning About the Spectrometric Methods and the Approach to Solving Spectral Problems

Part II is the *core* of the course where students are introduced to the basic principles of spectroscopy and to four of the primary spectral methods for structure elucidation (See Table II).

Table I. Outline of General Education Course

Lecture	Laboratory (one week each, or as noted)
First Section (4 weeks)	
• Periodic table, atomic theory, ionic and covalent bonding, intermolecular forces and solubility. • Organic structures, isomers, and nomenclature. • Chromatography	• Hazmat (using physical properties including solubility, mp, and flammability to identify compounds). • Organic Structures • Chromatography (TLC of OTC drugs, and Sep-pak separation of red and blue dyes in Kool-aid™).
Second Section (6 weeks)	
• Spectroscopy and light. • Basic theory, instrumentation, and interpretation of UV, IR, NMR spectroscopies, and mass spectrometry. • How to approach combination spectral problems.	• UV-Vis (Determine Extinction Coefficient for Benzophenone, and visible spectra of Kool-aid™ dyes.) • IR (Identification of liquid and solid unknown; calculation of C-H vs C-D stretch and bending frequencies using Hook's law, and confirmation by measurement.) • GC-MS (Identification of two components in a mixture; and identification of cinnamal using mass spec) • NMR (Collect and analyze ^{1}H NMR of ethylbenzene. Collect ^{1}H NMR of, and identify, one unknown given its molecular formula).
Third Section (4 weeks)	
• Applications of chemistry, including topics such as: Polymers and recycling; environmental; using library resources; forensics; drug testing; drug development; government regulations; other instrumentation; student initiated topics.	• Unknown- 2 week lab (Each group is given a vial with an unknown. They use solubility, mp/bp, in addition to NMR, IR and mass spectral data to identify.) • Projects –3 week lab– Isolation and Identification of an organic compound(s) from a "real world" source. • Class presentations on their project.

Table II. Summary of the Four Types of Spectral Methods Used

Type of Spectroscopy	Type of Light absorbed	Range we observe	Transition	Information obtained
UV	UV	200 - 400 nm	electronic	Extent of conjugated pi bonds
IR	IR	4000 - 400 cm^{-1}	vibrational	Types of functional groups
NMR	radiowaves	Depends on magnetic field strength	Nuclear spin	C, H structural units and framework
Mass spec	NA	NA	NA	Molecular mass and mass of fragments

UV-visible spectroscopy is an ideal way to introduce the concepts of electromagnetic radiation and the general principles of spectroscopy – in particular the idea of quantum states and quantized energy, and the absorption or emission of light being due to transitions between different levels of these quantized energy states. Students are already familiar with the concept of color and most know that the color we see is a complement to the color of light absorbed. In lab, we reinforce this by having the students separate the blue and red dyes in Kool-aid using a Sep-pak (mini-reverse phase C18 liquid chromatography column)(2) and obtain a visible spectrum of each solution. They observe the λ_{Max} and correlate this with the color of the dye. In particular, they observe that the λ_{Max} corresponds to the color which is complimentary to that of the dye. The two main concepts we stress for UV-Visible spectroscopy are: 1) that the absorption of UV-visible light is due to electronic transitions between n (lone pair), π, and π^* states, and 2) that Beer's law ($A_\lambda = \varepsilon_\lambda lc$) is an equation that quantitatively relates the "amount of light absorbed" with the identity and concentration of the molecule absorbing the light, the λ of the light being absorbed, and the pathlength of the light through the sample. Discussion of the value of UV-visible data to structure determination is limited to that of how λ_{Max} increases with increasing conjugation of π bonds.

Infrared spectroscopy further reinforces the idea of absorption of light, in this case correlating it with vibrational motion of bonds that are present. We use the ball and spring model (Hooke's law) to discuss the correlation between frequency absorbed and the bond strength and atom masses involved. Additionally we emphasize that for an IR photon to be absorbed it must have a mechanism to push and pull in synchronization with the vibrational motion. Thus, since IR photons are oscillating electric fields this means that only those vibrations whose oscillatory motion generates an oscillating dipole moment are capable of absorbing IR radiation. The animations in IR Tutor(3) offer a very capable of absorbing IR radiation. The animations in IR Tutor(3) offer a very powerful way of reinforcing these two points. IR Tutor also allows us to give students a visual picture of asymmetric vs symmetric stretching and bending motions. This leads to a discussion of the correlation table for the basic functional groups and how IR is used in structure determination.

In the mass spec section we cover basic instrumentation and interpretation. The students are asked to reason out how one might measure mass. Most have heard that mass is equal to force divided by acceleration, F=ma. This leads to a discussion of the need to have the particles (i.e. molecules) be able to move freely without friction when we "push" on them with a measured force. Then observing how they move in response to a precisely applied force, one can conceptually determine their mass. Electric and magnetic fields offer a way to "push" provided the particle is charged (i.e. ionized). Normally we cover electron impact as the ionization source and either magnetic sector or time of flight as the mass analyzer. The instrument the students use in lab employs a quadropole analyzer, so using the C-MoR Mass Spec Module (4) we often also show how that works. The C-MoR Mass Spec module offers animations which make these concepts come alive for the students. In terms of interpretation, we focus our attention on several key signatures often provided by mass spectral data. The molecular ion gives the most important piece of information, namely the molecular mass of the compound (which we later emphasize provides the ability to decide if all the pieces to the molecular puzzle have been identified). In addition, the molecular ion indicates whether an odd (or even) number of nitrogens is present (nitrogen rule) and provides a fingerprint for the presence of chlorine or bromine through the isotope patterns. The main fragment ions we discuss are phenyl(77), benzyl or tolyl(91), propyl or acetyl (43), and butyl (57). Of course, the mass of fragment differences such as loss of 15 for a methyl group, or loss of 29 (ethyl group), or loss of 28 (carbonyl or sometimes ethylene) also gives important fragment pieces. We also are able to show example spectra of various types of organic compounds with the C-MoR Mass Spec module.

Building on the concepts of IR and UV spectroscopies, we then introduce NMR theory. We use the very front end of NMR Tutor (5) to introduce the concept of nuclear spin and the alpha and beta state energy differences obtained when a proton or ^{13}C nucleus is placed in a magnetic field. The level of theory is kept simple, focusing on the simple (quantum model) view of α and β energy levels whose difference in energy depends on the B_{eff} experienced by the nucleus. Then drawing on analogies with the absorption spectroscopies discussed earlier, the NMR experiment is described as an absorption measurement in which the energy of radiation absorbed correlates to the α-β energy gap. Each signal therefore must correlate with a different B_{eff} being experienced by a proton nucleus. In this way, we can then introduce the different major factors which can contribute to the magnitude of B_{eff} experienced by each proton in a molecule, i.e. position-chemical shift and spin-spin coupling.

Thus, we do not discuss FT NMR in any detail but rather use the simpler concept of a CW instrument.

To focus on the main types of information available in a proton NMR spectrum, we use the acronym NIPS (this is SPIN backwards).

N	Number of equivalent hydrogens	How many types of hydrogens
I	Integration	Number of each type
P	Position or Chemical Shift	Environment of hydrogen
S	Splitting	Number of neighboring hydrogens

Using these four components and giving practice at each step, the students quickly become familiar with basic interpretation of proton NMR spectra. The key to the students' progress in this course is daily problem sets that focus on the concept that was covered that day. Specifically, as illustrated in the Table III, there is a problem set that focuses on the following topics in sequence: identifying equivalent hydrogens, predicting chemical shift and integration, and predicting splitting patterns, and integrating all four of these pieces of information that are available from the ^1H NMR spectrum.

This basic introduction to NMR takes about two weeks. The next two weeks the students continue to work on problem sets as additional topics, such as ^{13}C NMR, long range coupling, tree diagrams, coupling constants, etc. are discussed and demonstrated in class.

As they become comfortable with the types of information provided by each of the spectral data (UV, IR, NMR, and mass spec) they are assigned daily problem sets that contain increasing numbers of "combination spectral problems". Numerous problem sets are given and students are encouraged to work together.

General Approach to Structure Determination from Spectral Data

We have found that the nonmajors in this class, not surprisingly, do not have the same level of chemical intuition as science majors in sophomore organic where these topics are typically first introduced. Therefore, they needed help in establishing a systematic approach to using the spectral data to make decisions about structure. Outlined below is a method that has helped. Basically the three parts are: 1. Find all the pieces that the ^1H NMR spectrum gives you. 2. Compare the "molecular formula" of the pieces with the given molecular formula to ensure that you have all your pieces. 3. Put the pieces together (most often you will start with the end pieces i.e., fragments with only one open valence). Before introducing this, we had found that students had misled themselves by relying too heavily on chemical shift below $\delta5$ ppm. Whileas above $\delta5$ ppm, the

Table III: Sample Questions from NMR Problem Sets

NMR Problem Set #1	Find the number of equivalent hydrogens. Examples:
NMR Problem Set #2	Draw the predicted 1H NMR spectrum and relative integration. Examples:
NMR Problem Set #3	Draw the predicted ^{1}H NMR spectrum, relative integration, and splitting. Examples: On this problem set we also give some actual spectra where they match the compound to the spectrum.
NMR Problem Set #4	Give a reasonable structure for each of the following 1H NMR spectra. Examples: (Some are given in standard format as shown here, and some are given as the actual spectrum). Examples: a. $C_{10}H_{13}Cl$ b. $C_8H_{10}O$ δ 1.57 (s, 6H) δ 1.46 (t, 3H) δ 3.07 (s, 2H) δ 3.85 (q, 2H) δ 7.27 (s, 5H) δ 7.29 (s, 5H)

chemical shifts are fairly distinctive for different functional groups (such as –COOH, –CHO, phenyl, vinyl), below δ5 ppm it is really the integration that indicates the fragment. Whether the methyl group that integrates for 3H is next to a carbonyl or a phenyl can be easily determined once all the fragments are listed and one is deciding between isomers.

This general approach lends itself very well to integrating the spectral data from all the techniques. In particular IR and MS data can also be used to help accomplish step 1 and the molecular ion mass from mass spec can be used in lieu of a molecular formula to accomplish step 2.

Part III. Applications of the Tools in the Real World

During the last four weeks of the semester we cover applications of chemistry to the real world to expose them to the very real and wide variety of ways the material they are learning impacts the business world, the national economy, the healthcare industry, an political decisions, etc.. This is done with a combination of lectures, invited speakers, and student presentations. Topics that we typically try to cover include environmental chemistry and policy, polymers and recycling, government regulations and OSHA guidelines, natural product isolation and drug discovery (*e.g.*, the history of acetylsalicyclic acid and acetaminophen) . The students also prepare 10-15 minute presentations (alone or in pairs) on a topic of interest to them. Student topics in the past have included "Love Canal" (from a student who grew up near there), cosmetics and fragrances, fireworks, warfare agents, and drug testing. Often a student will present on a prescription drug that they or someone in their family uses. They are allowed, indeed encouraged, to discuss the topic in a broad sense as long as there is a substantial portion devoted to the chemistry behind the topic.

Lab Component

The lab is designed to correlate with the lecture part of this course. In the Table 2 above, we briefly outline the experiments students are doing each week in lab as they are covering related topics in lecture. While we are discussing intermolecular forces and polarity, students do boiling points and solubility tests in lab. The "Hazmat" lab involves a "10 bottle" experiment. In this case the students are given a list of what is in the ten bottles and they have to figure out which is which based on solubility in water, 10% HCl, 10% NaOH, and pentane and using flammability as an additional test. Another laboratory experience involves using organic model kits to build compounds and isomers. The students spend the lab period drawing line structures and naming organic compounds but this immersion gives them a level of comfort with organic functional group recognition and nomenclature that is necessary for the rest of the course. The next lab introduces chromatography (2,6).

While students are learning about each type of spectrometer and spectroscopy in lecture, they are learning to use each instrument individually in lab culminating in using all the instruments to determine the identity of an unknown sample. The main task each week is for the students to learn to collect data on each of the instruments and use it to identify an unknown. However they are also exposed to some quantitative aspects of chemistry when they use Beer's law to calculate an extinction coefficient after doing serial dilutions of benzophenone (UV) and when they use Hooke's law to predict the wavenumber shift for chloroform vs deuterated chloroform.

After identifying an unknown using all of the tools they have learned throughout the semester, each lab group chooses a project. Occasionally a group will devise their own project, but most often students choose from a list.

Examples of projects and how NMR spectroscopy is used for identifying their components is given below.

Real World Lab Projects

a.   Identification of the fragrant component in Sassafras:
Students find a sassafras plant near the campus lake, grind up a root sample, and by steam distillation and extraction isolate a few drops of saffrole which is then identified.   The NMR spectrum of one sample is shown below. Usually the students need to do a library search on their IR spectrum to determine the structure and then identify how the ^{1}H NMR data, shown in Figure 1, supports this structure.

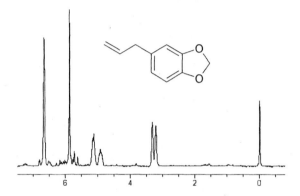

Figure 1. ^{1}H NMR spectrum of Saffrole isolated by steam distillation

b. Polymer sample:
This involved a destructive distillation (depolymerization) to generate monomers which were also identified. Typical examples were sample of plexiglass (poly methyl methacrylate) or plastic tableware (polystyrene). Students collect IR and ^{1}H NMR data on both the polymer and the monomer after thermal degradation. The ^{1}H NMR spectra of the monomers are shown in Figure 2.   A full description of this experiment has recently been published (8).

c. Drug Analysis:
Determination of the structure of one of several medicinal preparations. A particularly good sample we found is diphenylhydramine found in Dramamine. Using aqueous base to dissolve the pill, followed by extraction with diethylether provides a nice sample.   Mass spectral information can be

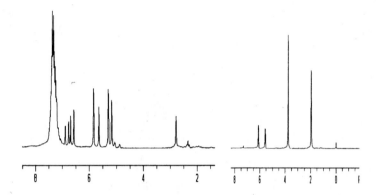

Figure 2. ^1H *NMR spectra of styrene and methyl methacrylate isolated by thermal degradation of a respective polymer sample*

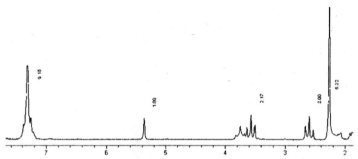

Figure 3. ^1H *NMR spectrum of diphenhydramine extracted from Dramamine tablets*

helpful in determining the final structure along with the ^{1}H NMR spectrum, shown in Figure 3.

d. Identification of the fragrant component in the spice Cumin:
This is a lab often done in organic classes (7). Steam distillation and extraction of ground cumin provides a sample of cuminaldehyde. The ^{1}H NMR data gives a clear analysis of this compound.

e. Waste material:
Several years ago, a 50 gallon drum of an unknown liquid was found on campus near our soccer stadium and track. It turned out to be polypropylene glycol, most likely used in refinishing the track. Students use IR and ^{1}H NMR spectra along with physical properties to identify this unknown.

The final week in lab, the students present their project to the class by discussing the experimental details, the spectra they obtained and how it fits with their determined structure, and finally they present additional information related to their project. For instance, if they did the polymer project, they could discuss recycling or new biopolymers. For drug analysis they might discuss the steps in developing new drugs or marketing of new drugs. The topic is up to the students and is meant for them to explore a topic of particular interest to them.

Instrumentation Needs

As a final note about the laboratory portion of this course, laboratory enrollment is limited to 16 students per section because of the need for instrument access. We currently use a Cary 50 UV-vis, a Nicolet Avatar 360 with a diamond coated HATR accessory, Anasazi FT upgraded 60 MHz ^{1}H NMR and 90MHz ^{1}H/^{13}C NMR spectrometers (both permanent magnets), and Shmidadzu gc-ms systems, all of which are dedicated for hands-on use by these students   Students generally work in groups of two - four to ease the instrument accessibility problem, and as a result practice team work in their problem solving. Working as a team seems to be perceived as an advantage as students puzzle things out together, assign each other specific tasks, and learn from each other. They tend to feed off each others' ideas and interests.

The choice of an Anazazi upgrade of our 60 MHz CW instrument originally hinged on four points: 1) Extending the useful life of the instrument. 2) Obtaining an FT based NMR for routine use which would require minimum personnel time and cost to maintain (i.e., one that could even sit for long periods unused without requiring effort or cost). 2. Obtaining an FT based NMR for routine use by introductory and nonmajor students for the benefit of sensitivity and therefore throughput (i.e., time required perspectrum). 4. To achieve all of the above at a reasonably low cost. After our experience with the Anasazo 60, we were so impressed at the increased throughput for 1D ^1H and the added capability of obtaining COSY's, that our department decided to upgrade a 90MHz permanent magnet instrument via Anasazi. This instrument also has ^{13}C capabilities and is used for our sophomore organic teaching laboratory and our nonmajors course.

Course Materials

This course is not a standard course for nonmajors found at other universities (9) and thus we have not found a suitable textbook in the market. As a result, we have developed our own "mini-text" which includes chapters on each of the topics covered, a lab manual, and an extensive set of problem sets. The text is sold to the students essentially at cost of printing, and is also available on-line to our students. The text is continuously being modified as the course evolves. Copies of the problem sets, lab manual, and mini-text are available upon request.

In addition, in the classroom we use a number of computer modules. This includes "NMR Tutor" and "IR Tutor" by Charlie Abrams, C-MoR modules on "Mass Spectrometry" and "Chromatrography" by René Kanters and the UR C-MoR team, and other commercially available programs with spectral data such as "SpectraDeck" and "SQUALOR".

The Problem Sets

The numerous problem sets, their design, and the method by which they are incorporated into the lecture, are critical to the success of the course. It is through these that the students actually learn the material. They stimulate interaction between and among the students as well as with the instructor. The students get frequent feedback which builds their confidence and adds to the intellectual satisfaction that comes from succeeding in solving the puzzles. Students are often given 10 minutes at the end of class to start on their problem set. This way they are able to still ask questions to get started and this prevents them from becoming frustrated when they work on them later.

Conclusion and Assessment

As the students gain confidence through the problem sets and their laboratory experience, their enthusiasm grows. They enjoy having a hands-on experience with instrumentation that is used by "real" chemists. The problem solving nature of interpretation is perceived as fun. By the time the students are working on the combination spectral problems, they are "hooked" and do not want to stop until they have come up with a structure that fits the spectral data. The objective is for the students to see that they are capable of using modern instrumentation and of solving real chemical problems. A few students that took this course went on to be science majors, but most of the students leave the course saying, "I can see why someone would enjoy being a chemist" and feeling it was a better experience than what they originally expected.

On the most recent set of student evaluations, students responded 77% strongly agree and 23% agree to the statement, "The course fulfilled its purpose as a general education requirement". One student wrote, "This course was a great way to fill my Gen Ed requirement. It was difficult but not impossible for non-majors. I definitely learned a great deal about Chemistry and even more about how chemists go about solving problems. I would certainly recommend this course to others who needed a Gen ed science class and who actually wanted to learn something." These sentiments were reiterated over and over by students who felt it was challenging but not impossible, they would recommend the course to their friends, and that it was different from their high school courses.

References

1. Clough, Stuart C.; Kanters, Rene P. F.; Goldman, Emma W. *J. Chem. Educ.* **2004**, *81*, 834-836.
2. Bidlingmeyer, B: Warren, Jr. F. *J. Chem. Educ.* **1984**, *61*, 716-720.
3. Abrams, Charlie "IR Tutor", Abrams Educational Software, see http://members.aol.com.charlie.abr/.
4. Kanters, Rene; Dominey, Raymond N. "C-MoR: Mass Spec Module", see http://oncampus.richmond.edu/academics/a&s/chemistry/CMoR/info/index. html.
5. Abrams, Charles "NMR Tutor", Abrams Educational Software, see http://members.aol.com.charlie.abr/.
6. Pavia, D. L.; Lampman, G. M.; Kriz, G. S.; Engel, R. G. *Organic Laboratory Techniques: Small Scale Approach,* 1st ed.; Brooks/Cole-Thomson Learning: Pacific Grove, CA, 1998; p 79.
7. Pavia, D. L.; Lampman, G. M.; Kriz, G. S.; Engel, R. G. *Organic Laboratory Techniques: Small Scale Approach,* 1st ed.; Brooks/Cole-Thomson Learning: Pacific Grove, CA, 1998; p 169.
8. Clough, Stuart C.; Goldman, Emma W. *J. Chem. Educ.* **2005**, *82*, 1378-1379.
9. Werner, T. C.; Hull, L. A. J. Chem. Educ. **1993**, *70*, 936-938.

Chapter 6

24/7 Dynamic NMR Spectroscopy: A New Paradigm for Undergraduate NMR Use

Robert M. Hanson

Department of Chemistry, St. Olaf College, 1520 St. Olaf Avenue, Northfield, MN 55057

In this chapter I present how we have used remote instrumentation in the sophomore organic chemistry laboratory at St. Olaf College to completely transform how NMR spectroscopy is carried out and how the concepts related to NMR spectroscopy are introduced. I will touch on the process we went through during the past few years to implement what we call 24/7 Dynamic NMR Spectroscopy, and I will show how the implementation we have put in place has enabled a completely new team-based approach to investigation by students at the sophomore level.

Introduction

At the outset of this program, back in the fall of 2000, we felt the need to improve our access to NMR spectroscopy at St. Olaf College. Our old NMR spectrometer, a 1980s-vintage 300-MHz instrument, had served us well, but we had continually been dogged by three intrinsic problems:

- Equipment: How does one make effective use of a very expensive ($400,000) piece of equipment at a small liberal arts college?

- Time: How can a substantial number (100+/year) of entry-level organic chemistry students be introduced to high-tech cutting-edge experimental techniques without overburdening the system?
- Pedagogy: How can students be encouraged to experiment on their own even at a relatively early point in their instruction?

Our feeling was that even with 13 years of experience using NMR spectroscopy in the organic lab, we just were not making good use of our investment. The basic problem is outlined in Figure 1. Basically, because of the large number of students and the limited amount of time involved, we needed the lab assistant (or lab professor) to run the NMR experiments. Students prepared their samples, handed them to the lab assistant, and either watched the process passively or went back to the lab, and that was it. The lab assistant did the rest.

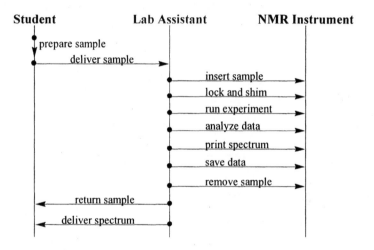

Figure 1. Prior to implementing automation, NMR spectra were run by lab assistants.

Our main problem was that with so many students and so little time (at least at the sophomore organic level), we just could not effectively introduce students to the immense power of NMR spectroscopy. The problem, of course, is intrinsic to the technique: In NMR spectroscopy, one needs to prepare and insert the sample, the instrument needs to be locked and shimmed, parameters need to be selected, data has to be gathered, and the data require sophisticated computer-based analysis prior to interpretation. Until just a very few years ago, "taking an NMR" had to be done manually, and this required substantial expertise.

Thus, the best we could do at the sophomore level was to show students the equipment and encourage them to watch as trained assistants carried out the

64

experiments. We knew that this was inadequate, particularly because students at this point in their learning appeared to be particularly interested in experimentation. Here we had the most sophisticated experimental apparatus on campus, and we didn't have a way of getting students involved with it in any really valid way unless they were already actively involved in research.

The 24/7 Dynamic[1] NMR Solution

With these ideas in mind, we set out in the fall of 2000 to find a solution to the three problems involving equipment, time, and pedagogy. Our goal was to transform the basic model of student/assistant interaction with the spectrometer, as shown in Figure 2.

Our goal was to design a system that would allow 100-120 students to individually be in control of the experiment – a system that would allow the

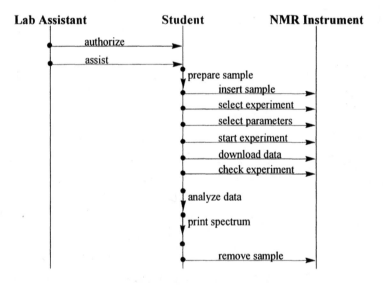

Figure 2. In our new paradigm, the lab assistant becomes the facilitator, and the student assumes responsibility for the NMR experiment.

---

[1] *Dynamic NMR* is not used here in the usual sense of samples undergoing chemical or physical changes over time(1) but, rather, in the sense of expressing action rather than inaction, which is appropriate for an instrument that can be acccessed 24 hours a day, 7 days a week.

assistant to be just that, an *assistant*. In order to achieve this goal, we recognized that automation would be a key principle. We identified five necessary elements of an automated system that would serve our undergraduate and faculty populations:

1. Choose an NMR instrument that is designed with automation in mind, with the components needed for unattended operation.
2. Design a simple yet effective web-based interface ("OleNMR") to distribute instrument access to any point on campus.
3. Extend the "user" system already in place for campus E-mail to a "team" system for NMR spectroscopy.
4. Make the instrument available to students and faculty 24 hours a day, 7 days a week.
5. Offer students the opportunity to try experiments that are not required but provide especially interesting and valuable information.

The first four of these elements are simply technological hurdles we had to overcome. Basically, they involved the purchase of a Bruker 400-MHz NMR spectrometer equipped with autotuning/automatching probe, deuterium gradient shims, and a 120-position autosampler, a bit of customization of the IconNMR software provided by Bruker, and the writing of a server/client interface.

The fifth point, involving student opportunity to experiment, is really the key to the pedagogical gains we have observed in implementing automation in the area of NMR spectroscopy at St. Olaf College and is illustrated in Figure 3.

Because the system is automated, because it does not require direct operator intervention, because the NMR experiment no longer needs to be carried out strictly during "laboratory hours" or by "designated experts," because the control of the experiment can now be in the hands of the student, students have access to a whole new experimental approach to NMR spectroscopy. The key innovation here, of course, is the ability to experiment. Once a sample is in the sample holder, the student is free to run as many experiments with it as he or she desires.

In practice, 24/7 Dynamic NMR works as follows:

- Student teams in the organic laboratory are assigned sample changer positions at the beginning of the semester. We generally make teams of two or three students, thus requiring about 50-60 of the available 120 sample positions. The remaining 60-70 positions are utilized by research groups and advanced courses such as biochemistry and physical chemistry.
- At any time, students may insert a sample in their designated holder in the sample changer.
- Once the sample is in the sample changer, the student may log on to the system from anywhere on campus anytime day or night using their

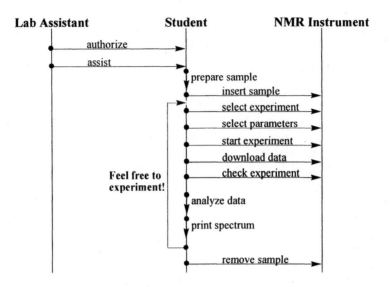

Figure 3. Automation allows for experimentation, because there is far less restriction in terms of instrument time.

favorite browser, set up an experiment, run it, analyze the data, and print the spectrum.

- Since the sample remains in the sample holder, students are free to experiment on their own, within limits.
- Students in advanced courses have more direct contact with the instrument and have a higher level of access to experiments.
- Authorized research assistants may have full instrument access.

Automation Enables Experimentation

The point here is that automation has enabled us to introduce a completely new way of using the technology of NMR spectroscopy in the undergraduate laboratory. More than that, what automation has allowed us to do is to transform the undergraduate laboratory itself into an opportunity for students to *experiment* in the authentic meaning of the word.

There are two fundamental ways in which automation has allowed students in organic lab to experiment. First, because the sample is in a robotic sample changer, students are encouraged to try more than just one "standard" NMR experiment. Thus, when I am supervising an organic lab, I ask that students show me their spectra. Even with absolutely no background in NMR spectroscopy, students know that there is information here, and we talk about what it means. "This is a very cool proton spectrum. Can you please take a carbon-13 spectrum and get back to me?" "Hmm, that didn't seem to give us much information. How much sample did you use? Could you add some more and try again?" "Oh, here's a beautiful spectrum. Good job! Say, would you be willing to try a COSY experiment? It only takes about 10 minutes. We can talk about what a COSY experiment is when it's done."

In addition, automation has allowed us to use the spectrometer exactly the way it is commonly used in research laboratories – for obtaining information relating to experiment progress, product identity, and sample purity – as a powerful diagnostic tool for the success of a synthetic reaction. "CONGRATULATIONS! This spectrum tells me that your experiment worked. Look here...." "Ah, I see you could have rotovapped that a little longer. This signal is for ethyl ether."

Automation Allows Earlier Introduction to NMR Spectroscopy

In our hands the use of NMR does not require any prior knowledge by the student about the theoretical basis of the experiment. In fact, what we have observed at St. Olaf is that by providing students with NMR data of their own early on in the course, long before NMR is introduced in class, many become highly motivated to learn more about it. Long before they know all about chemical shift, integration, and splitting, they have learned that the NMR technique answers the most practical of questions: What happened? Did I get the right product? Is it pure? Why is the melting point so low?

For example, because it is so easy to implement, we can now introduce NMR spectroscopy during the very first week of lab in organic chemistry. In this experiment, students are given an "unknown" solid and have to determine its identity. A list of approximately 20 possible compounds, with their known melting points, is posted in the laboratory. Depending upon the lab section, students might attempt this solely based on melting point and solubility properties, take proton spectra or, in some cases, obtain C-13 NMR data. With spectra in hand, students are given just a little in-lab introduction to the ideas of NMR spectroscopy, such as where aromatic protons appear and what an OCH_3 would look like. With just a very brief introduction, students lcan learn how to determine the number of nonequivalent sets of protons or carbons based on structure. Using their spectra, students then easily rule out compounds that have similar melting points. Students sometimes learn that taking good melting points requires more effort than they expected, since the listed compound with the nearest melting point is ruled out by the NMR data. The extra work of looking up compounds in the Aldrich catalog to determine their structure is rewarded with definitive correlation between structure and and NMR data.

Automation Encourages Teamwork

St. Olaf College has a long tradition of team-based pedagogy. (2 – 5) In this context, it is useful to consider the lab assistant and supervising professor as additional team resources. Overall, the team system consists of a network of teams involving supervisors, assistants, and experimentalists. There may be many supervisors, or there could be just one—the more there are, the more the responsibilities can be distributed. Thus, we designed our automated NMR spectrometer system to work within a team context and to reflect that approach. The teams involved with NMR spectroscopy at St. Olaf College are organized based on two broad categories: course and research. Each of these "teams" is subdivided, providing a whole series of working groups (Figure 4).

In this way, the concept of a team need not be restricted to laboratory courses. A professor doing research that involves NMR spectroscopy could set up any number of teams related to their work. We implemented three classifications of team members:

1. **Experimentalist** (students) Experimentalists design and submit experiments, monitor instrument status, and analyze data. Generally experimentalists work in teams of two or three students. The idea is that each member of a team has equal access to the instrument, and no two teams have access to each others' work. All team members see the same files, join in the responsibility for designing and submitting experiments, and are free to analyze the data either collectively or independently.

2. **Assistant** (laboratory or research assistants) Assistants maintain the team member lists, assign and identify samples, and help experimentalists with instrument operation. Each lab section at St. Olaf is facilitated by an undergraduate lab assistant. The primary responsibilities of the lab assistant includes (a) assigning students to teams and (b) helping students interact with the instrument and interpret the data. (Note that it is expressly *not* the lab assistant's job to carry out the NMR experiment. That job falls squarely on the student experimentalists.)

3. **Supervisor** (professors) Supervisors create teams, set experiment availability and options, allocate sample changer slots, help students and assistants interact with the instrument, and oversee the server file system.

This division of labor has proven effective. One of the goals of the team system is to distribute management functions so that no single person is overburdened. There may be a principal supervisor who sets up the teams and adjusts them periodically, but other supervisors can certainly set up teams of

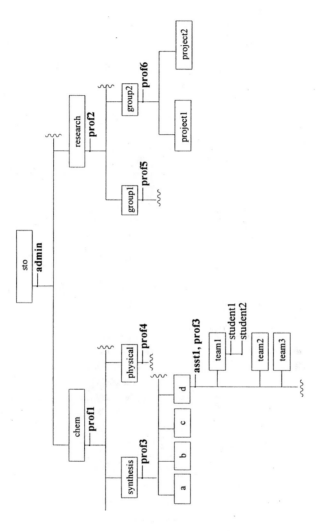

Figure 4. Teams are set up either in terms of courses and sections or research groups and projects.

their own. Management of team member lists is the responsibility of the assistant-level management team. These players use very simple tools to add and remove users from teams. The username base is drawn from a campus-wide database and needs no particular management.

The Importance of Context

One of the interesting benefits of our team model is that context is everything. It is most important to understand that a real person might play more than one of these roles. When a team member logs in, they also select a team in order to provide a context for their work (Figure 5).

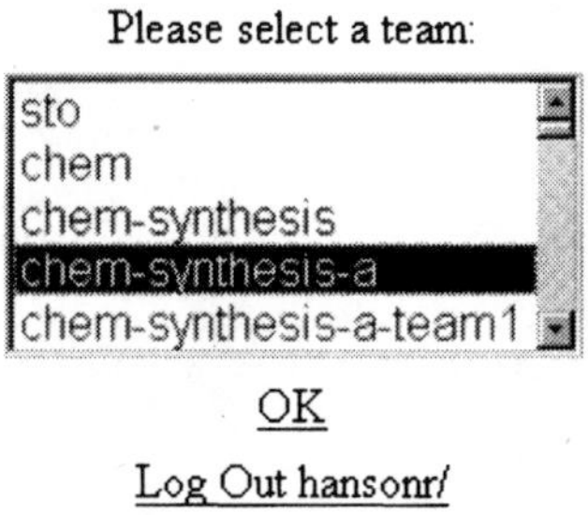

For example, a supervisor might play the role of an assistant (assigning samples) or an experimentalist (running experiments). A junior chemistry major may be an assistant for chem-synthesis-d, for example, and also a general user in chem-physical-team2.

Access to spectrometer function is based on the team that is designated at log in. A person logging in as a member of chem-synthesis-d-team1 sees only the data that their team has produced (or is given access to) and has access only to the sample changer slots assigned to their team. Likewise, a supervisor or assistant logging in as that team sees what the student sees, though they would have certain additional rights.

Please select a team:

sto
chem
chem-synthesis
chem-synthesis-a
chem-synthesis-a-team1

OK

Log Out hansonr/

Figure 5. After logging in, students choose a team context within which to work.

As a supervisor or assistant, I can log in as any of the teams under my supervision. When I do, I see what that level of user would see. I can log in as chem-synthesis-d and see all the spectra recently acquired by section D; I can log in as chem-synthesis-d-team3 and see just what that team has done. And, of course, I can do this from my office, not just at the spectrometer. Similarly, an assistant can log in either at his or her default level or as one of the teams under supervision.

Typical user systems (such as E-mail), even though they may involve working groups (aliases), don't really support group work. The 24/7 team concept fosters group work, because all users, once logged in as individuals, work exclusively in the context of a team. Although I may be a member of more than one team, by logging in as a specific member of a specific team, the system automatically helps me focus my activities. As a supervisor, this is particularly useful, because I can log in as one of the teams under my supervision and focus specifically on what the members of that team are working on. By allowing all team members equal access to the original data (not just copies), the system supports true team work. One student might start an experiment, but another might be the one to analyze the data and suggest improvements. The division of labor is up to the individual teams.

Automation Requires "Fair" Queuing

One aspect of the system that took considerable thought was the implementation of a method by which requested experiments could be ordered in a way that a majority of users would consider "fair." This was necessary because an automated system, at least in this context, is a multi-user environment. Unlike our old spectrometer, with one seat, our new spectrometer has hundreds. "First-come, first-served" just didn't seem like the right approach.

In addition, the "factory default" queuing routine built into Bruker's IconNMR system was designed for a completely different use of the sample changer. Designed with high throughput in mind, this routine would cycle through all samples sequentially, running all designated experiments for each sample in turn. The premise is that one would load up the sample changer, designate the experiments to run using IconNMR, set it going, and walk away. While this certainly makes for efficient use of the sample changer, it was not sufficient for our needs because it is sample-oriented rather than team-oriented.

For these reasons we implemented the following rules:

1. Number of experiments: You may submit as many experiments as you like. If another experiment is running, your experiment will enter a queue with the status pending. When it is your turn, your experiment will automatically run.

2. Order of Execution: The queue is not a simple first-in-first-out queue, because if it were, one team could submit a large number of experiments and tie up the whole system. Rather, if other teams have experiments queued and you have more than one experiment in the queue, only your first experiment will advance. Once it has run, then your next experiment will advance. Thus, if one team submits five experiments and another team submits five experiments after that, the experiments will run in an order that alternates between teams. Similarly, if three teams submit experiments, then there will be a three-way rotation as experiments are run.

3. Composite experiments: Some experiments may involve more than one spectrum. These composite experiments are run as a set.

4. Canceling experiments: You may cancel any of your experiments at any time, even after it starts running, if possible. If an experiment is canceled before it starts running, it will be removed from the queue and its status will change to canceled. If an experiment is canceled after it has started running, its status will change to aborted, and data may be lost.

5. Changing the order of jobs: Supervisors may at their discretion expedite a job, putting it to the top of the queue, or suspend/reinstate a job, which will place it at the end of the queue.

The implementation of these rules primarily amounted to rewriting a single TCL (tool command language) function in a single module of Bruker's IconNMR code. In addition, we developed a JavaScript-based monitoring system that allows users to check the current instrument and queue status.

Obtaining and Analyzing Spectra

Once an experiment is complete, the spectrum needs to be obtained, adjusted, and printed. We have set our IconNMR system to print one paper copy of the transformed 1D or 2D NMR spectrum automatically on the lab printer. For most students, this is sufficient. However, if the student has left the laboratory and needs to work with the spectrum, OleNMR provides an online tool that can facilitate this work. After logging in and checking that an experiment is complete, a student can call up and manipulate a "thumbnail" spectrum (Figure 6).

Manipulations are limited to horizontal and vertical expansions and setting of integration scale and regions. Clicking on "Create PDF" sends the integration/scaling data back to the server, which in less than a second turns it into a template, runs the standard plotting program provided by Bruker, deposits the PDF file on the web server, and returns a link to that document. Students simply click on the link, which opens the PDF document, and print the spectrum on a printer somewhere in their vicinity (Figure 7).

Implementation Details and Challenges

This chapter is not intended to explain all of the implementation details of our system. Basically, though, we have a Bruker Avance 400 MHz spectrometer with automatic tuning/matching Z-gradient broadband probe with deuterium gradient shims, and a BACS-120 sample changer, all standard issue from Bruker. We use Bruker's standard "IconNMR" front end to the XWin-NMR system. On the Windows 2000-based PC that runs the spectrometer, we made some minor

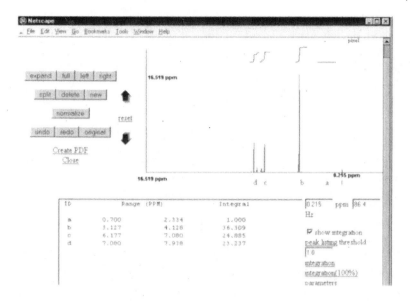

Figure 6. The "thumbnail" spectrum, for expansion and setting integration.

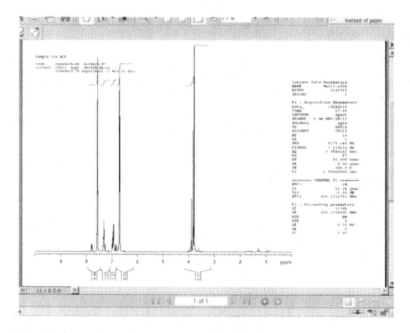

Figure 7. The final PDF spectrum returned, equivalent in all respects to that produced sitting at the instrument.

modifications to the Bruker open-code TCL language routines and wrote two Visual Basic 6.0-based communications software programs (DDESERVE, OLENMR).

In addition, on a separate campus server, a set of client JavaScript files gets loaded onto the user's machine as the web site is accessed. Development of this code was by far the most complicated aspect of the project and involved two full summers of work and five research assistants.

One of the greatest challenges was the design and implementation of a system by which spectral data could be manipulated remotely -- integrated, enlarged, expanded, and printed. After considerable work we settled on the method described above, which allows real-time manipulation of a moderately precise spectrum followed by delivery of a full-precision spectrum in the form of a PDF document.

Remote spectral phasing has not been implemented, but it also has not been necessary. (For research groups with higher demands, the system offers delivery of the FID in JCAMP-DX format for full remote processing.)

In all, the operation is totally transparent to the user and is relatively easy to implement and manage.

Assessment

Although we have done no formal assessment of the use of our spectro-meter, the OleNMR system saves detailed use logs, and they are informative. Over the 22-month period of August, 2003, to May, 2005, 2848 jobs were submitted. Of these, 21% were research and 79% course-related. Most (2048, 72%) were standard proton experiments. Almost all (2834, >99%) were remote jobs using the OleNMR system; only 14 (0.5%) were taken seated at instrument. Specifically during the 2004-2005 academic year 912 jobs were run. If one were to compare this with single-seat use at about 15 minutes per experiment, this use represents about six 40-hour weeks of laboratory assistant time that were spent more productively. As of September 2006, the system has been in continuous operation for three full years, with minimal maintenance and supervisory work.

Summary: Advantages of Automation

In summary, advantages of 24/7 Dynamic NMR are several:

- For students:
    o opportunity to experiment
    o better integration of class/lab experiences

- For lab assistants:
  o more effective use of assistants as lab assistants
  o more interaction with students in and out of lab

- For faculty:
  o more time for teaching during lab rather than playing the technician role
  o allows for a focused "as needed" instruction in spectroscopy
- For staff:
  o more efficient use of the spectrometer
  o fewer problems with down time

From a pedagogical perspective, automated NMR has completely changed our use of technology in the organic laboratory. Students have had far more access to NMR than ever before and can experiment with techniques not specifically called for in the laboratory manual. NMR can be used in lab in the way it is used in research: as a diagnostic tool for answering practical questions related to how an experiment is going or how successful it was. In addition, because we are now able to introduce NMR spectroscopy so early in the sequence, when it comes time to discuss NMR in class, students already have a context and even a fair amount of experience with real problem solving using NMR.

From a practical perspective, automated NMR completely transformed the role of the laboratory assistant. No longer are they sitting in a room running spectra or explaining over and over to students how to lock, shim, acquire, and analyze spectra. Instead, they assist in the assignment of teams and sample positions, and they get students started with sample preparation. They are *assisting*.

For faculty, who often were heavily involved with training of assistants and students on the detailed technicalities of getting "good" spectra, automated NMR has released them to do what they far prefer to do: assist in the selection of appropriate NMR experiments and in the interpretation of spectral data.

Overall, being able to have the instrument in active use any time day or night is far better than trying to fit all student use into blocks of time determined by the registrar. In addition, one of the most interesting results of automation has been the effect of "down time." With the traditional approach to NMR spectroscopy, when the machine went down, it was a disaster. Samples backed up; whole sections went without data. We have had our share of times when the machine was not available. For example, when the hard disk crashed it took a full day to put in a new drive and reinstall all the software. While this was happening, amazingly, nobody cared. Students loaded the sample changer as usual during lab and, upon getting an email from the supervisor that the system was back up, students started submitting experiments from wherever they were on campus at that time.

Acknowledgments

This project could not have been possible without the overwhelming support of the administration of St. Olaf College, which, after two unsuccessful attempts at NSF CCLI funding ("too ambitious"/"not hands-on"), awarded the Chemistry Department $500,000 for the acquisition of a new NMR instrument and the necessary building renovations. Clemens Anklin, Bruker Biospin, was the first to say that our seemingly crazy idea was conceivable on a Bruker system. His confidence in our plan is greatly appreciated. Chemistry majors Mike Purnell, Gregg Sydow, and Stephanie Skladzien (summer 2002) and physics majors Jared Irwin and Bryan Anderson (summer 2003) contributed tremendously to the design and implementation of the client-side JavaScript coding that has made the interface almost trivial for use by novice undergraduates. Finally, I thank my colleagues in the Chemistry Department at St. Olaf College for their patience and enthusiasm during the development of this novel way of utilizing NMR spectroscopy. Portions of this text were presented at the 88th Canadian Chemistry Conference, Saskatoon, Saskatchewan, May 31, 2005 and the online conference, *Trends and New Ideas in Chemical Education*, Jan-Feb, 2005, sponsored by the American Chemical Society Committee for Computers in Chemical Education. Thanks are given to the Howard Hughes Medical Foundation for a generous grant that paid for research assistant summer salaries.

References

1. Jackman, L. M., Cotton, F. A. eds. "Dynamic Nuclear Magnetic Resonance Spectroscopy", Academic Press, New York , NY, 1975.
2. Walters, J. P. *Analyt. Chem.* 1991, *63*, 977A.
3. Walters, J. P. *Analyt. Chem.* 1991, *63*, 1077A.
4. Walters, J. P. *Analyt. Chem.* 1991, *63*, 1179A.
5. Jackson, P.T. and Walters, J.P. *J. Chem. Ed.* 2000, *77*, 1019.

Supplemental Material

The St. Olaf College 24/7 Dynamic NMR site is restricted to on-campus computers only. However, a limited-functionality demonstration site can be accessed at http://www.stolaf.edu/people/hansonr/nmr/24-7/index.htm?DEMO.

Chapter 7

Inclusion of NMR Spectroscopy in High School Chemistry: Two Approaches

Brian Esselman[1] and Donald E. Mencer[2,*]

[1]Science Department, Whitewater High School, 401 South Elizabeth Street, Whitewater, WI 53190
[2]Department of Chemistry, Wilkes University, 84 West South Street, Wilkes-Barre, PA 18201

More high schools are introducing organic chemistry into their curricula for a variety of reasons. Since NMR spectroscopy is a powerful tool for understanding molecular level properties, such as molecular connectivity and inductive effects, the introduction of NMR spectroscopy into the high school chemistry curriculum is a natural step. In this chapter, the authors examine two programs, one in a high school and the other a university outreach program, that expose high school students to the use of NMR as a tool to probe the structure of organic molecules.

Chemists recognize the enormous power of Nuclear Magnetic Resonance (NMR) spectroscopy as a tool for probing the structure, bonding, and chemical environments present in gases, solids, and liquids. NMR spectroscopy can be employed in the study of any magnetically active nuclei.

Historically the two most studied nuclei have been hydrogen-1 (^{1}H) and carbon-13 (^{13}C). Conceptually NMR spectroscopy of these nuclei is rather simple. When exposed to a magnetic field, ^{1}H (and ^{13}C nuclei), which can have a nuclear spin of either ±½, align with or against the applied magnetic filed and exhibit two energy states. The gap between the two energy states for any particular nuclei increases linearly with the strength of the applied magnetic field and also depends on the identity of the nuclei. For example, when exposed to the same applied magnetic field, the energy gap for ^{1}H is about four times greater than for ^{13}C. However, the local magnetic environment generated by neighboring atoms of different electronegativities generates small differences in the energy gap. Additionally, the presence of neighboring atoms with their own nuclear spins also creates very small changes in the energy gap. NMR spectroscopy probes these subtle changes in the energy gap and thus reveals information about the structure of the molecule.

It would be difficult to overstate the significance of the role played by NMR spectroscopy in advancing chemical knowledge to its present state. It is also clear with the advent of *in-vivo* biochemical studies, along with 3-D, 4-D, and imaging techniques, that the full potential of NMR spectroscopy has not yet been exploited. These facts support the argument that NMR spectroscopy will play a critical role in the future exploration of problems in biochemistry, chemistry, medical imaging, and biology. Given the range of research problems that will be tackled using NMR techniques, it is safe to say that NMR will likely impact the lives of most people living in developed countries. Incorporation of critical technologies such as NMR spectroscopy into the chemistry curriculum at the high school level should not be overlooked as we move forward.

^{1}H NMR Spectroscopy – for High School Students

Rationale

The rationale for teaching the interpretation of ^{1}H NMR spectra is very similar to the rationale for teaching organic chemistry topics to high school students. Many arguments have been put forth calling for an increased role of organic chemistry in introductory chemistry courses and high school courses. These arguments tend to center around the sheer percentage of chemists who work in the organic field (*1*). While this argument may have value, it is not the central justification for including almost a full semester of organic chemistry in a high school chemistry course. A better argument can be made based on the skills and preparation of the students. Some high school students are ill-equipped to handle several of the topics that are included in introductory college chemistry courses or the AP-Chemistry curriculum. In general, the mathematical background of high school students is quite poor in comparison to students in an introductory college chemistry course at a college or university (2). Yet these same students are well-equipped to deal with the concepts of organic chemistry such as the interpretation of ^{1}H NMR spectra. Specifically, learning to interpret ^{1}H NMR spectra requires that students utilize several topics of general and organic chemistry and improves problem solving skills. Also, about nine out of ten teachers of high school chemistry have completed a course in college level organic chemistry (3) and, as a result, are well-acquainted with the material associated with organic chemistry.

Even in the second year high school chemistry course, less than half of the students have completed or are currently enrolled in calculus and the prerequisite for Chemistry II is only a second year algebra course and Chemistry I. Students at this stage struggle with the a variety of the skills, such as dimensional analysis and solving word problems, needed for a traditional AP or college chemistry course (4, 5). Perhaps more time and effort should be directed toward introducing topics for which the students are well prepared. The focus would be on the synthesis of several chemistry concepts, improvement in real problem-solving skills, and preparation for chemical thinking that could be employed when confronted with chemical information in future settings.

Introductory organic chemistry, while having a reputation as being very difficult, places very little emphasis placed upon mathematics and calculations. The problems and thinking involved in organic chemistry are far more conceptual and qualitative. While there may be an assumption that this level of thinking is beyond the grasp of high school students, this is not the case (6). Even though they may not be as capable as their college-level peers, students of this level are able to understand a variety of concepts like isomerism, polarity, resonance, torsional and steric strain, and delocalization. As well, they are able to draw and explain simple electron pushing mechanisms and interpret IR and ^{1}H NMR spectra (7). In fact, the idea of using NMR as a tool for high school education has actually been explored in the past in Germany. (8)

The Personal Perspective of a High School Teacher – Brian Esselman

Part of my rationale for teaching ^{1}H NMR spectroscopy comes from my own experiences as a high school chemistry student; as a learner in my high school course, I was bothered by how much chemistry I was supposed to accept without solid justifications. I wanted to understand the how and why of chemical reactions. I wanted to know how we knew the structure of molecules and atoms given the limitations of our observations. Given the questions that I had, I remember that I found the organic chemistry of my second year college course very exciting and also comforting as we covered reaction mechanisms and ^{1}H

NMR spectroscopy. Finally, I had some of the answers to my questions regarding the justifications for the structures and reactions that I had been shown. Although all students may not have the same questions, many of the students that I have taught have felt the same excitement interpreting ^{1}H NMR spectra and seeing a reaction mechanism.

The interpretation of ^{1}H NMR spectra is an excellent example of a topic in organic chemistry that can provide a wealth of chemical understanding without the use of mathematically complex explanations. Additionally, ^{1}H NMR spectroscopy helps resolve one of the persistent questions that many beginning chemistry students have; it offers experimental information to help explain the structure of molecules. Despite the initial appearance of ^{1}H NMR spectroscopy being very difficult, the interpretation of ^{1}H NMR spectra is more like a logic puzzle that involves the application of some simple rules related to molecular structures. High school students are well prepared to solve the puzzle of the ^{1}H NMR spectra and build the molecular structure that it fits. I have found that for the more creative students who enjoy solving puzzles, interpretation of ^{1}H NMR spectra quickly became their favorite topic of the year.

The Whitewater High School Experience

Whitewater, Wisconsin is a small mid-western city with a population of approximately 13,400. The city has strong ties to agriculture and is also a university town that is located within commuting distance of Madison and Milwaukee. The Whitewater School District has long served the children of university professors, local business people, and local farmers. During the past ten years there has been an increase in the diversity of the community as many non-native English speakers immigrate into the area. The high school

Table I. Relevant demographic information describing the learning environment and size of Whitewater High School 2004-2005.

Total Students	653
% of Students who attend a 4 year college	60 %
% of Students who attend a 2 or 4 year college	80 %
% of Students on Free or Reduced Lunch	14 %
% of Students designated English Language Learners	11 %
Average ACT Score	22.3

community reflects this changing population seeing a dramatic increase in the number of students for whom English is a second language (See Table I).

Whitewater High School has a typical eight period day where each class period is approximately 45 minutes. Each year there are approximately 120 -

130 students enrolled in Chemistry I and approximately 25 – 30 of those students continue on to take Chemistry II. While the Chemistry I course is very similar to courses taught in many other high schools, the Chemistry II course is a bit unusual. Though it is of Advanced Placement Chemistry difficulty, a little less than half of the year is devoted on organic chemistry as summarized in Table II. A complete description of the two courses can be found on the course websites (*9, 10*). Additionally, at least two days each week are devoted to laboratory experiments placing a heavy emphasis on chemical experimentation.

Table II. Whitewater High School Course Outlines

Chemistry I*	*Chemistry II***
Atomic and Molecular Structure of Matter (Ch. 1, 3, 7)	Organic Nomenclature
	Hybridization
The Mole and Conservation of Matter (Ch. 8)	Organic Functional Groups
Stoichiometry (Ch. 9)	Polarity
Light and Atomic Structure	Constitutional & Geometric
Periodic Trends (Ch. 5)	Isomerism
Molecular Structure (Ch. 6)	Ring, Torsional, and Steric
Gas Behavior (Ch. 10-11)	Strain
Solutions and Equilibria (Ch. 13, 18)	NMR
Acids and Bases (Ch. 15-16)	
Thermochemistry (Ch. 12, 17)	

*Chapters refer to Holt – Modern Chemistry

**No textbook used

Wilkes University Summer Workshop

The idea of a College or University providing as a location for outreach programs for high school teachers and students is well-accepted. However, the literature on outreach programs employing NMR spectroscopy is very limited. (*11-13*) The focus of the Wilkes University workshop is to create an innovative project oriented summer workshop for teams of high school students (completing their sophomore or junior year) and their teachers. The program consists of intensive full-time immersion in problems involving chemical synthesis and instrumental analysis. The teams of high school students and chemistry teachers work with college students and faculty members on projects of substantive chemical interest. The workshop aims to serve the broad interest of the chemical discipline by exposing highly motivated students to a cross-section of a college chemistry curriculum. Funding for the workshop is provided by the Science in Motion Program, a Pennsylvania Department of Education funded project (*14*).

The projects were developed to include a mixture of synthetic work, computational work, and instrumental characterization. The organic synthesis involves the oxidation of an alcohol to an acid as well as the experimental evaluation of the yield, purity, and the acidity (pK_a) of the resulting acid. The instruments used to investigate the starting materials and the products formed include GC-MS, FT-IR, and NMR spectroscopy (^1H and ^{13}C). The inorganic project, involves the creation of complex ions and/or compounds and the investigation of the wavelength of maximum absorption (λ_{max}) for coordination complexes using UV-Vis spectroscopy. Participants are also introduced to computational methods for evaluating properties of molecules. The computational power now available to chemists makes it possible to compute the expected IR and NMR spectra as well as other useful molecular features.

The workshop does not provide a *crash course* in organic synthesis, but the rationale for the synthesis of an organic acid from a primary alcohol starting material is described as an example of functional group transformation. This provides the necessary background for a hands-on experience in organic synthesis. Time constraints and pedagogical considerations limit the coverage of the fundamental operating principles of, or detailed data analysis for, UV-Vis, NMR, IR, and GC-MS. However, the participants are introduced to the physical basis of each technique along with directions on the operation of each instrument. Participants also are trained to interpret their data with the aid of structure libraries on the GC-MS and FT-IR and the use of standard charts and tables for each method. No attempt is made to provide instruction in the details of quantum mechanics or molecular mechanics calculations. However, the participants do receive an introduction to the general types of molecular modeling along with examples of typical applications, strengths, weaknesses and limitations of each type of modeling. The participants perform a series of hands-on computational activities that demonstrate the capabilities of these methods.

This workshop has been run as a pilot with limited enrollments for three years and, although this will further complicate the logistical challenges created by limitations in equipment availability, outside funding is being sought to increase the scale of the workshop to include more participants. The course content is ambitious, but the high school students are able to process most of the content, especially with the assistance of their high school teacher/mentors. Although there are topics in the workshop unrelated to organic chemistry and NMR spectroscopy, about eighty-five percent of the time spent and the content in the workshop has an organic chemistry focus.

Common Modalities and Methods

In comparing notes on the curricula of the Whitewater High School Chemistry II program and the Wilkes University Summer Workshop program, common topics and concepts emerged. Both programs attempt to focus on common themes of molecular structure and function, relationships between common functional groups, and the use of spectroscopic tools to probe molecular structures. In these two very different settings, the instructional methods for the introduction to NMR spectra are very similar.

To introduce NMR spectroscopy, some time is devoted to lecture on how spectroscopy works and very generally on how an NMR spectrometer works. Little time is devoted to discussing the details of the operation of the spectrometer; the ideas of time domain spectroscopy, the details of the collection of the data, and of pulse sequences are beyond high school students and not particularly important to interpreting the spectra. When tackling ^{1}H NMR, both programs discuss that each hydrogen atom has a property that creates a tiny magnetic field and that it is necessary to apply an external magnetic field to detect the differences in the energy states of the protons in the sample. This is the extent of the general theory behind ^{1}H NMR spectroscopy that is discussed. However, the relationships of the various pieces of information that can be observed in spectra, such as chemical shift and splitting patterns, are related back to the magnetic field generated by each hydrogen atom in the molecule and the local environment that it is experiencing.

In both of these programs, example spectra are employed to discuss shielding, signal splitting, signal integration, and purity of sample. Within a time period equivalent of two or three days of high school class time, most of the students are able to start working through solving the spectra on their own. In the Whitewater High School class, each year several of the students get so enthused that they start ignoring the discussion and rush to solve the practice spectra. Not unlike college students first exposed to NMR spectroscopy, high school students must devote time to practice. Additionally, very few high school students are able to get proficient enough at solving the spectra that they do not require chemical shift correlation charts. For the practice components the Whitewater High School students receive about fifty spectra. These spectra tend to be relatively simple and do not include many examples of long-range coupling or complex splitting that result from two different coupling constants, or molecules containing halogens, oxygen, and nitrogen. They do include molecules with symmetry, aromatic rings, and double and triple bonds. Students who spend the necessary time to work through practice spectra generally find that they understand ^{1}H NMR spectroscopy much better and the spectra become easier to interpret. Students are asked not only to determine the structure of the molecule but also to explain how that structure gives rise to its ^{1}H NMR spectrum. A sample ^{1}H spectrum is shown in Figure 1. The typical features of chemical shift from the electronegative functional group and the aromatic ring can be seen. In addition, typical splitting is seen for the methyl and methylenes and the integrated peak areas accurately represent the number of protons present.

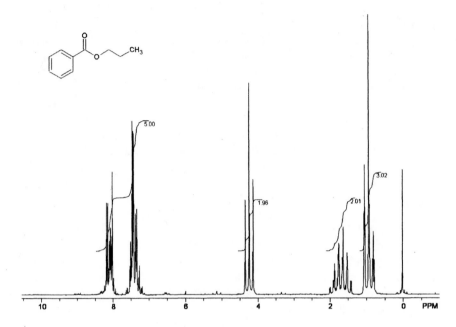

Figure 1. ¹H NMR spectrum for n-propyl benzoate. Aromatic protons centered at ~δ=7.7 are visible. Typical splitting is evident for the methyl (triplet at δ=0.95) and methylenes (triplet at δ=4.3 for the O-CH₂- and mutiplet for the - CH₂- split by both the methyl and the methylene neighbors seen at δ=1.7).

In the Wilkes University Workshop, students are able to examine sample spectra and then make predictions about what spectral features should be present for their molecules of interest. The students sit at the NMR console and acquire data (both ¹H and ¹³C) using a Varian EM 360 upgraded with the Anasazi dual channel EFT capabilities. The students can see chemical shifts and splitting for molecules that they also synthesize in the lab. In order to incorporate the discussion of ¹³C NMR spectroscopy during the Wilkes University workshop, the only additional information needed consists of the difference between the isotopic abundances of ¹H vs. ¹³C and the introduction of the idea that ¹³C spectra can be collected in a manner that eliminates coupling to ¹H atoms. An example ¹³C spectrum is shown in Figure 2. Some who teach NMR spectroscopy in college level organic courses argue that instruction on ¹H decoupled ¹³C NMR should actually precede instruction in ¹H NMR. In that manner, students can focus on the concept of chemical shift, employing all the structural features that determine extent of shielding and desheilding, prior to the introduction of the complicating features of splitting patterns.

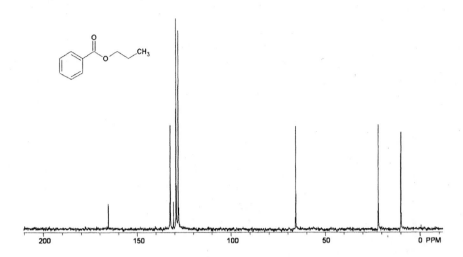

Figure 2. ^{13}C NMR spectrum for n-propyl benzoate. The carbonyl C is located at ~ δ=165. Four magnetically "unique" aromatic carbons are seen from ~δ=125-135. Finally, carbons in the n-propyl group are located at –OCH$_2$- (~δ=67), -CH$_2$- (~δ=22), and CH$_3$ (~δ=10).

Access, Resources, and Cost

Access to NMR instrumentation is very limited for high schools, but creative educators can find community resources that afford access to NMR spectrometers. For a high school with an approximate budget of $3,400 per year for seven to eight sections of Chemistry I & II, a discussion of ^{1}H NMR is by nature very theoretical. During the first three years of teaching ^{1}H NMR at Whitewater High School, ^{1}H NMR remained a purely theoretical discussion; however during the 2004-2005 school year, WHS students gained access to NMR spectrometers utilizing external resources.

The first access to a spectrometer was through the local university. One student who had finished Chemistry II as a junior wished to continue studying chemistry during her senior year. She completed several organic syntheses at the high school and then spent a couple of hours analyzing her products and starting material using the NMR spectrometer with the assistance of Steven Anderson, one of the Organic Chemistry professors at the University of Wisconsin-Whitewater. With some training, the student and teacher were able to use the spectrometer to analyze several the products and starting materials of her syntheses. The student was also allowed to return to the university at a later date run the remainder of her samples.

Secondly, Frank Contratto of Anasazi Instruments contacted Whitewater High School after learning through a former student that ^{1}H NMR spectroscopy was included in the curriculum. He offered to lend a demonstration spectrometer to the school for a couple of weeks. The presence of this 60 MHz spectrometer moved the discussion of ^{1}H NMR spectroscopy out of a theoretical context and allowed the students to collect spectra. The students prepared samples and dealt with the difficulty of getting enough solid to dissolve in the limited solvents available. They worked with computer software to collect the spectrum and analyze it. These tasks provided them with a hands-on connection to the spectroscopy that had been discussed in previous weeks. The students were excited to have an opportunity to use the spectrometer while several former students complained that they did not have the same opportunity.

The hands-on access afforded students during the week-long workshop at Wilkes University is also clearly a positive aspect of the workshop. Nearly immediate feedback can be obtained using an FT instrument. If a permanent magnet (60 or 90 MHz) spectrometer has been upgraded for FT data collection of both ^{1}H and ^{13}C spectra, students can obtain spectra on many compounds in a very short period of time. These instruments are low maintenance and have low operating budgets. Also, any organic compounds can be run as neat liquids, eliminating the need for deuterated solvents and long runs for ^{13}C data collection. Perhaps someday an NMR spectrometer will be as commonplace in high schools as electronic balances are today. This is not a likely short-term objective however, as the cost of these instruments is still prohibitive for most high school programs.

Assessment of NMR Spectroscopy Instruction

Whitewater High School

Students are assessed formally in three separate methods: an experiment that included ^{1}H NMR analysis, an in-class quiz, and a group take-home exam. First, students convert maleic acid to both fumaric acid and succinic acid. They analyze their starting material and products by pH titration, melting point, ^{1}H NMR, and ^{13}C NMR. The spectroscopic data was used to confirm formation of the desired product and to assess relative product purity. Since these molecules all have very simple spectra, the students are able to interpret the spectra despite the limited resolution of the spectrometer or the presence of impurities. As part of the laboratory assignment, students assign all of the signals in the spectra to atoms in the molecules. One of the added benefits of having the spectrometer for these experiments is that students are able to see in the ^{1}H NMR spectra how much maleic acid starting material is left in their converted products. Often left out of textbook discussions of ^{1}H NMR spectroscopy is its usefulness for determining the purity of a sample, and the experiment allows students to see this application directly.

In addition to the practical use of ^{1}H NMR, each student is assessed with a quiz on ^{1}H NMR that includes both interpreting a spectrum to determine a molecular structure and predicting a spectrum for a given structure. Students were permitted the use of any NMR correlation data they wished to use in completing the quiz.

Two sample quiz problems:

1. Predict and draw the ^{1}H NMR spectrum for 3-[(1E)-1-(bromomethylene)propyl]phenol and identify which signals are caused by each hydrogen environment.

2. Determine the structure and identify which signals are caused by each hydrogen environment for a formula of $C_{10}H_{14}O_2$ and a spectrum of 2.9 δ (doublet, 2H), 3.3 δ (singlet, 6H), 4.5 δ (triplet, 1H), and 7.2 δ (multiplet, 5H). (An image of the spectrum was also included.)

Following the individual in-class quiz, students are given a take-home exam of more complex spectra to be completed in student groups of three. Each group is given a set of ^{1}H NMR spectra unique to that group's exam, which allows the instructor to tailor the challenge that each group will face to the strength of its members. For each spectrum, students assign signals to the structure that they determine and explain their reasoning in writing. The students typically report that they have a very difficult time on the exam, but understand their spectra very well after completing it.

Whitewater High School's Student Responses:

Though WHS is in only its 5th year of teaching ^{1}H NMR spectroscopy, and there is not a large population of alumni who have taken college organic chemistry courses, very positive feedback has been collected from those students who have taken those courses involving ^{1}H NMR spectroscopy. A few of those comments are presented below.

"I fell in love with organic while studying it in the first semester of my Chemistry II class my junior year of High School, and it was in that class that I was introduced to ^{1}H NMR. As a senior I had an exciting opportunity to work with the UW-Whitewater's ^{1}H NMR Spectrometer under the observation and supervision of the organic chemistry Professor Steven Anderson. With Professor Anderson's help, I was able to run each of the organic substances I had synthesized earlier in the semester. Compounds that I synthesized included: aspirin, salicylic acid, dulcin, phenacetin, ferrocene, etc. Using the ^{1}H NMR [spectra] I obtained from my samples, I was able to determine relative purity by comparing them to the substances' known spectrum." - Jamie

"By learning about NMR in high school, I was more prepared for my college Chemistry classes. Not only did it increase my knowledge of chemistry before heading into challenging college courses, but it also helped me to improve my skills for analysis and problem solving. The concepts involved in understanding NMR are concepts that appear frequently in every branch of chemistry and so it is useful to understand them. For example, through learning about NMR I learned more about molecular structure and electronegativity. The process of learning how to assign a molecule to an NMR spectrum was challenging but I feel that being exposed to it in high school made me much more comfortable with it when I got to college." - Kari

Wilkes University Workshop

High school teachers and students complete a pre and post-test of the topics incorporated into the workshop. To date the most popular portion of our workshop has been the computational chemistry component. The second most highly rated portions of the workshop, in terms of popularity, is actually the inorganic synthesis of a coordination compound. However, the feedback on the spectroscopy components has also been favorable. The students and teachers pay close attention during the lecture, hands-on, and discussion portions of each topic and are particularly impressed that they can compute the expected infrared (IR) and Nuclear Magnetic Resonance (NMR) spectra expected for some of the molecules with which they are working.

In addition, all of the post-test results demonstrate improvement in knowledge after the workshop when compared with pre-test results prior to the workshop. Two example questions (a structure of ethyl acetate is provided) from the post-test that relate to NMR spectroscopy are:

1. In a molecule of ethyl acetate, how many magnetically unique types of hydrogen are observed in the ^1H NMR spectrum?
   a. One   b. Two   c. Three   d. Four   e. I do not know
2. In a molecule of ethyl acetate, into how many peaks will the resonance for the protons on the carbon labeled with a "*" be split?
   a. One   b. Two   c. Three   d. Four   e. I do not know

In the pre-test, out of the ten students tested to date, all have answered, "I do not know" to both of these questions. In the post test, 90% answered question number 14 correctly and 80% answered question number 15 correctly. Given that NMR spectroscopy is covered by some of these students three days prior to the post-test, the results demonstrate a clear mastery of two of the critical concepts in NMR.

Limitations and Possibilities for the High School

Understandably, there are significant barriers to teaching ^{1}H NMR spectroscopy to high school students, but the students' intellectual ability is not one of them. The most significant barrier, the lack of sufficient financial resources for public or private high schools to gain access to a ^{1}H NMR spectrometer for student use, will need to be addressed. Additionally, it is likely that only a minority of certified high school chemistry teachers feel confident enough in their own knowledge of the subject matter to attempt to teach it to high school students. Even if an instructor feels confident with his or her ^{1}H NMR knowledge, with the increased pressure from administrators, parents, school boards, and the community to offer Advanced Placement courses, few second year courses are able to devote the necessary time to studying organic chemistry to reach the topic of ^{1}H NMR spectroscopy. Yet some states have incorporated NMR into high school curriculum requirements. For example, the academic standards in the Pennsylvania Code (*15*) require for students in the tenth grade the following: "Describe and demonstrate the operation and use of advanced instrumentation in evaluating material and chemical properties (e.g., scanning electron microscope, nuclear magnetic resonance machines)."

The strengths of NMR spectroscopy make it a unique tool for understanding key concepts of molecular structure and connectivity. Teaching NMR spectral interpretation as part of the high school curriculum enhances student understanding of key molecular concepts and promotes the problem solving skills that many view as critical to life-long learning. Incorporation of this important spectroscopic tool into the high school curriculum provides a chance for a different kind of high school chemistry course: a course that fosters molecular level thinking instead of algorithmic computation. This is a possibility that deserves further consideration.

References

1. Annual Employment Survey *Chem. Eng. News*, annually.
2. *Trends in International Mathematics and Science Study (TIMSS)* 2003, URL http://nces.ed.gov/TIMSS/index.asp
3. Smith, S.P. *2000 National Survey of Science and Mathematics Education: Status of High School Chemistry Teaching*, pages 4 and 7, URL http://2000survey.horizon-research.com/reports/high_chemistry/high_chemistry.pdf
4. McFate, C.; Olmsted III, J. *J. Chem. Ed.* **1999**, *76(4)*, 562-565.
5. Legg, M.J.; Legg, J.C.; Greenbowe, T.J. *J. Chem. Ed.* **2001**, *78(8)*, 1117-1121.
6. Dori, Y. J.; Barak, M. *Educational Technology & Society* **2001**, *4(1)*, 61-74.

7. Alonso, D.E. *Organic Synthesis and NMR Characterization for High School Students* Abstracts, 40th Midwest Regional Meeting of the ACS, Joplin, MO, United States, October 26-29, **2005.**

8. Hallpop, P.; Schuetz, H. *Pocket Text, Vol. 31: Use of Proton NMR Spectroscopy. Instruction Program for High Schools* Verlag Chemie-Physik Verlag, Weinheim, Ger. **1975**, 103 pp.

9. Esselman, B.; Kachel, K. Whitewater High School Chemistry I website, URL http://www.wwusd.org/whs/chem/index.html

10. Esselman, B. Whitewater High School Chemistry II website, URL http://www.wwusd.org/whs/advchem/index.html

11. Lessmann, J.J.; Benefiel, C.; Newton, R.B. *Link to the future: a partnership program connecting high school and university chemistry.* Abstracts, 58th Northwest Regional Meeting of the American Chemical Society, Bozeman, MT, United States, June 12-14, **2003.**

12. Scully, S.M.; Hannay, E.L.; Skiver, R.L.; Vodhanel, D.A.; Blankenburg, M.A. *University of Toledo summer camp for the integrated activities grant* Abstracts of Papers, 225th ACS National Meeting, New Orleans, LA, United States, March 23-27, **2003.**

13. Mencer, D.E. *Permanent magnet FT-NMR spectrometers: Versatile tools for undergraduate instruction and high school outreach.* Abstracts of Papers, 230th ACS National Meeting, Washington, DC, United States, Aug. 28-Sept. 1, **2005.**

14. The Pennsylvania Science in Motion & Advancing Science Outreach Program, URL http://services.juniata.edu/ScienceInMotion/

15. 22 The Pennsylvania Code, § 4.83, **2006**, URL http://www.pacode.com/secure/data/022/chapter4/s4.83.html.

Teaching NMR
with Technology

Chapter 8

C-MoR: Computer Modules That Assist in Teaching NMR Theory and Interpretation

Raymond N. Dominey[1], Charles B. Abrams[2], René P. F. Kanters[1], and Emma W. Goldman[1]

[1]Department of Chemistry, University of Richmond, 28 West Hampton Way, Richmond, VA 23173
[2]Department of Physical Sciences and Engineering, Truman College, Chicago, IL 60640

The C-MoR project leverages the 3-D visualization and animation abilities of modern computing to help students better visualize, and thus understand, ideas that are based on spatial/conceptual relationships (such as Symmetry), as well as ideas that are based on a time-ordered process, such as the development of an FT-ICR or FT-NMR signal. We describe four such computer modules (C-MoR Symmetry, C-MoR MS, NMR Tutor, and C-MoR NMR) that we use extensively in our chemistry majors Spectroscopy and Instrumentation course to teach the theory, instrument design and function, and spectral interpretation skills that are essential to appreciating the full power of modern NMR Spectroscopy.

Introduction

Computer use in the classroom has grown exponentially in the past 15-20 years (1). All the advances in computer technology that have occurred not withstanding, there is still uncertainty about how to use the technology effectively in the classroom (2). This article discusses the use of a number of animation-based computer programs to assist in teaching both NMR theory and NMR interpretation.

In the early 1990's, a group of us at the University of Richmond started to envision how computers could be used to increase students' understanding of chemistry by incorporating three-dimensional visualization and animation methods throughout the four-year chemistry curriculum. We found that many entering college students find it difficult to visualize and animate three-dimensional images "in their heads". Yet this ability is essential to being able to succeed and enjoy chemistry, physics, and even calculus. Out of our assessments and discussions, many of which were stimulated by regular contact with Prof. Nicholas Turro (Columbia University; chair of a committee/task force studying the use of technology in science education for the National Academy of Sciences), we started a project that we now call **C-MoR** (Chemistry Modules of Richmond) (3) with the explicit goal: to develop and integrate computer visualization and animation throughout the Chemistry curriculum. We wanted to help students understand the space-filling and dynamic nature of molecules, promote a habit of using 3-D visualization and animation to gain a more thorough and intimate understanding of chemical systems, and thereby help all students to think of chemistry as tangible, dynamic processes. All C-MoR modules are delivered through a common web-browser interface with appropriate plugins to make the materials as versatile and widely usable as possible.

A synergy developed between our C-MoR efforts at Richmond and Prof. Nick Turro and the Edison Chemistry Education Project (4) at Columbia. One of us (Charlie Abrams), while working as a key participant in the effort at Columbia University developed the highly effective animation-based computer program, IR Tutor (5). We came to view IR Tutor as an important model of how new technology can support and enhance teaching. Specifically, we saw the effectiveness of IR Tutor's use of animation to communicate abstract ideas within chemistry as "*the*" model to be emulated. After completing IR Tutor, Abrams developed animations that would help in teaching concepts central to an introduction to NMR theory, resulting in NMR Tutor (6).

The C-MoR project identified a myriad of topics, all central to the teaching and/or practice of chemistry, in which the 3-dimensional nature of molecules or ideas associated with the time progression of a process are critical elements of a proper understanding of the topic. Since NMR Tutor focuses on introductory NMR theory, the C-MoR project developed materials to assist teaching ideas fundamental to the design and function of modern FT instrumentation and to assist the teaching of NMR spectral interpretation. As will be discussed, one module was developed specifically to complement NMR Tutor. Several others, while focused on ideas that underlie seemingly unrelated areas within the chemistry curriculum, also have fundamental utility in teaching NMR. In this way, we hoped to build bridges between seemingly disconnected topics within the chemistry curriculum via underlying connections that are fundamentally common to each (*e.g.*, Fourier Transform methodologies in chemistry instrumentation, and symmetry).

We will describe four computer modules and how they are useful: in explaining concepts of central importance when teaching introductory NMR theory, in recognizing aspects of FT-NMR that can be connected in a generalized way to other FT methods, in explaining the conceptual underpinnings of the design of FT-NMR instrumentation, and in teaching principles of NMR data interpretation. There are a number of programs available that predict NMR spectra (7), but the ones discussed here help understand the theory and instrumentation appropriate for use in an undergraduate chemistry curriculum. One benefit of these programs is that they are useful in a variety of levels of courses where NMR is discussed—from non-majors courses, to sophomore organic courses, and in upper level spectroscopy and instrumentation courses. The four computer modules we will discuss are: C-MoR Symmetry, C-MoR Mass Spectrometry (C-MoR MS), NMR Tutor, and C-MoR NMR. The first two were specifically developed for other topics, but are helpful in discussing ideas related to chemical shift equivalency and in understanding principles associated with FT methods in general.

The focus of this paper is a description of how we use these resources in our advanced analytical chemistry course, *Spectroscopy & Instrumentation*. In this course we intentionally cover Mass Spectrometry before NMR specifically because the animations in the C-MoR MS module provide the opportunity to make powerful conceptual connections, directly and by analogy, to several ideas that are of central importance in NMR.

Contexts for Teaching Some NMR fundamentals

Symmetry

In discussing the part of NMR theory relevant to developing interpretation skills, one of the topics that is discussed is *chemical shift*. It is important to discuss it from the perspective of a theoretical model by which we can understand the origins of, and therefore differences in, *chemical shift* and also from the perspective of deducing which nuclei in a particular structure will be *chemical shift equivalent* versus *chemical shift inequivalent*. It is also important to recognize the distinction between *accidental* and *symmetry* based chemical shift degeneracy. In this section we will focus on the latter perspective, and then return to the former in the next section.

The C-MoR Symmetry module is very useful in explaining the criteria for deducing which nuclei in a particular structure are chemical shift equivalent. As is explained in several standard texts (8, 9), the requirements for chemical shift equivalency or non-equivalency is based on specific symmetry elements and associated symmetry operations. Given an achiral environment, the requirements for chemical shift equivalency are either: 1) the positions of the nuclei in question are interchanged through symmetry, or 2) the nuclei are physically exchanged by some rapid process which is fast compared to the NMR time scale. In either case, the nuclei in question are indistinguishable by NMR. For the former, they are indistinguishable because they occupy fundamentally indistinguishable points in space. For the latter, they are indistinguishable because the physical exchange process produces nuclei that have an "intermediate" resonance frequency, i.e. the concentration weight average of the frequencies associated with the exchanged positions.

In terms of chemical shift equivalency, the three relevant symmetry elements/operations are: 1) rotational symmetry (C_n); 2) reflection symmetry (σ, or S_1 or C_s); and 3) inversion symmetry (i, or S_2). Simply stated, if a symmetry operation transforms a molecule's structure into one that is superposable with the original, then the symmetry operation has "interchanged" nuclei that are in structurally indistinguishable positions. Such nuclei give resonances which are said to be "isochronous" or, more simply, chemical shift equivalent.

To properly appreciate the symmetry criterion, it is important for students to have a clear understanding of the fundamental distinctions between types of symmetry elements and operations, i.e. how they each relate points in a 3-D molecular space to each other. Furthermore, this needs to be understood both relative to an intrinsically achiral environment (solvent or host environment surrounding the molecule) as well as relative to a chiral environment. In particular, nuclei positions related by rotational symmetry are referred to as "homotopic", and thus would be chemical shift equivalent in chiral as well as achiral (hosts) environments. On the other hand, reflection or inversion related nuclei positions are referred to as "enantiotopic", and thus would be chemical shift equivalent only in achiral (hosts) environments. Finally, if there are no symmetry elements/operations that interchange two nuclei positions then they are referred to as "diastereotopic" positions, and would be chemical shift inequivalent under either achiral or chiral (hosts) environments. This language dovetails nicely with the standard language associated with stereochemistry covered in sophomore organic courses, so this additional contextual exposure helps by further connecting these ideas of symmetry and equivalency, to the related but more familiar ideas of stereochemistry.

C-MoR Symmetry

As an aid to teaching these ideas, the C-MoR Symmetry module provides an interactive 3-dimensional visualization tool for demonstrating symmetry operations as animations in a virtual 3D space by using VRML objects whose 3D structures have representations of the associated symmetry elements embedded within them. The instructor can "pick-up" (so-to-speak) a VRML object, and rotate it around at will; thus providing to the spectator(s) an intuitive appreciation for the 3-dimensional arrangement of the components of the object being viewed. This is very helpful in illustrating what a *symmetry element* is and how it is oriented with respect to the overall structure of the object. It also helps the spectator develop an appreciation of the fact that the *symmetry element* itself is also a representation of a particular symmetry characteristic of the points in space occupied by an object to which it applies.

The most powerful feature of this module, however, is the animations themselves. Animations of a symmetry operation can be activated from any orientation of the object being displayed by "clicking" on the *representation* of the embedded symmetry element. In the animation, a *ghosted* version of the atoms is subjected to the particular symmetry operation via a VRML animation that shows these more transparent atoms move through space. Thus, it shows symmetrically equivalent atoms being interchanged by movement through space. For rotational symmetry operations, this is movement around the rotational symmetry axis. For the reflection symmetry operation, this is movement to, and then through, the reflection plane in trajectories perpendicular to the plane. For improper rotational symmetry operations, this is movement first by rotation around the symmetry axis immediately followed by reflection perpendicular to, and then through, the associated reflection plane. The powerful impact on pedagogy from this module derives directly from the fact that its animations essentially instantaneously communicate to the spectator precisely what a particular "symmetry operation" is, does, or rather, represents. It shows that the visual representations of the object before and after the animation are exactly the same even though the indistinguishable parts (atoms) were ones that moved during the animation.

Figure 1A below shows a series of images captured during the animation that illustrates the C_3 proper rotation operation. Note that the C_3 symmetry element is depicted as an axis that is embedded within the structure of the object and whose cross-sectional shape matches the associated crystallographic symmetry element symbol. Figure 1B similarly shows a series of images captured during the animation that illustrates the reflection operation. Again, it is very important to note that the power of this module is intrinsic to the animations themselves. Thus, these static 2-D images only hint at the effectiveness of using the actual module itself.

We should also point out that one can accomplish this pedagogical task with physical hand-held models. However, the beauty of using the C-MoR module is how rapidly one can explore the 3-dimensional relationships for a number of symmetry operations, and that the whole class can see clearly the operation that is being executed by the instructor while he/she is executing it. Additionally, students enjoy exploring these symmetry animations on their own using the same computer module available to them via the university's network.

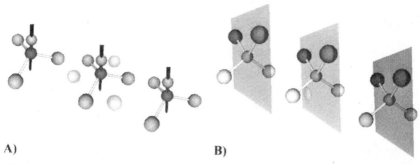

*Figure 1: <u>C-MoR Symmetry:</u> Three points (frames) each in the animation of:
A) rotation around a three-fold axis; B) reflection through a symmetry plane.
In both examples, the lighter colored balls in the second frame are moving in
the animation. (See page 1 of color inserts.)*

This particular module was first developed for use in other courses. It provides animations that cover all the different types of symmetry operations as well as animations that connect these to the idea of *point groups* (and eventually to *space groups*). We have found that the use of even a small subset of animations from the symmetry module, even for a few minutes, is as effective, if not more so, than spending much more time using hand-held models in getting the concepts of symmetry equivalent atoms, and by extension their chemical shift equivalence, across to students.

Concepts Transferable Either Directly, or by Analogy, Between Teaching FT-ICR and Teaching FT-NMR

Teaching Mass Spectrometry immediately prior to teaching NMR, and in particular using the C-MoR MS module to do so, provides a real (almost tangible) added value to the teaching of NMR. This added value derives principally from the FT-ICR animations within the C-MoR MS module; the animations instantaneously create important mental images for each student that visually illustrate several fundamental and generalizable concepts common in the contexts of both FT-ICR and FT-NMR. Furthermore, when the FT-ICR animations are considered along side the animations from NMR Tutor (to be discussed in the next section) several fundamental underlying similarities of all Fourier Transform (FT) methods become intuitively obvious. When teaching NMR, we typically try to make direct connections/analogies with ideas initially introduced via the FT-ICR animations in explaining: 1) the model most commonly invoked to rationalize the origins of chemical shift, *i.e.* cyclotron motion of charged objects, and thus factors that influence it; 2) the semi-classical model of precession, specifically resonance as a source of precessional motion and how the precessional model provides an intuitive understanding of the requirements for detection, *i.e.* correlation or coherence of precessional motion; and 3) the basic design and function of a modern pulsed FT-NMR instrument.

1) Cyclotron motion of charged objects

A key feature of the FT-ICR section of the C-MoR MS module is the animation of cyclotron motion for ions with translational movement in a plane perpendicular to the magnetic field. When generalized, such ion cyclotron motion is a concrete example of the more fundamental concepts expressed by Faraday's and Lenz's Laws (10a), which describe the force that exists on charges moving in a magnetic field. This animation therefore affords an opportunity to discuss the phenomenon described by these laws and the equally important counterpart, the Biot-Savart law, which describes the magnetic field that is induced by circulating charge density (10b).

These two principles find applications in both the theoretical framework of NMR (*e.g.*, explaining shielding) and the more practical framework of NMR instrument design (*e.g.*, explaining how a coil can be used either as transmitter or detector for oscillating magnetic fields). Therefore, these animations allow one to make the point that other charged objects, such as electrons, will act similarly under the influence of a strong magnetic field, *i.e.* a cyclotron (periodic) motion of electron density will be induced. The extra step needed to transfer these concepts to shielding in NMR is provided by the fact that the induced current, according to the Biot-Savart law, induces a magnetic field with associated toroidal field lines. Most undergraduate texts present *diamagnetic shielding* and *diamagnetic shielding anisotropy* as two essentially disconnected phenomena, but the powerful, analogy to the ICR animations helps one understand that the distinction between *diamagnetic shielding* and *diamagnetic shielding anisotropy* stems from the degree to which the magnetic fields lines extend away from the axis of the electron density's induced cyclotron motion. The degree of extension into space of the magnetic field lines then is connected logically to the extent of space over which the associated electron density's cyclotron motion occurs. Thus, "simple" *diamagnetic shielding* generally refers to the situation where the particular electron density involved is associated with (or localized in) the atomic or bonding orbital spaces of a particular atom; in this case the induced toroidal magnetic field lines are generally limited to that particular atom's own atomic space. However, it is a natural extension of these ideas to conclude that electron density delocalized over more than one atomic space (*e.g.*, π systems) will generate magnetic field lines that extend further into space. Thus, these magnetic field lines can affect nuclei further away from the axis of the circulating electron density and also affect nuclei differently in different regions of space surrounding the electron density reservoirs in question. This anisotropic effect results from the anisotropic nature of the toroidal field lines relative to the orientation of the inducing magnetic field and to the rest of the molecule's structure. From this perspective the large effects of so-called *ring currents* associated with aromatic systems is a trivial extension, and even σ-bonding electrons can be seen to have the potential to contribute.

2) Periodic (time-dependent) Motion, the Precessional Model, and the FT-NMR Experiment

There are also strong analogies that can be drawn between cyclotron motion of charged objects (ions) and precessional motion of magnetic dipoles (nuclei). For example, the fact that an externally applied magnetic field generates a torque on both moving charges and magnetic moments provides a fundamental source of analogous behavior. Additionally, the differences in individual versus group behavior of particles that undergo periodic motion are directly analogous and lead to the important idea of *phase coherence* for both FT-ICR and FT-NMR. Furthermore, many of the same conceptual processes associated with implementing an FT-ICR experiment are also associated with implementing an FT-NMR experiment. Thus, there are numerous points in the development of the semi-classical model of precession and of the basic FT-NMR experiment at which direct connections can be made to ideas first introduced in the FT-ICR animations. An important pedagogical point is that by introducing the ideas in the more concrete and intuitive context of translationally revolving ions, the essence of the ideas/processes involved are more easily recognized when they appear again in the more abstract context of precessing magnetic moments where the time-dependent motion is initially less intuitive. Additionally, the appearance in these two different contexts of the time evolution of an interference pattern resulting from the superposition of phase coherent periodic motion for subpopulations with different frequencies provides an opportunity to focus on some of the intrinsic features associated with time dependent signals and Fourier transform techniques in general.

We make several such connections and analogies between FT-ICR and FT-NMR when describing and discussing the following points: a) the individual behavior of objects undergoing periodic motion; b) the ensemble (superposed) behavior of a collection of objects undergoing periodic motion c) the generality of the Fourier transformation applied to such time dependent interference patterns; and d) design and implementation of a simple FT-NMR experiment and instrument.

a) The individual behavior of an object undergoing periodic (time-dependent) motion and the fact that the frequency of such motion can be directly related to a property of interest is very nicely illustrated by the FT-ICR animations. Features such as frequency, phase of the motion in the x-y plane, and a method for observing the periodic motion are all concepts that are intuitively obvious from the animations. All of these ideas are easily discussed based on the FT-ICR animations, and then quickly transferred to a discussion focused on the precessional model of magnetic nuclei in a strong magnetic field when using the animations in NMR Tutor.

b) If these ideas are extended via animations to "ensembles" of many such objects undergoing periodic motion, two critically important ideas can be very effectively developed: 1) the importance of the relative phases of periodic motion of an ensemble of objects that have the same frequency of periodic motion; and 2) the importance of interference patterns that result from the superposition of periodic motion of objects with different frequencies of periodic motion. The first point is made obvious by animation of a collection of objects having the same frequency of motion, but where the primary difference is the point in space and/or the time at which each enters the motion. The "collective" difference between in-phase motion and out-of-phase motion for a collection of precessing vectors of the same frequency is made visually clear, and intuitively direct, by the corresponding animations in NMR Tutor. Thus, the importance of phase coherence in detecting periodic motion appears as a natural outcome of the ensemble behavior of a collection of objects undergoing independent periodic motion. Furthermore, application of the Faraday's and Lenz's Laws, discussed earlier, can then be used to amplify the importance of such phase coherence by suggesting a method to detect coherent precessional motion.he second point is also made obvious by animation, but this time of a collection of objects having different frequencies of motion. The periodic, or repeating, pattern of the collective motion observed in the ICR animation after the dipolar excitation of several ion of different cyclotron frequency (*i.e.*, different masses) is visually striking and intuitively clear. Likewise, correlation of this repetitive pattern in the ion motion with that of the signal induced in the receiver circuit is also visually obvious. A direct analogy with that of precessional motion of a collection of magnetic moments of different precessional frequency is clearly appropriate. Again, both of these distinctions are most easily made, first in the context of the FT-ICR animations, and then transferred by analogy to that of ensembles of precessing nuclei with the help of the animations in NMR Tutor.

c) and d) The animations depicting implementation of an FT-ICR experiment are clearly analogous to the animations in NMR Tutor depicting implementation of the basic FT-NMR experiment. Implementation an FT-ICR experiment is shown by animation of the following sequence of events: 1) application of a dipolar excitation pulse to generate phase coherent cyclotron motion; 2) observation of an oscillating signal induced in the receiver plates as the interference pattern of the ensemble's magnetron motions and phase coherent cyclotron motions evolve with time; and 3) the application of a quadrupolar excitation pulse to prepare the sample for a repeat of the experiment. Implementation of the basic FT-NMR experiment is similarly shown in NMR Tutor by animation of the following sequence of events (*i.e.* pulse sequence): application of a 90° pulse to generate phase coherent precessional motion, followed by observation of an oscillating signal induced in the receiver coil as the interference pattern of the ensemble's phase coherent precessional motions evolve with time. The delay time typically at the beginning of a NMR pulse sequence to allow for complete relaxation serves as the preparation period that is analogous to the quadrupolar excitation in FT-ICR experiment. Furthermore, loss of coherence in the cyclotron motion via collisional dampening in ICR is directly analogous to loss of phase coherence in the precessional motion via relaxation effects in NMR. So we see that FT-ICR and FT-NMR both have several common requirements: 1) to generate observable motion by generating phase coherence in the respective periodic motions; and 2) to detect the phase coherent periodic motion of the ensemble (sample). Additionally both FT-ICR and FT-NMR require Fourier transformation to convert the time dependent signal into a plot of a frequency distribution.

C-MoR FT-ICR

It is important to note that all the animations contained in the C-MoR MS module for the Quadrupole, Ion Trap, and ICR are accurate simulations of the ion trajectories. The animations are 3-dimensional renderings of the time dependent ion trajectories that are obtained by solution of the coupled differential equations of motion for any specific set of operational conditions of the device. The module allows interactive changes of the operational conditions followed by "on-the-fly" solution of the associated equations of motion to explore how the ion trajectories are affected by changes in operational conditions of the mass analyzer.

To give a sense of the content and interactive nature of the ICR animations, we present here a series of figures showing images captured during some of the animations. As with the symmetry animations, the power of this module is intrinsic to the animations themselves. Thus, these static 2-D images only hint at the effectiveness of the actual animations. We have chosen ones that illustrate some of the concepts we discuss above, but also some that highlight other analogies with FT-NMR that are only briefly mentioned, or not mentioned at all in the above discussion. In particular, Figures 2-4 show images taken from animations that show, respectively: 1) the build-up of coherent cyclotron motion for several ions of different cyclotron frequency and the resulting interference pattern (signal) that becomes observable from the superposition of their motion; 2) the effect of collisional dampening on the coherent cyclotron motion in the ICR and the similarity to the effect of relaxation in NMR; and 3) the correlation between duration of acquisition time and signal resolution.

Figure 2 shows two frames in the animation that illustrates the effect of a dipolar excitation pulse. The two frames show charged particles starting to precess around the magnetic field lines in response to kinetic energy provided by the dipolar excitation pulse. This animation also illustrates a major difference in

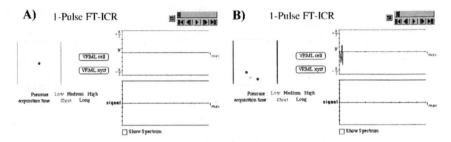

Figure 2: <u>C-MoR MS FT-ICR Excitation:</u> The effect of a dipolar excitation on three ions with slightly different masses: A) The frame before the dipolar excitation is activated; B) A later frame taken after the dipolar excitation is activated. The adjacent animations show the motion for each ion and on the right side in B the build-up of detectable signal (interference pattern) resulting from superposition of periodic motion all ions present.

the way phase coherent periodic motion is developed in FT-ICR versus in FT-NMR. In FT-ICR, the longer the dipolar excitation field is applied, the larger the amplitude of observable signal that will build-up. In NMR however, the amplitude of observable signal varies sinusoidally with the duration of the applied excitation field.

Figure 3 shows frames from two animations that illustrate the effects of different rates of coherence loss due to differing rates of relaxation. In the ICR, pressure and thus collisions cause dampening (i.e., relaxation) of the amplitude of motion and thus reduction in signal amplitude. This dampening has as its direct equivalent in NMR in the relaxation time of the NMR excited state. A shorter NMR relaxation time is analogous to higher rates of collisional dampening due to higher pressures in the ICR chamber. The spectra resulting from Fourier transformation of the detected signal highlights the effect in terms of signal intensity and width. Note that both the cyclotron and magnetron motion result in peaks in the spectrum, a complication that NMR does not have. However, discussion of the use of high frequency band pass filtering in FT-ICR to remove the signal contribution from magnetron motion has its useful analogies to their use in FT-NMR hardware design.

Figure 4 shows frames from two animations that illustrate the effect of the total measuring time on the resolution of the final spectrum. The duration of the acquisition time in Figure 4A corresponds to a fraction of the acquisition time in Figure 4B. The difference in spectral resolution of the Fourier transformed signal is obvious.

After using the C-MoR MS animations to illustrate many of these principles, we use NMR Tutor to formalize their connection directly to introductory level NMR theory.

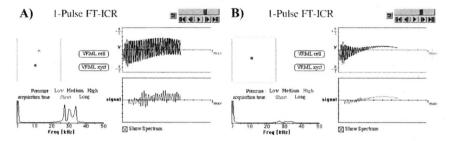

Figure 3: C-MoR MS FT-ICR Relaxation: Damping of ion motion in the ICR chamber and its effect on the signal intensity. A) Simulation for low pressure conditions; B) Simulation for high pressure conditions. Note the longer lifetime of coherent cyclotron motion in A than in B, and the corresponding greater signal intensity and narrower peaks in the Fourier transformed spectrum. This is analogous to long and short relaxation times in NMR, (A and B respectively). (See page 1 of color inserts.)

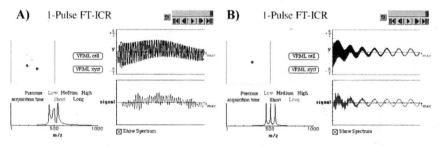

Figure 4: C-MoR MS FT-ICR Acquisition Time: The effect of acquisition time on spectral resolution. The acquisition time for B is 8 times that for A. (See page 1 of color inserts.)

NMR Tutor

NMR Tutor is a package of animations that very efficiently presents many of the basic ideas critical to an introduction to the theory of pulsed FT-NMR. Among the NMR concepts that may be explained more easily with interactive, 3-dimensional animations are: 1) the concept of precessional motion itself together with the magnetic field dependence of its frequency, 2) the correlation of the semi-classical precessional model of resonance frequencies with the quantum model of spin states and transition energies, 3) the difference between phase-incoherent precession (before a 90 degree pulse) and phase–coherent precession

104

(after a 90), 4) the concept of the rotating frame of reference, 5) the behavior of the net magnetization during a simple pulse sequence, and 6) the basic design of more complex pulse sequences. The program NMR Tutor (6) was created to help explain these concepts. The following series of figures, Figure 5-10, are snapshot images taken from various frames in the NMR Tutor animations to try to illustrate the power and effectiveness that NMR Tutor offers for communicating these types of ideas. But again, as was pointed out earlier, much of the real power lies in the animations themselves; they provide clear, almost intuitive communication of ideas that are intrinsically about a process—either a progression of changes as a function of time or as a function of changes in some other parameter.

Several techniques are incorporated within the animations to allow the user to explore the effects of changing relevant parameters. For example in Figure 6, clicking either on "Increase H_o" or "Decrease H_o" initiates an animation that illustrates the corresponding effect, $e.g.$ clicking on "Increase H_o" in Fig. 6A shows a faster precessional motion in the animation associated with Fig. 6B.

The concept of phase coherent motion is clearly illustrated in Figure 7 and dovetails nicely with the ideas discussed in the context of FT-ICR.

The several stages of a modern pulsed FT-NMR experiment, along with the associated concepts, are integrated very effectively in animations of several pulse sequences. These animations, as illustrated by Figure 8 for a typical one-pulse experiment, show a clear correlation between the time course of the pulse sequence and the time evolution of the net magnetization as well as the correlation of both to the time evolution of the observed signal.

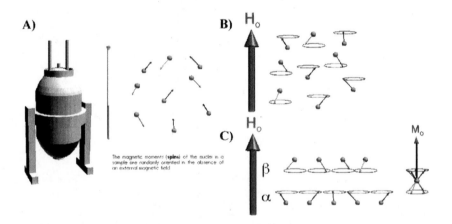

Figure 5: The Precessional Model: A) Image showing the random orientation of nuclear magnet dipoles and the lack of precessional motion in the absence of a magnetic field; B) animation showing, for a collection of spins of two distinct orientations, the precessional motion induced by a magnetic field; C) animation showing the correlation of the quantum model with the precessional model.

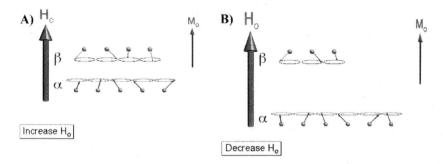

Figure 6: <u>The Effect of Magnetic Field Strength</u>: Two frames illustrating that stronger applied magnetic fields result in high precessional frequencies, larger α-β state energy gaps, and increased α-β state population differences.

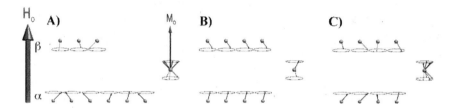

Figure 7: <u>The Effect of a 90° Pulsed Excitation</u>: A) Animation depicting phase in-coherent precession that exists before excitation. B) Frame immediately after an idealized 90° excitation in which the animation depicts phase coherent precession as well as "net" α → β transitions to give equal α,β state populations. C) As the animation continues it depicts the loss of phase coherence that evolves over time.

A clear understanding of not only what is meant by the *laboratory* and *rotating frames of reference*, but also the significance of the *rotating frame* as a particular item of utility for thinking of NMR, is always an abstract and difficult concept. These too are made almost immediately intuitively obvious by the animations that illustrate them (Figure 9).

Beyond the above basic concepts, NMR Tutor also has interactive animations that begin to explore more advance topics such as phasing effects in the processing of Fourier transformed signals both in 1-D and 2-D spectra, the basic structure of heteronuclear multipulse sequences (*e.g.*, animation of the Bilinear Rotation Decoupling, BIRD, pulse sequence), and essential features of a 2-D NMR pulse sequence.

In all of these animations, innovative use of methods to allow the user to interact with the animation gives the user the chance to explore the phenomenon

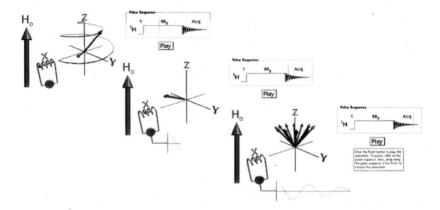

Figure 8: *A One-pulse Sequence:* A series of images taken from the animation depicting the behavior of the net magnetization during a simple 1-D pulse NMR experiment. A) Animation depicting the time evolution of the net magnetization during the progression of a 90° pulse; B) and C) two frames taken from the animation during the acquisition time period showing the time evolution of the net magnetization, including a depiction of loss of phase coherence, and the associated time dependent signal induced in the receiver coil.

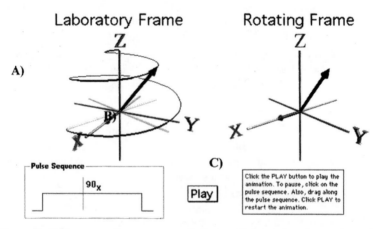

Figure 9: *Laboratory Frame versus Rotating Frame:* One frame from the middle of the animation depicting the evolution of the net magnetization, simultaneously from the viewpoint of the laboratory and rotating frames of reference.

being illustrated in the animation. For example, in order to help explain the second dimension in 2D NMR, it is helpful to have an interactive model of the pulse sequence. The central problem of animating a 2D NMR experiment is that it is actually a series of 1D experiments, repeated many times with small, incremental changes in the sequence for each repeat. An animation of the net magnetization during a 2D experiment would be very repetitive, with only slight changes in each repetition. Furthermore, the relationship between the delay parameter and the net magnetization would be lost in such a one-dimensional animation. Typically, a one-dimensional animation is controlled by a drag bar that represents the progression of time through the pulse sequence. By adding a second drag bar as the delay parameter in the pulse sequence, the student can directly interact with the second time dimension in the same way she interacts with the first time dimension (Figure10). After viewing the pulse sequence in the first time dimension, the student can position the drag bar at any point in the sequence, for example, immediately before acquisition, and then change the delay parameter via the second drag bar. The resulting animation of the net magnetization and of the FID signal will then illustrate the changes in net magnetization that results from the change in the delay. As an additional aid to the student, a caption underneath the pulse sequence changes to describe what is happening at the point in the pulse sequence indicated by the time indicator. This feature is even more helpful with the HETCOR sequence.

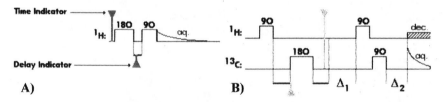

Figure 10: Multiple drag bars in NMR Tutor A) Experiments with multiple time variables (delays) are illustrated with two drag bars as shown in the T_1 inverse recovery pulse sequence. B) The HETCOR pulse sequence, showing one possible position of the first time dimension drag bar (upper triangle) and the variable delay (lower triangle).

As powerful as NMR Tutor's animations are in conveying many of the essential ideas of basic NMR theory, there are some fundamental limitations to using such animations for many advanced topics in NMR. As appealing as the semi-classical model is for understanding basic pulse sequences, it is not helpful for explaining many intrinsically quantum mechanical phenomena, *e.g.* double quantum coherence. It can also potentially lead the student to believe that the quantum mechanical underpinnings are simply a more exact form of the animations, rather than correctly understanding that the animations are misleading for all but the most basic pulse sequences. Thus, when using NMR Tutor as a teaching tool it is important to communicate these facts concerning the semi-classical model and use of such animations.

C-MoR NMR Module

In order to produce a module that complements the strengths of NMR Tutor, the C-MoR NMR module was designed specifically as a tool to assist in teaching NMR spectral interpretation skills. It provides a versatile, high quality interface through which one can efficiently display a wide variety of modern NMR spectral data within a common interface. The display features embedded make it easy to highlight correlations between signals and structural features, as well as correlations between the same or complementary information contained within different NMR experiments on the same sample. It is developed to illustrate: 1) the specific types of information contained in different NMR experiments, and 2) the different interpretational strategies used in practice to extract that information. Implementing these objectives in the classroom with this module, however, will be left up to the particular pathway that a particular instructor using it decides to take; we have intentionally _not_ scripted any particular order in which most of the features can be accessed. Rather we have merely organized them so that those that are logically connected are correspondingly available to the user. At this point we have primarily focused on developing the graphical user interface, embedding within it the types of display features that would support broad exploration of the correlation of information that is contained in NMR data; and so at this point, we have only included a small set of compounds in the module's associated database for illustration purposes. We hope to gather, and will welcome, input from colleagues across the country (or world) especially in the form of suggestions of pedagogically useful examples to add to the set of spectra.

Now for some particulars: The NMR module provides menus from which one can select interactive spectra. The viewer itself is a Flash movie that takes its information regarding the menus and the associated spectrum Flash file from an XML file that can be modified to add features or to tailor the module to a new way of using it. Figure 11 shows the "expanded" menus that are currently available. By selecting an entry from one of these submenus, the menu system available to the user is automatically updated in order to provide ready access to other spectra relevant to the option selected. The spectral data are correlated by compound and by experiment, and logical spectrum overlays are made available. Thus, the complementary nature of the data available between 2-D experiments can be illustrated by selectively overlaying spectra. In addition to being used by an instructor in a targeted teaching mode, as Figure 11 reveals, this module can also be used by the student in "quiz" mode—a mode in which the student can access any of the spectra but without displaying the corresponding structure itself. This allows the student to self-test their ability to determine the structure of a compound based on its NMR spectra.

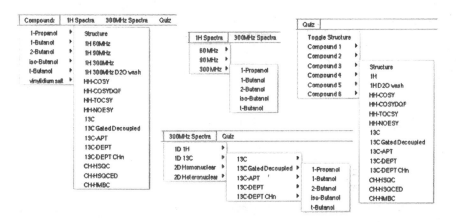

· *Figure 11: Some expanded menus in the NMR spectrum viewer. On the far left top is the top-level menu bar showing the first expandable menu with compound selected. It shows the spectra that are available for display for that compound. Expansions of each successive top-level menu bar item together with one of the sub-options selected are display in the middle and right portions of the figure.*

One form of interactivity consists of being able to click on a peak in the spectrum, or on a structural feature in the molecule's structure, and to show the underlying correlation between the two, *e.g.* a proton in a structure and signal in the 1H-NMR (see Figure 12). This form of interactivity is also extended to the two-dimensional spectra as well. For example, clicking on an atom in the structure, or its corresponding signal in the F1 dimension, signifies the source of the magnetization that can be detected. The convention adopted for color coding 2-D spectra are red for the magnetization source, blue for the observed magnetization, purple for cross peaks, and green for diagonal peaks. Spectra that rely on correlation of magnetization through bonds (HH-COSYs and HH-TOCSY) also allow for the selection of all signals that propagate through bonds to a heteroatom by clicking on the heteroatom (see Figure 13). It goes without saying that one can also click on a particular cross peak to see the originating nucleus (in red) and the observed nucleus (in blue), as shown in Figure 14A.

Finally some spectra can be overlaid, as shown in Figure 14B. The overlaid spectrum has the same color as the text buttons on the side. In this particular case

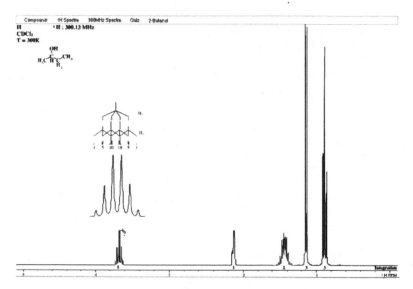

Figure 12: a 1D proton NMR of 2-butanol after clicking on the peak with the expanded coupling pattern. This same image would show after clicking the C-2 hydrogen. Note the additional 2-Butanol menu item.
(See page 2 of color inserts.)

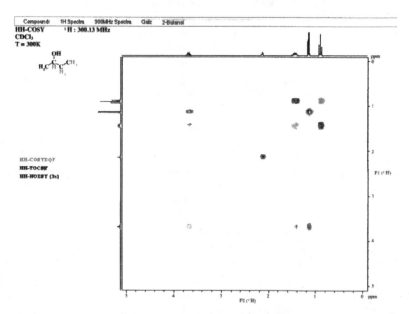

· Figure 13: An HH-COSY of 2-butanol after clicking the C-3 carbon, showing the signals detectable due to HH-coupling going 'through' this atom.
(See page 2 of color inserts.)

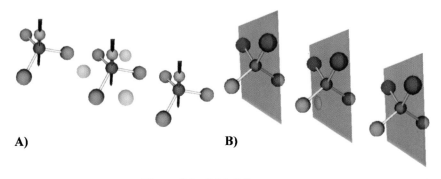

Figure 8.1. C-MoR Symmetry.

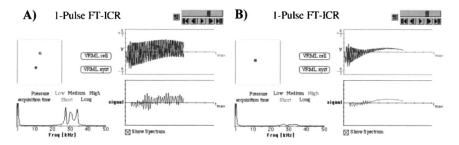

Figure 8.3. C-MoR MS FT-ICR Relaxation.

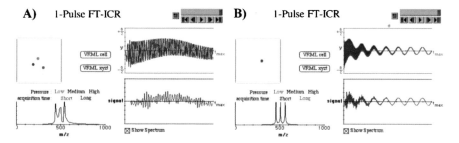

Figure 8.4. C-MoR MS FT-ICR Acquisition Time.

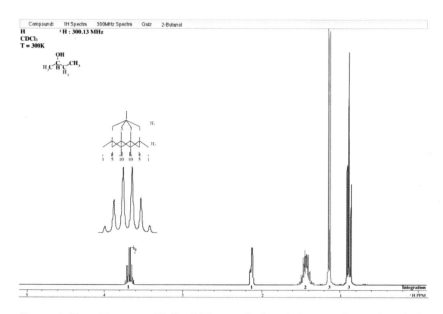

Figure 8.12. a 1D proton NMR of 2-butanol after clicking on the peak with the expanded coupling pattern (blue). This same image would show after clicking the C-2 hydrogen (blue). Note the additional 2-Butanol menu item.

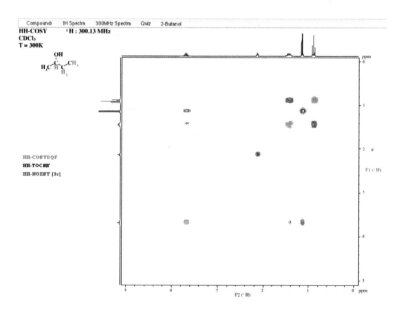

Figure 8.13. An HH-COSY of 2-butanol after clicking the C-3 carbon, showing the signals detectable due to HH-coupling going 'through' this atom.

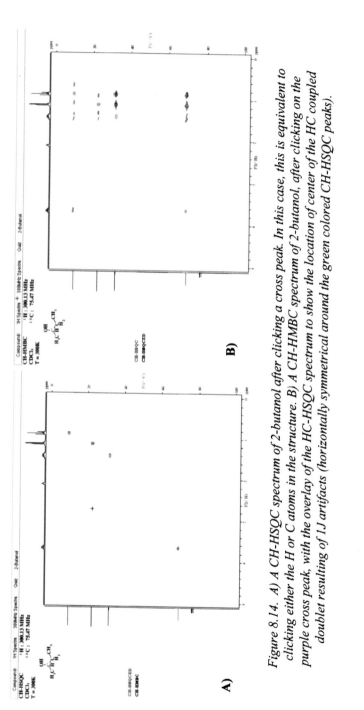

Figure 8.14. A) A CH-HSQC spectrum of 2-butanol after clicking a cross peak. In this case, this is equivalent to clicking either the H or C atoms in the structure. B) A CH-HMBC spectrum of 2-butanol, after clicking on the purple cross peak, with the overlay of the HC-HSQC spectrum to show the location of center of the HC coupled doublet resulting of 1J artifacts (horizontally symmetrical around the green colored CH-HSQC peaks).

the spectral overlay helps in showing the, sometimes strong, artifacts caused by the ^{1}J HC couplings in the CH-HMBC spectra by overlaying the CH-HSQC spectrum.

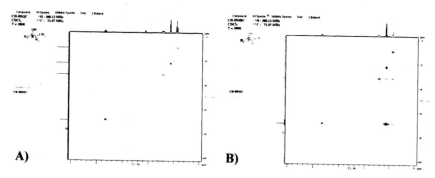

A)   B)

Figure 14: A) A CH-HSQC spectrum of 2-butanol after clicking a cross peak. In this case, this is equivalent to clicking either the H or C atoms in the structure. B) A CH-HMBC spectrum of 2-butanol, after clicking on the dark cross peak, with the overlay of the HC-HSQC spectrum to show the location of center of the HC coupled doublet resulting of 1J artifacts (horizontally symmetrical around the medium grey colored CH-HSQC peaks). (See page 3 of color inserts.)

Conclusion

We have found these computer programs and animations to be extremely valuable in providing students with an intuition and understanding of NMR theory, instrumentation, and interpretation. The C-MoR MS module, in particular its FT-ICR component, provides a concrete example through which one can think of, and understand, the essential characteristics of physical objects that are undergoing periodic motion, so that it is easier for students to then understand the more abstract concept of nuclear magnetic moments undergoing periodic motion. Likewise, an understanding of basic symmetry gives an easy way to look at and determine chemical and magnetic equivalency, which are important in NMR interpretation.

The power of animations lies in their ability to rapidly communicate the time evolution aspect of things that are intrinsically a process, *i.e.* changes that take place as a function of time. The important aspects of any FT technique are intrinsically dynamic in nature and thus animations offer a powerful vehicle for explaining these phenomena. Just as it is true that "a picture is worth a thousand words" when describing a static object, "an animation is worth a million pictures" when describing a process. This benefit is even more dramatic for more abstract processes. This is why the ICR animations are so efficient in getting someone to understand what cyclotron motion is and what it means for objects to have different frequencies of cyclotron motion. The animations are even more effective because they clearly illustrate the essential features of such abstract concepts as phase coherent motion versus phase in-coherent motion. These ideas are directly analogous to corresponding concepts in NMR, and thus the effectiveness of NMR Tutor in helping to explain NMR.

We are continuing to develop both NMR Tutor and the C-MoR NMR module and plan to merge them into one package that covers both theory and interpretative skills.

Acknowledgments

We wish to acknowledge the University of Richmond, the Jesse Ball DuPont Fund, the FIPSE program of the U.S Department of Education, and the Henry and Camille Dreyfus Foundation, and the Virginia Foundation for Independent Colleges for the financial support over a ten-year period that made the C-MoR work and its dissemination possible.

References

1. a) Freeman, A. W .; *Journal of Allied Health* **1987**, *16*, 177-183; b) Sanger, Michael J.; Phelps, Amy J.; Fienhold, Jason; *Journal of Chemical Education* **2000**, *77*, 1517-1520; c) Lloyd Jan; Moni Karen B; Jobling Anne; *Down's Syndrome, Research and Practice : the journal of the Sarah Duffen Centre /* University of Portsmouth **2006**, *9*, 68-74; d) Hasselbring, T. S.; Glaser, C. H *The Future of children /* Center for the Future of Children, the David and Lucile Packard Foundation **2000**, *10*, 102-122.

2. a) Kiboss, Joel K.; Ndirangu, Mwangi; Wekesa, Eric W. *Journal of Science Education and Technology* **2004**, *13*, 207-213; b) Roschelle J M; Pea R D; Hoadley C M; Gordin D N; Means B M *The Future of children /* Center for the Future of Children, the David and Lucile Packard Foundation **2000**, *10*, 76-101; c) *Anatomical record. Part B, New anatomist* **2005**, *286*, 8-14.

3. Further information on the collection of web deliverable C-MoR modules that are is available at http://oncampus.richmond.edu/academics/a&s /chemistry/CMoR/info/index.html.

4. Edison Project, Chemistry Department, Columbia University, New York City, NY; http://www.columbia.edu/cu/chemistry/ugrad/multimedia.html.

5. Abrams, Charles; "IR Tutor"; Abrams Educational Software; see http://members.aol.com/charlieabr/

6. a) Abrams, Charles; "NMR Tutor"; Abrams Educational Software; see http://members.aol.com/charlieabr/; b) Also NMR Tutor is available in Shockwave format at http://faculty.ccc.edu/cabrams/projects/nmrtutor.

7. a) Advanced Chemistry Development ACD 1H-NMR Predictor; http://www.acdlabs.com/products/spec_lab/predict_nmr/hnmr/ b) Bruker NMR, Billerica, MA

8. Some good introductory reference texts on introductory level NMR include: a) Harris, R. K. *Nuclear Magnetic Resonance Spectroscopy*; Pitman Publishing Inc, Marshfield, MA, 1983. b) Friebolin, H. *Basic One-and Two-Dimensional NMR Spectroscopy*, 4th ed.; Wiley-VCH, Weinheim, Germany, 2005. c) *One-dimensional and Two-dimensional NMR Spectra by Modern Pulse Techniques*; Nakanishi, K., Ed.; University Science Books, Mill Valley, CA; d) Günther, H. *NMR Spectroscopy*, 2nd ed.; John Wiley and Sons, New York, NY, 1995.

9. The texts we use for our Spectroscopy & Instrumentation course are: a) reference 8b; b) Silverstein, R. M.; Webster, F. X.; Kiemle, D. J. *Spectrometric Identification of Organic Compounds*, 7th ed.; John Wiley & Sons, New York, NY, 2005.

10. Serway, R.A.; Jewett, J. W., Jr. *Physics for Scientists and Engineers with Modern Physics,* 6th ed.; Brooks/Cole–Thomson Learning; Belmont, CA, 2004. a) Chapter 31. b) Chapter 30.

Chapter 9

The Ubiquitin NMR Resource

Richard Harris[1] and Paul C. Driscoll[1,2]

[1]Department of Biochemistry and Molecular Biology, University College London, Gower Street, London WC1E 6BT, United Kingdom
[2]National Institute for Medical Research, The Ridgeway, Mill Hill, London NW7 1AA, United Kingdom

We describe the ubiquitin NMR resource (http://www.biochem.ucl.ac.uk/bsm/nmr/ubq), which evolved from the need to teach postgraduate students, within our laboratory, the process of obtaining sequence specific resonance assignments. Ubiquitin was chosen as a model protein, because it is commercially available and high quality NMR spectra can be acquired in a relatively short period of time. The data in this resource can be used as a practical demonstration of obtaining sequence specific resonance assignments, which provides an introduction to triple resonance NMR experiments and how the structures of the individual amino acid types give rise to different ranges of $C\alpha$ and $C\beta$ chemical shifts. From the assignment of the backbone resonances a comparison of the chemical shifts of the $C\alpha$, $C\beta$ and CO resonances to random coil values allows for a straightforward prediction of the secondary structure content. All the software used in the resource for processing and spectral analysis is freely available from the Internet.

Introduction

Ubiquitin is a small protein (76 amino acid residues) that derives its name from its occurrence throughout the plant and animal kingdoms (1, 2). The three-dimensional structure of ubiquitin has been extensively characterized by both X-ray crystallography and NMR spectroscopy (3, 4) and comprises a combination of alpha helices and beta strands in a ββαββαβ motif, with a chain topology now known as the ubiquitin fold. Over-expression of ^{15}N- and ^{13}C/^{15}N-isotope labeled ubiquitin in *Escherichia coli* regularly yields tens of milligrams per litre of media and, as such, represents an economic method to provide a stable, low molecular weight, non-aggregating, highly soluble protein that is ideal for testing new NMR pulse sequences and other aspects of the application of NMR methodology.

The Ubiquitin NMR Resource, described here, comprises a series of NMR spectra acquired on commercially available samples of ^{15}N-labeled and double-labeled ^{13}C/^{15}N ubiquitin (5mg in *ca.* 300uL 20mM potassium phosphate, pH 5.8, VLI Research Inc.). The spectra are divided into three archives: (i) triple resonance experiments for assignment of the backbone resonances; (ii) experiments to obtain side chain assignments; and (iii) spectra for assignment of inter-proton distance restraints (Table 1). Each archive contains the raw Varian FID data, made available to be processed according to taste, an NMRpipe (5) script for referencing information and for outputting the data into a format for the user's preferred NMR analysis software, and pre-processed files ready to be analyzed using the program ANSIG (6). A discussion of NMR data processing is beyond the scope of this chapter, however most general NMR texts review this area (for example Rule and Hitchens (7)); for a more in-depth treatise on data processing see Hoch and Stern (8).

Table 1. Experiments acquired on ^{15}N or ^{13}C/^{15}N ubiquitin at 25°C (mixing times given in parenthesis)

Archive 1	*Archive 2*	*Archive 3*
2D ^{1}H,^{15}N HSQC	2D ^{1}H,^{13}C HSQC	2D ^{1}H,^{15}N HSQC
3D HNCA	2D (HC)CH-TOCSY	3D ^{15}N TOCSY-HSQC (80ms)
3D HN(CO)CA	3D H(C)CH-TOCSY (7ms)	3D ^{15}N NOESY-HSQC (100ms)
3D HNCACB	3D H(C)CH-TOCSY (21ms)	2D ^{1}H,^{13}C CT-HSQC
3D HN(COCA)CB		3D ^{13}C NOESY-HSQC (100ms)
3D HNCO		
3D HN(CA)CO		

For an explanation of the nomenclature see (9, 10)

The NMR study of any protein, whether aimed at the investigation of ligand binding, dynamics or determination of the three-dimensional solution structure, will require the cross peaks in the 2D ^{1}H,^{15}N-heteronuclear single quantum correlation spectroscopy (HSQC) spectrum to be assigned to their individual amino acid residues. The ^{1}H,^{15}N-HSQC is often described as a "fingerprint" of

116

the protein because every amino acid (except proline) contains a backbone amide group that should give rise to a single cross peak that correlates the amide proton and amide nitrogen. Figure 1 shows the (already-assigned) $^{1}H,^{15}N$-HSQC spectrum of human ubiquitin acquired on a Varian UnityPlus spectrometer operating at a nominal ^{1}H frequency of 600MHz. For small proteins, such as ubiquitin, the assignment process is relatively straightforward and may by achieved with good accuracy by "traditional" analysis of three-dimensional ^{15}N-edited TOCSY-HSQC and NOESY-HSQC spectra (10).

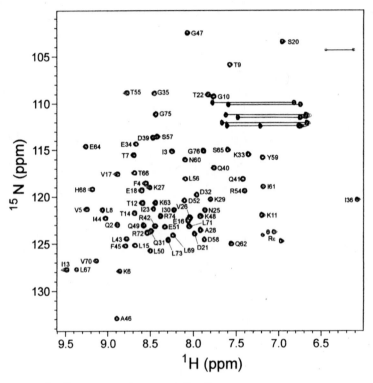

Figure1. 2D $^{1}H,^{15}N$ HSQC spectrum of $^{13}C/^{15}N$-labelled ubiquitin recorded on a 600 MHz Varian UnityPlus spectrometer at 298K, with final assignment of backbone amide NH groups.

As an example of how the sequential assignment can be made using the ^{15}N-edited TOCSY-HSQC and NOESY-HSQC spectra, a series of strip plots for residues S57 to N60 is shown in Figure 2. The first requirement of the strategy is the identification of the amino acid type from the through bond $^{1}H-^{1}H$ TOCSY correlations (11). Then the sequential residue can be identified via the $H^{N}(i)$-$H\alpha(i-1)$, $H^{N}(i)$-$H^{N}(i-1)$ or $H^{N}(i)$-$H^{N}(i+1)$ NOE correlations and maybe confirmed with $H^{N}(i)$-$H\beta(i-1)$ NOEs.

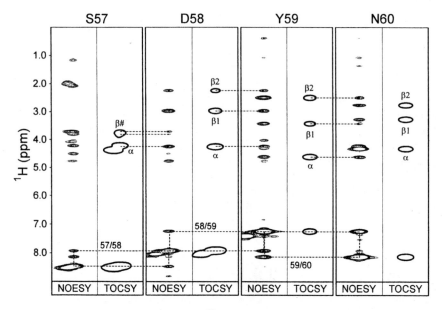

Figure 2. 2D Strip plots from the 3D ^{15}N edited NOESY-HSQC (multiple contour levels) and TOCSY-HSQC (single contour level) spectra for the backbone amide resonances of Ser57, Asp 58, Tyr 59 and Asn 60. The $H^N(i)$- $H^N(i-1)/H^N(i+1)$ NOEs are indicated with dotted line boxes and labelled with residue number and dotted lines show the $H^N(i)$-$\alpha/\beta_1/\beta_2(i-1)$ inter-residue NOEs. Mixing times for the ^{15}N NOESY-HSQC and TOCSY-HSQC are 100ms and 60ms, respectively.

Even for small proteins the identification of amino acid through spin system is complicated because of the degeneracy of the pattern of side chain proton correlations (for example differentiating Gln and Glu; Asp and Asn residues) and it can be difficult to delineate long-chain amino acids, such as Lys and Arg, because relayed connectivities do not often travel completely down the side chain. For example, in Figure 2 the pattern of TOCSY cross peaks for residues D58, Y59 and N60 are similar and the unambiguous determination of amino acid type for each of these NH cross peaks would not be possible. If we consider the through-bond correlations for residue S57, the chemical shift degeneracy of the beta protons results in a pattern of cross peaks that is ambiguous because at first glance it could result from either Ser (α, $\beta\#$) or Gly (α_1, α_2). However, assuming that the set of TOCSY correlations is complete for the i+1 residue (H^N to α, β_1, β_2), the residue is most likely Asn, Asp, Cys, His, Phe, Trp or Tyr. Inspection of the primary amino acid sequence of ubiquitin reveals that of the six glycines and three serines only S20 and S57 precede a residue that would give rise to a α, β_1 and β_2 pattern of correlations (D21 and D58). Considering the next amino acids in the sequence, T22 and Y59, the α, β_1

118

and β_2 set of cross peaks observed for the i+2 residue would eliminate T22 as a possible assignment and so these three NH cross peaks can be assigned as S57-D58-Y59.

However, the ^{1}H-directed general approach to resonance assignment is non-ideal as it relies upon the correct identification of potentially ambiguous through-space NOE connectivities. There is the potential for error because of chemical shift degeneracy and difficulties in amino acid identification because of inefficient magnetization transfer in TOCSY spectra; for larger proteins the ^{1}H-^{1}H J-correlation techniques frequently fail because of the increased proton resonance linewidth. To alleviate these problems, an alternative approach, employing triple resonance NMR experiments that rely on through-bond connectivities via well-resolved one- and two-bond heteronuclear couplings, has been developed (for reviews see references (7, 9, 10, 12)). This approach, whilst being arguably more complex and often requiring more time on the NMR spectrometer, has the advantage of being less ambiguous and more reliable. A combination of some (or all) of the following triple resonance NMR experiments, HNCO, HN(CA)CO, HNCA, HN(CO)CA, HNCACB, HN(CO)CACB (Figure 3) can be used to obtain the backbone amide ^{1}H and ^{15}N assignments. The nomenclature of the triple resonance NMR experiments describes, in a codified manner, the magnetization pathway for that experiment. For example, the HN(CO)CA experiment correlates the amide ^{1}H and ^{15}N resonances of one residue (i) with the Cα chemical shift of

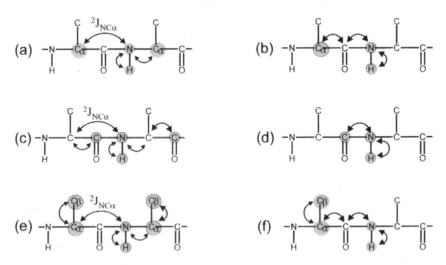

Figure 3. Schematic diagrams of the nuclei that are correlated in the triple resonance NMR experiments (a) HNCA, (b) HN(CO)CA, (c) HN(CA)CO, (d) HNCO, (e) HNCACB and (f) HN(CO)CACB. Nuclei for which the chemical shift is mapped in the experiment are shown in dark circles. Magnetization transfer in these experiments is indicated by solid lines, with the direction of transfer shown by arrows.

the preceding residue (i-1). Chemical shifts of nuclei in parenthesis - here the carbonyl group - are not mapped.

Analysis of the triple resonance NMR experiments in the ubiquitin resource allows identification and sequential assignments for 70 out of the 73 (76 less 3 prolines) backbone amide nitrogen and proton resonances. The absence of resonances corresponding to residues M1, E24 and G53 is attributed to exchange broadening. A cross peak is observed for every correlation that is anticipated from the primary structure of the ubiquitin polypeptide chain. Therefore, the NMR data available on ubiquitin represents an ideal model system for demonstrating the basic aspects of the analysis of triple resonance NMR experiments.

Sequence Specific Resonance Assignments using the Resource

By way of a typical example, we describe here one way of using the ANSIG session containing the triple resonance NMR spectra to teach students the rudiments of sequential assignments for ubiquitin. ANSIG (Assignment of NMR Spectra by Interactive Graphics) is a graphics tool designed by Per Kraulis for the analysis of NMR spectra (6) that has been widely adopted in the protein NMR community and has provided the model for the recent development of a new package with similar 'philosophy', CcpNmr Analysis (www.ccpn.ac.uk). A complete description of ANSIG can be found at www-ccmr-nmr.bioc.cam.ac.uk.

Briefly, each ANSIG session within the resource contains a series of files required by the program: a control file (.ctr) contains the pathname of the ANSIG program and a listing of the other required files; a spectrum description file (.spd) containing a list of all the types of spectrum that will be loaded into the ANSIG session; a sequence file (.seq) which is the primary amino acid structure for the protein being studied; an initialisation script (.ini) which contains information about any user-defined ANSIG macros, the number of spectrum windows to be displayed, and which of the spectra defined in the .spd file format to display in each of the windows; and a cross peaks file (.cpk) which is a binary file containing the spectral coordinates and intensities of the loaded cross peaks and any assignment of those cross peaks. The spectral contour maps and cross peak lists, pre-calculated using AZARA(13) prior to loading in ANSIG, are contained in a separate directory.

When the ANSIG session contained in archive 1 is accessed, by using the command ansig ubq in a shell window, seven windows will be initiated (Figure 4). The windows labeled ANSIG w1 through ANSIG w6 display contours for the NMR spectra (described briefly below), whilst the window titled ANSIG v3.3 contains a menu of the standard ANSIG macros, list of spectra (defined in the .spd file), ubiquitin amino acid sequence (from the .seq file), and a list of user-defined macros (from the .ini file).

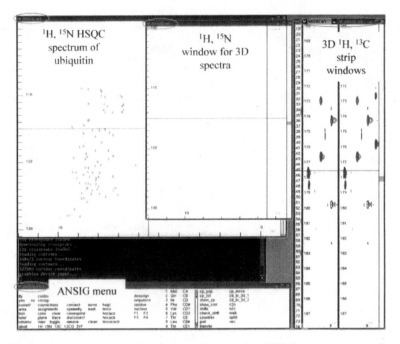

Figure 4. A screen snapshot image showing the ANSIG sequence specific resonance assignment session obtained upon loading.

Within ANSIG all commands are executed using the middle mouse button, whilst moving around and zooming in and out of spectra are operated using the left and right mouse buttons, respectively.

- ANSIG w1 – $^1H^N,^{15}N$-HSQC spectrum of ubiquitin
- ANSIG w2 – first $^1H^N,^{13}C$ strip window for the 3D HNCA, HN(CO)CA, HNCACB and HN(COCA)CB spectra.
- ANSIG w3 – second $^1H^N,^{13}C$ strip window for the 3D HNCA, HN(CO)CA, HNCACB and HN(COCA)CB spectra
- ANSIG w4 – $^1H^N,^{15}N$ window displaying cross peaks from 3D spectra.
- ANSIG w5 – first $^1H^N,^{13}CO$ strip window for the 3D HNCO and HN(CA)CO spectra
- ANSIG w6 – second $^1H^N,^{13}CO$ strip window for the 3D HNCO and HN(CA)CO spectra

The data can be used to "walk" either forwards or backwards along the protein backbone chain (until either a proline or a residue with a missing amide resonance is reached). The following describes the process of starting from an HSQC cross peak for residue i to find the next i+1 residue, which is shown diagrammatically in Figure 5.

Displaying 2D ^1H, ^{13}C slices

By clicking on the 2d_to_3d_1 macro with the middle mouse button and then selecting a cross peak in the ^1H, ^{15}N HSQC spectrum a slice through all the 3D experiments at the corresponding ^{15}N chemical shift is displayed in windows w2 (Cα/Cβ) and w5 (CO). Dotted lines in the HSQC spectrum (Figure 5a) highlight the cross peak selected and, in the strip windows, the NH ^1H chemical shift of the given HSQC cross peak.

Four spectra corresponding to the HNCA, HN(CO)CA, HNCACB, and HN(COCA)CB experiments are superimposed in window w2. The spectra can be toggled on or off using the toggle command or by typing [tab]-<n>, where <n> represents the number key of the experiment (the order is given by the definition in the .spd file). For example, the HNCA experiment is spectrum 1, HN(CO)CA is spectrum 2, *etc.* Sampling this operation enables the user to establish the colour 'code' that connects the contours with the spectrum labels.

Finding the next residue along

To start the assignment, the main cross bar (solid line) is placed, using the middle mouse button to drag it into position, on the Cβ and CO correlations for residue i in windows w2 and w5, respectively (Figure 5b). The c2n macro is used to take a slice through the 3D data set at the specific carbon frequency and the result is displayed in window w4. By selecting the Cα of residue i (a dotted line through the cross peak is then displayed) the cross peaks displayed in window w4 represent all the possible amide NH signals that have a correlation at a similar Cα chemical shift and so provides a readout of all the candidates for residue i+1 (Figure 5c). For clarity of the subsequent discussion the candidates for the i+1 residue will be denoted 'j' and where we are searching for j=i+1.

To determine which of the cross peaks arise from the next residue along is now a process of "trial-and-error". The scope of this testing can be sensibly limited to encompass only those signals that give Cα correlations in both the HNCA and HN(CO)CA spectra. Any NHs that only have cross peaks in the HNCA experiment alone can be reliably eliminated, as these amide signals cannot correspond to the following residue. In the example shown (Figure 5c), there are four candidates for the i+1 residues (j=1,2,3,4). By using the 2d_to_3d_2 macro to display the 2D ^1H,^{13}C strips for each possibility in turn in window w3, a comparison can be made to determine whether the Cβ(j-1) correlation in the HN(COCA)CB experiment of the "new" residue has the same chemical shift as the Cβ(i) cross peak of the starting residue. If the shifts do not match then we can move onto the next candidate amide NH cross peak. However, if the Cβ correlations do match (Cβ(j-1) ≈ Cβ(i)) then this possibility could be the next residue. Sometimes the non-degeneracy of the Cα and Cβ shifts will be sufficient to make an unambiguous selection of j=i+1. However, in

122

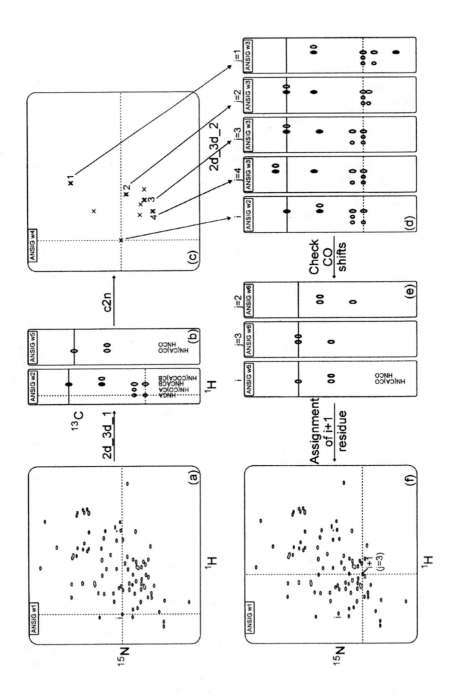

Figure 5. Representation of the process involved in assigning the (i+1) residue using triple resonance NMR spectra in an ANSIG session starting from the NH cross peak for residue 'i'; (a) selecting an amide cross peak in the $^1H,^{15}N$ HSQC spectrum, (b) displaying $^1H,^{13}C$ strips for that amide, (c) taking a slice through the $^{13}C\alpha(i)$ correlation, (d) comparison of the $^1H,^{13}C$ strips for all possible next-residue amide NH groups, (e) comparison of the ^{13}CO strips for the two amide NHs that have matching $^{13}C\alpha(j-1)$ and $^{13}C\beta(j-1)$ correlations with residue i (see text), (f) unambiguous assignment of residue i+1.

the particular example shown, both amides j=2 and j=3 appear to have both Cα(j-1) and Cβ(j-1) chemical shifts at the correct shift (Figure 5d). Therefore, it is not possible to determine which is the i+1 residue on the basis of these shifts alone. An analysis of the carbonyl group correlations can be used to determine which spin system is indeed the correct sequential assignment (Figure 5e). In this case the pattern of NH, CO correlations unambiguously identifies amide j=3 as the correct i+1 residue (Figure 4f). Generally, assuming good quality NMR spectra, only one combination of the Cα, Cβ and CO chemical shifts will provide a good match. This reflects the very small likelihood that 2 residues simultaneously have degenerate NH, N, Cα, Cβ and CO shifts. The process is then repeated for the new i+1 residue to find the i+2 residue, and so on.

Sequence specific resonance assignments

In "sequential assignment" a good place to start is an amino acid with a highly characteristic combination of Cα and Cβ chemical shifts. Glycine (Cα *ca.* 45ppm, no Cβ), alanine (Cα *ca.* 53ppm, Cβ *ca.* 19ppm), serine and threonine (Cβ 60-75ppm) are the generally preferred starting points. For example, an easy way to search for the NH groups that have a correlation to an alanine is to look for cross peaks in window w4 that have Cβ cross peaks with shifts upfield of (i.e. lower than) 20ppm. This range of carbon chemical shifts is readily selected by opening up the scroll bar (red bar on the right hand side of all the windows), using the right mouse button. The two amino acid sequences that contain an alanine in ubiquitin are Lys-Ala-Lys (residues 27-29) and Phe-Ala-Gly (residues 45-47). The cross peaks for residue G47 can be identified straightforwardly through the unique Cα chemical shift range for glycine residues and the Cβ(i-1) correlation corresponding to an alanine.

Aside from Gly, Ala, Ser and Thr, the other amino acid types do not have specifically characteristic combinations of Cα and Cβ chemical shifts. However, these residues segregate into distinct sub-groups that have similar chemical shifts. For example, Ile, Phe and Tyr residues have chemical shifts for Cα in the range 58-61ppm and for Cβ at ~39ppm; Asn, Asn and Leu have Cα 53-55ppm and Cβ 39-41ppm; whilst for Arg, Cys, Glu, Gln, His, Lys, Met and Trp the Cα resonances reside in the region of 55-57ppm and Cβ at 28-33ppm. Two C programs, which are accessed through macros in the ANSIG session, can be used to predict the amino acid type based upon the differences in the combination of Cα and Cβ chemical shifts (9). The onecacb macro calculates the probability of an amino acid type using a single combination of Cα and Cβ chemical shifts. The fewcacb macro takes multiple combinations of Cα and Cβ chemical shifts from sequential residues and calculates the probabilities of each amino acid type for all possible positions within the particular primary sequence provided.

Secondary structure prediction

The Cα, Cβ and CO chemical shifts relative to standard values have a clear correlation with the polypeptide torsion backbone angles φ and ψ (14, 15). For example, for the Cβ resonance a downfield shift from the random coil position is observed for extended (β-) structure, with ψ ~ 130°, whereas for helical structures a small upfield shift is observed (ψ ~ -50°). On this basis the local secondary structure in ubiquitin can be predicted using the Cα, Cβ and CO chemical shifts by either plotting the secondary shifts, corresponding to the difference between the observed and random coil chemical shifts (calculated using the ANSIG macro sec), or by using the chemical shift index (CSI) program (15) available elsewhere.

Concluding Remarks

We have presented a resource that is suitable to guiding graduate students in the application of NMR to proteins, and in particular the sequence specific resonance assignment of proteins using triple resonance heteronuclear spectroscopy. The data is freely available in a downloadable form on the Internet to empower undergraduate educators to teach protein NMR using the real NMR spectra acquired on ubiquitin. We regard the relatively simplicity and excellent quality of the ubiquitin data as providing close to the optimum in what budding spectroscopists could possibly encounter in later projects. In addition, in an era when we can expect more and more computer-aided or automated assignment tools that will remove the need for the spectroscopists to engage directly in the assignment process, the resource can provide a valuable and readily available hands-on example of the underlining principles of such tools - especially useful when these more sophisticated approaches fail. The ultimate aim of the resource is to provide data on a model system for determining a three dimensional solution structure of a protein. Thus further developments of the resource would provide a test sample dataset to be followed through the process of assigning all the proton resonances in a protein, specific assignment of interproton nuclear Overhauser effects (NOEs) - whether manually or through automatic packages - and 3D solution structure calculations. Additional data, such as spectra used in end-stage structure refinement (e.g. for the measurement of residual dipolar couplings), could also be included.

Acknowledgements

We would like to thank Dr Michael Plevin for the original inspiration for setting up the Ubiquitin NMR Resource and Dr Acely Garza for insightful discussions.

NMRpipe - http://spin.niddk.nih.gov/bax/software/NMRPipe
ANSIG - (windows) - http://www.csb.ki.se/nmr/AFW.html
  (irix) http://www.bio.cam.ac.uk/nmr/ccmr/public/ANSIG/obtain_info.html
  (linux) - http://vivace.bi.a.u-tokyo.ac.jp/_takeshi/ansig3opengl/
Azara - http://www.bio.cam.ac.uk/azara/

References

1. Hershko, A. A. Ciechanover *Annu. Rev. Biochem.* **1982**, *51*, 335-364.
2. Ciechanover, A., D. Finley A. Varchavsky *J. Cell. Biol.* **1984**, *23*, 27-53.
3. Vijay-Kumar, S., C. E. Bugg W. J. Cook *J. Mol. Biol.* **1987**, *194*, 531-544.
4. Cornilescu, G., J. L. Marquardt, M. Ottinger A. Bax *J. Am. Chem. Soc.* **1998**, *120*, 6836-6837.
5. Delaglio, F., S. Grzesiek, G. W. Vuister, G. Zhu, J. Pfeifer A. Bax *J. Biomol. NMR* **1995**, *6*, 277-293.
6. Kraulis, P. J. *J. Magn. Reson.* **1989**, *24*, 627-633.
7. Rule, G. S. T. K. Hitchens, *Focus on Structural Biology Volume 5: Fundamentals of Protein NMR Spectroscopy*, Springer:2006.
8. Hoch, J. S. A. S. Sterm, *NMR Data Processing*, Wiley:London, UK, 1996.
9. Grzesiek, S. A. Bax *J. Biomol. NMR* **1993**, *3*, 185-204.
10. Clore, G. A. M. Gronenborn *Methods Enzym.* **1994**, *239*, 349-363.
11. Wüthrich, K., *NMR of Proteins and Nucleic Acids*, Wiley:London,UK,1986.
12. Evans, J. N. S., *Biomolecular NMR Spectroscopy*, Oxford University Press:Oxford, UK, 1995.
13. Boucher, W., "Azara v2.0". Department of Biochemistry, University of Cambridge, UK. http://www.bio.cam.ac.uk/azara/
14. Spera, A. B. A. *J. Am. Chem. Soc.* **1991**, *113*, 5490-5492.
15. Wishart, D. S. B. D. Sykes *J. Biomol. NMR* **1994**, *4*, 171-180.

Modern NMR in Laboratory Development

Physical Chemistry

Chapter 10

More Than Just a Characterization Tool: Learning Physical Chemistry Concepts Using NMR

Alexander Grushow and John E. Sheats

Department of Chemistry, Biochemistry and Physics, Rider University, 2083 Lawrenceville Road, Lawrenceville, NJ 08648

Four NMR experiments which emphasize the teaching of physical chemistry concepts will be described. Each has the characteristic of using NMR as a means to measure and quantify rather than just characterize qualitatively the system under investigation. The experiments include a modification of the classic keto-enol equilibrium experiment, a kinetic study of benzenediazonium ion decomposition, observation of a metastable intermediate during free radical decomposition, and the measurement of internal rotation barriers. In each experiment the student observes a physical chemical phenomenon taking place. The experiments are also designed such that students can develop a chemical understanding of the system even if they are unable to obtain quantitatively accurate data.

Introduction

At our institution a medium field NMR (~300 MHz) is used in all parts of our curriculum after the General Chemistry experience. However, we find it particularly important to emphasize that NMR is more than just a tool for identifying compounds. That chemical instrumentation can serve more than one purpose is a valuable idea for our students. In the area of physical chemistry we have successfully modified or adapted previously published experiments which utilize NMR for its ability to quantify physical chemical phenomena which are not readily characterized with other techniques. In the experiments described below, we discuss both the experimental details and our philosophy behind utilizing these experiments as a means for students to learn physical chemistry.

Keto-Enol Tautomerism

NMR investigation of the tautomerism of 2,4-pentanedione is a "classic" experiment in physical chemistry in that it is included in one of the standard physical chemistry laboratory textbooks (1). The experiment as published, however, was designed for use on a 60 MHz CW-NMR instrument. To adapt this experiment for use with a FT-NMR we have made some modifications that have been outlined previously (2). Briefly, the goal of the experiment is the measurement of the equilibrium constant of the following tautomerism reaction:

In solution, the keto form, **I**, is found in equilibrium with the enol forms **II** and **III**. The enol forms are characterized as having an intramolecular hydrogen bond between ketone oxygen and the hydroxyl proton. The rate of interchange between forms **II** and **III** is much faster than the time scale of the NMR experiment and thus a single set of NMR signals is observed for the average of these two structures. In a 60 MHz NMR spectrum, the vinyl and methylene protons are well resolved. A spectrum of 2,4-pentanedione is given in Figure 1. The vinyl proton is at 5.6 ppm and the methylene protons are at 3.6 ppm. From the integration of the vinyl and methylene peaks the relative concentrations of the keto and enol forms can be determined and an equilibrium constant can be calculated. From this information students can develop an understanding of which is the favored form of the molecule in a given solvent. Structure **I** is favored in polar solvents such as DMSO-d_6 and CD_3CN; whereas the enol form is favored in the nonpolar solvents C_6D_{12} and CCl_4. The beauty of the experiment lies in the instructor's ability to modify the experiment for a number of different student groups in a given class section, each perhaps performing the experiment with a different solvent. By collecting the class data on equilibrium constant as a function of solvent, students can gain an understanding of how a chemical environment (solvent polarity) affects chemical structure (tautomerization). This idea is important to convey to students since it has applications from organic synthesis to biochemical structure.

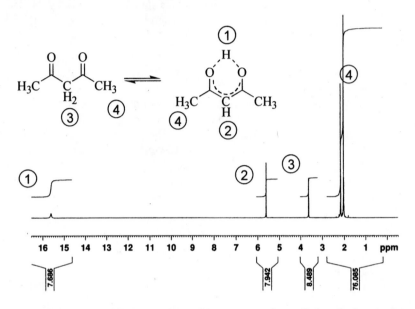

Figure 1. Proton NMR spectrum of 2,4-pentanedione (0.2 mole fraction) in CD₃CN solvent obtained at 300 MHz. (Reproduced with permission from reference 2. Copyright © 2002, Division of Chemical Education, Inc.)

Experimental Details

This basic experiment (*1*) and a number of modifications (*2*) designed for use with a medium field FT-NMR have been outlined previously. In our laboratory students prepare three solutions of 0.2 mole fraction of 2,4-pentanedione in solvents ranging from non-polar (C_6D_6) to polar (CD_3CN). NMR spectra are recorded at a fixed temperature. Students integrate the appropriate peaks in the spectrum and determine an equilibrium constant for the tautomerism reaction in each solvent.

We have observed that students obtain equilibrium constant values that are within 20% of the literature values. However, given the size of the polarity effect on the equilibrium constant, a student can observe how polarity affects the solution equilibria even if the data are not perfect. While obtaining valid data is an important student exercise, understanding chemical phenomena is a more important goal for this physical chemistry laboratory. After experience with this experiment we find that students are strongly aware of the effect of solvent on an equilibrium system.

Kinetics of Benzenediazonium Ions

The acid catalyzed decomposition of benzenediazonium ions is a useful means to develop *in situ* phenyl cations for use in further synthesis. Early work on the generation of phenyl cations was performed by one of the authors (*3*) as a Ph.D. thesis and was subsequently incorporated into the literature of physical chemistry experiments (*1,4*). However, research on the mechanism of decomposition of this class of compounds continues still (*5*). NMR has been extremely useful in our laboratories in helping to elucidate details not explored in earlier research. More importantly, it has created a plethora of student research projects allowing students to explore a kinetics problem in a quantifiable way. The decomposition occurs as follows:

$$\overset{+}{N}\equiv N \text{—} \underset{R_3}{\overset{R_1 \quad R_2}{\bigcirc}} \xrightarrow{\text{slow}} \quad + \quad \underset{R_3}{\overset{R_1 \quad R_2}{\bigcirc}} \quad + \quad N_2(g) \xrightarrow{H_2O} HO\text{—}\underset{R_3}{\overset{R_1 \quad R_2}{\bigcirc}}$$

By changing the substituent at the *ortho*, *meta* and *para* positions, one sees a change in the rate of decomposition occurring in the first step: generation of the phenyl cation. In a water solution the phenyl cation reacts quickly with the solvent to form a phenol. The early kinetic experiments on this system relied on the UV absorption of benzenediazonium ion between 295 and 325 nm. The phenol product is transparent in this region, but will absorb at wavelengths below 290 nm. However, when substituents are added, the overlapping UV spectra of the reactant and product can create difficulties. NMR allows examination of both reactant and product concentrations in comparison to an internal standard (DSS: 2,2-Dimethyl-2-silapentane-5-sulfonate) and has the added advantage of being able to detect impurities in the starting material. We have had students use both UV and NMR spectroscopies to examine a number of substituted benzenediazonium ions as a means to compare the utility of the techniques and as a means to examine the kinetic effect of a substituent on the formation of the phenyl cation. Phenyl cations will typically be stabilized by an electron-donating substituent such as *m*-methoxy or *p*-methyl or destabilized by an electron-withdrawing group such as *p*-carboxyl. The stabilization or destabilization can be observed from the half-life of the reactant ion.

Experimental Details

Benzenediazonium fluoroborate is prepared using the procedure of Dunker, Starkey, and Jenkins (6) by adding aniline to 48% fluoboric acid at reduced temperature followed by the addition of a sodium nitrite solution. The resulting product can be purified by recrystalization from 5% fluoboric acid. The product crystals can be stored in a refrigerator for several weeks before a noticeable amount of decomposition occurs. We also use this method for synthesizing substituted benzenediazonium ions by replacing aniline with an appropriately substituted phenyl amine to generate the appropriate substitution. For example, m-anisidine is used to synthesize m-methoxybenzenediazonium fluorborate. Deuteration of the phenyl ring is achieved by selectively deuterating the corresponding phenyl amine (7). For example, anilinium chloride is formed by bubbling HCl through an ether solution of aniline. The resulting insoluble crystals are collected and refluxed in D_2O for 24 hours. The deuterated compound is then used in the same manner as above to provide the deuterated benzenediazonium ion.

Purified diazonium fluorborate samples are dissolved in a 0.2 M H_2SO_4 solution with a fixed quantity of DSS, used as a fixed internal standard, and placed in an NMR tube. We have tried two measures to avoid tube breakage due to excess N_2 pressure generated by the reaction. Early on, we would freeze the solution in the NMR tube. The tube was then evacuated and sealed prior to the run. This process was time consuming, and by the time the sample was warmed and ready for the NMR experiment a significant amount of the reactant was found to have decomposed. More recently, we have taken a more simple approach and used an NMR cap with a pin-hole punched in it. We also run solutions with a very low concentrations of diazonium ion, typically 10^{-4} M. In 1 mL of NMR solution this generates only minimal partial pressure of N_2 in the space above the solution. The NMR sample is placed in the magnet and the instrument is programmed to maintain a constant temperature over the course of the kinetic run. Data is acquired at set intervals during the run. The temperature of the sample is calibrated using neat ethylene glycol or methanol depending upon the temperature of the experiment(8). The time length of the run and interval between acquisitions are dependent upon the substituent on the phenyl ring and how it will affect the half-life of the decomposition.

Resulting spectra are analyzed by comparing the relative integrations for the reactant and product to the DSS which will be constant throughout the run. The NMR spectrum of p-methylbenzenediazonium fluorborate during the course of a run is shown in Figure 2. In this spectrum phenyl peaks for the reactant are found between 7.7 and 8.2 ppm and the phenolic peaks for the product are found between 6.5 and 7.2 ppm. The peak at 4.8 ppm is due to H_2O in the D_2O solvent. Additionally for p-methylbenzenediazonium one could compare the reactant and product methyl peaks at 2.4 and 2.0 ppm respectively.

Results

Our students have performed experiments on a number of substituted benzenediazonium salts, also examining the effect of selective deuteration of different positions on the phenyl ring on the kinetics of formation of the phenyl cation. Previous investigations had shown that deuteration of benzenediazonium ion (*9*) slows the rate approximately 25% for an *ortho*-deuterium, 10% for a *meta*-deuterium and 2% for *para*-deuterium. This reduction was attributed (*9*) to reduction in hyperconjugation when D was substituted for H. The *ortho* deuterium isotope effect of 1.25 is larger than the value of 1.10-1.15 commonly observed for β-deuterium isotope effects in aliphatic carbocations. Since these are both teaching experiments and part of on-going research into isotope effects, students are more engaged in the work, given the lack of a "known" outcome.

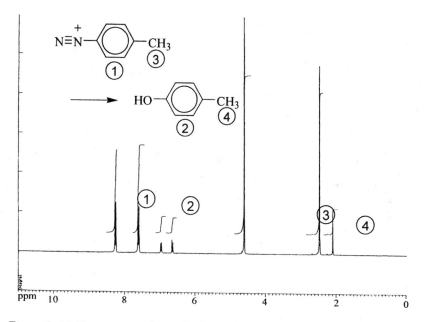

Figure 2. NMR spectrum of p-methylbenzenediazonium fluorborate during the course of a kinetics run.

Free Radical Decomposition with a Metastable Intermediate

2,2'-Azo-bis-isobutyronitrile (AIBN) is a commonly used free radical polymerization initiator. The kinetics of decomposition of this compound were initially studied fifty years ago, and the resulting mechanism was the subject of some controversy because of the difficulty in determining the intermediate steps

(*10,11*). With NMR, however, the kinetics are easily followed. The initial formation of the free radicals has a half-life of about 80 minutes at 80°C.

$$H_3C \overset{CH_3}{\underset{CN}{\rule[-1ex]{0.4pt}{3ex}}} N=N \overset{CH_3}{\underset{CN}{\rule[-1ex]{0.4pt}{3ex}}} CH_3 \xrightarrow{\text{60-80°C}} 2 \cdot \overset{CH_3}{\underset{CN}{\rule[-1ex]{0.4pt}{3ex}}} C\text{-}CH_3 + N_3$$

The radical products are stabilized by the presence of the cyano group and the odd electron is shared approximately equally by the tertiary carbon and the nitrogen.

$$\cdot\ \overset{CH_3}{\underset{\underset{N}{\overset{|||}{C}}}{\underset{|}{C}}}\text{-}CH_3 \quad \longleftrightarrow \quad \overset{CH_3}{\underset{\underset{\cdot\,N}{\overset{||}{C}}}{\underset{|}{C}}}\text{-}CH_3$$

IV   **V**

If no other reactive species are present, the free radicals will recombine in two ways. Two radicals of structure **IV** will combine carbon to carbon to form tetramethylsuccinonitrile (TSN), structure **VI**, or the nitrogen of structure **V** will combine with the carbon on **IV** to form structure **VII**, dimethyl-N-(2-cyano-2-propyl)-ketenimine (DKI). The combination of two radicals in form **V** leads to a much weaker nitrogen to nitrogen bond and is not observed.

$$2 \cdot \overset{CH_3}{\underset{\underset{N}{\overset{|||}{C}}}{\underset{|}{C}}}\text{-}CH_3$$

$$H_3C\text{-}\overset{CH_3}{\underset{CN}{\overset{|}{C}}}\text{—}\overset{CH_3}{\underset{CN}{\overset{|}{C}}}\text{-}CH_3 \quad \textbf{VI}$$

$$H_3C\text{-}\overset{CH_3}{\underset{CN}{\overset{|}{C}}}\text{—}N{=}C{=}\overset{CH_3}{\underset{CH_3}{\overset{|}{C}}}\text{-}CH_3 \quad \textbf{VII}$$

Other side products arise from the abstraction of a proton from one radical by another forming isobutyronitrile and methylacrylonitrile. Methylacrylonitrile will react further with radicals, if their concentration is high enough, to produce a polymeric material. In all, the side products will be approximately 12% of final product. Of the two main products, TSN is the more thermodynamically stable species. If the reaction were quenched at any point DKI would remain (*12,13*). However, during a kinetic run at 60-80°C DKI will decompose back into free radicals with a half-life only slightly shorter than that of AIBN. Thus, over time, NMR signals for DKI in the reaction rise quickly to an equilibrium level of approximately 10% of the starting AIBN concentration. As the reaction nears completion, the DKI signals slowly disappear. In the NMR spectrum with benzene-d_6 as the solvent, both AIBN and TSN give rise to sharp singlets (integration of 12 protons) at 1.0 and 0.8 ppm, respectively. Two peaks (integration of 6 protons each) at about 1.1 and 1.3 ppm are observed for DKI. In our laboratory exercise we have students integrate peaks from AIBN, TSN and DKI in comparison to an internal standard (anisole) for several runs and plot the resulting "concentrations" as a function of time.

Experimental Details

Commercial AIBN is available from a number of sources. Before use, AIBN is recrystalized by dissolving 5 g in a minimum amount of ethanol at 45 °C, filtering and quickly cooling to 0 °C. White needle-like crystals are a typical form of the purified AIBN. The recrystalized material is dried *in vacuo* for several hours to remove the solvent. A 0.25 M solution of AIBN is prepared by dissolving 1.03 g in 25 mL benzene-d_6. About 0.25 g of anisole is added as the internal standard. The decomposition is conducted in a Schlenk tube under N_2 atmosphere with a water-cooled condenser positioned above the tube. The condenser is then connected to a bubbler. The Schlenk tube is immersed in an oil bath which is maintained at 100 °C, well above the boiling point of benzene. Once the benzene starts to reflux, 1 mL aliquots of the reflux solution are removed every 20 minutes for about 4 hours. Each aliquot is placed in a NMR tube and stored in an ice-bath to quench the reaction until it is time to take the spectrum. The reaction is allowed to reflux over night and a final sample is removed at the "end" of the reaction. Similarly a student can prepare between ten and twenty NMR tubes with 0.5 mL of solution each. These tubes can be placed in a constant temperature bath, removing them at appropriate intervals. Either procedure will allow a student to collect enough data points to follow the reaction and determine the rate of decomposition of AIBN and the rate of DKI decomposition and reformation into TSN. The derivation of the rate equations is given by Smith and Carbone (*11*).

After our department acquired a more automated NMR spectrometer, we began to explore the possibility of performing the reaction in the NMR spectrometer. This was initially a difficult endeavor since we were concerned both with bringing a sample to near reflux inside the probe and that the reaction also generates nitrogen gas. The problems are solved by reducing the concentration of AIBN to about 0.01 M and using a solvent with a higher boiling point such as toluene-d_8. We again employed an NMR cap with a small pin hole in the top as a safety measure. The student can then set up a series of programmed acquisitions at 10 or 20 minute intervals for four to six hours. The NMR temperature is set prior to the run. The resulting kinetic data from this

136

procedure are very similar, however, the chemical shifts in the different solvent are also a little more difficult to resolve since the AIBN peak is very close to the upfield peak of DKI. As a result, some deconvolution of the integration values is necessary to determine the true relative populations of each. An NMR spectrum of a reaction sample is given in Figure 3. In this spectrum, other peaks are observed as a result of some of the secondary side products as well. In particular, the doublet at about 0.6 ppm is attributed to isobutyronitrile. Other side products can be identified and quantified by the students in this experiment.

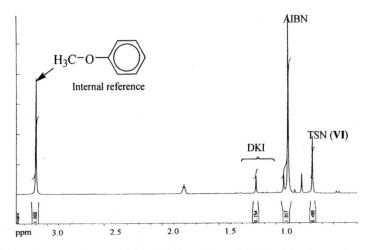

Figure 3. NMR spectrum of the decomposition of AIBN with toluene-d$_6$ as the solvent. The reaction is taking place in the NMR at 80°C. This spectrum was taken 61 minutes into the run.

Results

A plot of relative population of the three main species in the reaction as a function of time is given in Figure 4. According to the derivation of Smith and Carbone the decomposition of AIBN is first order with

$$\ln \frac{\left[AIBN \right]}{\left[AIBN \right]_0} = -k_1 t \qquad (1)$$

providing a simple means to determine the decomposition rate constant. The decomposition of DKI is a little more complex but a rate equation can be approximated by

$$\ln \frac{[\text{DKI}]}{t} = \frac{-\left(k_1 + k_4 U\right)}{2} t + \ln \left(k_1 V [\text{AIBN}]_0\right) \tag{2}$$

where k_4 is rate constant for the decomposition of DKI into the cyanopropyl radicals, U is the fraction of DKI that goes to form TSN and other side products, and V is the fraction of radicals from AIBN that recombine to form DKI. Figures 5 and 6 are plots of Equations 1 and 2 based on the data given in Figure 4.

This particular experiment is more than just a kinetic experiment followed by NMR. Students are both examining a slightly more complex reaction mechanism and quantifying the existence of a metastable intermediate, DKI, which actually achieves a steady-state concentration that is large enough to be determined readily. There are scant few experiments accessible to an undergraduate in which a student can experimentally observe such an intermediate and follow its subsequent decomposition. It should be noted that a subsequent student research project (currently in progress) on this system has shown that when AIBN undergoes photolytic decomposition at room temperature the NMR shows a 50:50 mixture of DKI and TSN as products with little evidence of other side products. This a further indication that DKI is only metastable in the reaction mixture because of the elevated temperature of the thermolytic decomposition.

Barriers to Internal Rotation: Amides

We have recently revisited another NMR experiment with roots in physical chemical processes. The barrier to internal rotation about the carbon-nitrogen bond of an amide group has been well studied for its role in biochemical peptide structure. While several NMR experiments using the enhanced methods available to FT-NMR instruments are capable of measuring the barrier to this internal rotation, we have opted to follow the experiment outlined in the paper by Gasparro and Kolodny (14). One dimensional NMR spectra are acquired at different temperatures. Of particular interest to students in this experiment is the unusual effect that temperature has on the observed spectrum. Temperature effects for most samples that students see are typically small changes in chemical shift. However, a rise in temperature in this system produces a collapsing of peaks. Given the simplicity of the inversion-recovery experiment it is also possible (15) to measure some of the same kinetic parameters at room temperature. While this is a simple acquisition and data manipulation, the observations made in the experiment are qualitatively different and we have found they have less impact on student understanding than the experiment where the effect is observed directly in the spectrum. It might be useful to perform an inversion recovery experiment, however, as a complementary method of

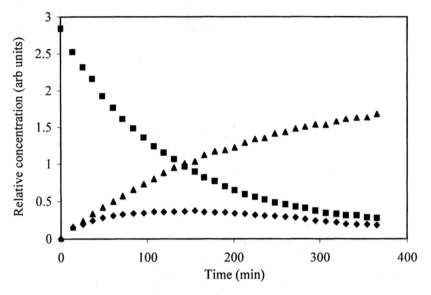

Figure 4. Relative population of the reactant of AIBN (■) and its two main decomposition products, DKI (◆) and TSN (▲).

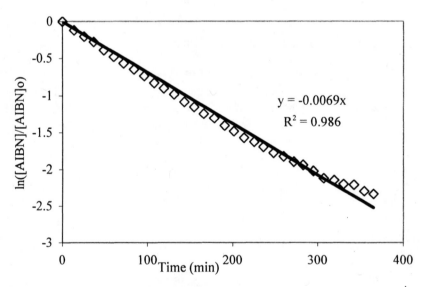

Figure 5. Plot of data to eq 1. The resulting slope gives a $k_1 = 0.0069$ min^{-1}.

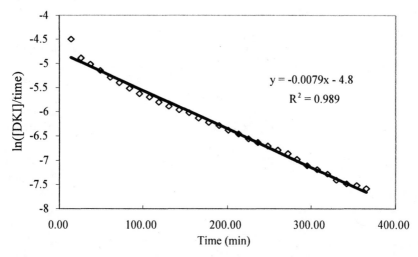

Figure 6. Plot of data to eq 2. The resulting plot provides a value of $V = 0.43$ *and* $k_4U = 0.0089$ *min^{-1}.*

confirming the result from the temperature programmed experiment. Since the details of the experiment have already been outlined, we will discuss the modifications we have made and the reasons we advocate use of this experiment in the curriculum.

Students usually have some experience with the concept of internal rotation from their organic chemistry courses. However, this example gives them a concrete notion that molecules can and do undergo internal rotation and that these internal rotations can be hindered by the chemistry of the system. In the case of dimethyl acetamide (DMA), the A and B methyl groups will undergo interchange only when the molecule has enough energy to overcome the pseudo

conjugation between the carbonyl and the non-bonding electrons on the nitrogen. Formation of this weak π bonding orbital "freezes" the methyls in position, such that methyl A and B are in chemically different environments at room temperature. However, since the barrier to this rotation is small (less than 100 kJ/mol) it is possible for internal rotation about the carbon-nitrogen bond to occur with reasonable temperature increases. Through the use of an appropriate introduction, some of our students are fascinated with the idea that at two different temperatures the observed spectra have a different number of peaks. After a suitable discussion of the origin of this phenomenon, we have found students keen to pursue the line of inquiry to determine the temperature of coalescence and the thermodynamic parameters of activation for the internal rotation process.

In the earlier reference (*14*) the solute DMA was dissolved in CCl_4 since the CW NMR instrumentation and the formerly high cost of deuterated solvents dictated this practice. In our experiments students dissolve DMA in pyridine-d_5 or benzene-d_6. These solvents were initially chosen for students to use given their structural similarity, yet measurable polarity differences. Samples are degassed using several cycles of the freeze-pump-thaw method. The NMR tubes are flame-sealed before the final thaw. To assure that we will not damage the NMR when the temperature is raised above the boiling point of the solvent, flame sealed tubes were heated in an oil bath inside a fume hood with the sash down to about 400 K for at least 30 minutes. During this time we carefully inspected the tubes for signs of vapor leakage; excess bubbling of the solvent (indicative of boiling in an open system) was a sure sign that the tube was not completely sealed.

Spectra for both solvents are shown in Figure 7. Of particular interest in these two spectra is the affect that solvent plays on chemical shifts. It can be indicated to students that the difference in chemical environment of the N-methyls is much greater in the non-polar solvent. Such observations are a central part of our physical chemistry curriculum. As with the tautomerism experiment students show in their lab reports a strong understanding of the effect of solvent polarity on the chemical environment.

Concluding Remarks

We have described four different experiments which utilize the quantitative aspects of NMR spectroscopy as a means to familiarize students with physical chemical concepts that are not easily elucidated using other experimental methods. In each case we emphasize the role that changes in NMR spectra play through the course of each experiment. This also reinforces the idea that NMR spectra are not just static characterization tools. In addition we use the ability of the NMR experiment to observe differences in chemical environment as a means to examine the physical effect that solvent plays on the compounds under study.

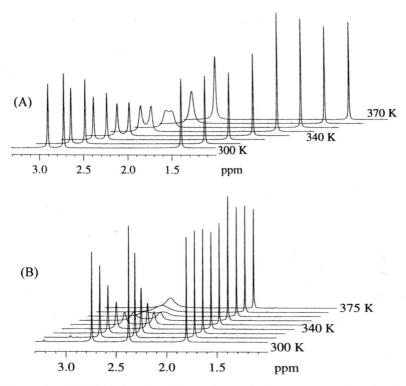

Figure 7. (A) DMA in pyridine-d₅ as a function of temperature. Spectra are shown at 10 K intervals. The coalescence temperature is about 350 K. (B) DMA in benzene-d₆ as a function of temperature. Spectra are shown in 10 K intervals up to 370 K. A final spectrum was done at 375K to observe a single peak above the coalescence temperature which is approximately 365 K.

Acknowledgments

We would also like to thank the many students in the Rider Chemistry program who have contributed to this work over the years. The NMR was purchased on a grant from the NSF (DUE-9952369).

References

1. Garland, C. W.; Nibler, J. W.; Shoemaker, D. P. *Experiments in Physical Chemistry,* 7th ed.; McGraw-Hill: New York, NY, 2002.
2. Grushow, A.; Zielinski, T. J. *J. Chem. Educ.*, **2002**, *79*, 707-714.

142

3. Sheats, J. E. Ph.D. thesis, Massachusetts Institute of Technology, Cambridge, MA 1965.
4. Sheats, J. E.; Harbison, K. G. *J. Chem. Educ.*, **1970**, *47*, 779-780.
5. Ussing, B. R.; Singleton, D. A. *J. Am. Chem. Soc.*, **2005**, *127*, 2888-2899.
6. Dunker, M. F. W.; Starkey, E. B.; Jenkins, G. L. *J. Am. Chem. Soc.*, **1936**, *58*, 2308.
7. Crossley, M. L.; Kienle, R. H.; Benbrook, C. E. *J. Am. Chem. Soc.*, **1940**, *62*, 1400.
8. A tool is available on the web which does the calculation and also provides reference to each calculation: http://www.spectroscopynow.com/FCKeditor/ UserFiles/File/specNOW/HTML%20files/NMR_temperature_measurement. htm
9. Swain, C. G.; Sheats, J. E.; Gorenstein, D. G.; Harbison, K. G. *J. Am. Chem. Soc.*, **1975**, *97*, 791-795.
10. Talat-Erben, M.; Bywater, S. *J. Am. Chem. Soc.*, **1955**, *77*, 3712-3714.
11. Smith, P.; Carbone, S. *J. Am. Chem. Soc.*, **1959**, *81*, 6174-6177.
12. Smith, P.; Rosenberg, J. *J. Am. Chem. Soc.*, **1959**, *81*, 2037.
13. Smith, P.; Sheats, J. E. *J. Org. Chem.*, **1962**, *27*, 4053.
14. Gasparro, F. P.; Kolodny, N. H. *J. Chem. Educ.*, **1977**, *54*, 258-261.
15. Williams, K. R.; King, R. W. In *Physical Chemistry: Creating a Dynamic Curriculum;* Editors, Schwenz, R. W.; Moore, R. J. Eds.; American Chemical Society: Washington, DC, 1993, p 315.

Chapter 11

Gas Phase ^{1}H NMR

C. Drahus, T. N. Gallaher, and T. C. DeVore[*]

Department of Chemistry MSC 4501, James Madison University,
Harrisonburg, VA 22807

Gas phase ^{1}H NMR spectra of water, acetone, methanol,
ethanol, heptane, and 2,4-pentanedione were obtained by
placing less than 0.3 ml of the compound into a standard NMR
tube and using a Bruker Spectrospin 400 NMR equipped with
a variable temperature probe to obtain the spectra. These
spectra can be used to determine gas phase chemical shifts and
to measure the enthalpy of vaporization. The spectrum of 2,4-
pentanedione can also be used to measure the gas phase
equilibrium constant between the keto and enol tautomers.

Introduction

Although gas phase ^{1}H NMR was first done in 1939, poor signal to noise
ratios limited its use until the advent of FT-NMR *(1-4)*. Since 1980, gas phase
NMR has been used to measure equilibrium constants, internal rotation rates, to
explore catalyst sites, to determine shielding constants, and to monitor high
temperature thermal decompositions *(4)*. In spite of the growing number of
applications, there is still little or no mention of this technique in most organic
or physical chemistry textbooks *(5)* and the general perception that exotic
procedures are needed to obtain good quality spectra still exists. The sensitivity
of modern high field NMR is sufficient to obtain high quality gas phase ^{1}H
NMR for many small molecules using the equilibrium between a small amount
of the liquid and the vapor in a conventional NMR sample tube as the sample.

Laboratory exercises that use gas phase NMR to measure the enthalpy of vaporization and the equilibrium constant for 2,4-pentanedione have been used in the applied physical chemistry laboratory at JMU for the past three years. Since the gas phase spectra of each of these molecules shows significantly different chemical shifts relative to the chemical shifts observed for the neat liquid, the gas phase spectra could also be used to illustrate the effect of the solvent on the chemical shift in the neat liquid and in solution. However, this experiment has not been done in any teaching lab at JMU.

Experimental

Proton NMR spectra of water, methanol, ethanol, acetone, 2,4-pentanedione, and heptane vapor molecules were obtained by placing a small volume of the neat liquid in a 5 mm OD NMR tube. The only restriction on the volume of liquid used is that it must be small enough to leave "empty space" in the sampling region of the NMR. We have been able to obtain good quality spectra with added liquid volumes ranging from less than 0.05 ml to approximately 0.5 ml. There are advantages to sealing the tubes (we us a glass blowers torch to seal them under a rough pump vacuum) since the sample is then contained in the tube, but the capped tubes will also work if slightly larger volumes of liquid are used. By adjusting the sample temperature and the amount of liquid initially placed in the tube, it is possible to obtain NMR spectra of only the neat liquid, of only the vapor, or of both.

A Bruker Spectrospin 400 NMR equipped with a variable temperature probe was used to collect the NMR spectra. Most of the laboratory exercises were done using eight (8) scans with a receiver gain of 1300 and a screen width set between 25 ppm to 30 ppm. Since a lock solvent was not used, the FID was maximized manually before collecting each spectrum. The spectra were manually phase corrected and automatically baseline corrected before measuring the chemical shifts and the peak intensities. Since the spectra were manually adjusted, the chemical shifts and the absolute intensities vary somewhat between sets of scans. The chemical shifts were determined by shifting the whole spectrum by the amount needed to bring the liquid CH_3 peak into agreement with the chemical shift for the CH_3 peak relative to liquid TMS reported in the literature. *(6,7)* The intensities were adjusted either by using an internal standard or by assuming that the sum of the intensities of the liquid and the vapor was constant as the sample was heated and cooled. If desired, it should be possible to use a double walled NMR tube to hold the lock solvent while obtaining the gas phase spectrum. The liquid portion of the sample could be used as the internal standard. However, we have not tried this procedure.

The NMR spectra observed for water are presented in Figure 1. Only two peaks were expected (one for the liquid and one for the vapor molecule), but three peaks were observed when less than 0.15 ml of water was placed in some sample tubes. As the sample was heated causing more of it to vaporize, the peak assigned to the liquid (4.7 ppm) decreased in intensity and the intensity of the vapor signal (3.4 ppm) increased (see figure 1). The intensity of the weak peak near 8 ppm remained approximately constant and was assigned as water molecules bonded to the surface of the NMR tube based upon reports in the literature *(8,9)*. Extra peaks arising from unexpected species can complicate the analysis of the spectrum for the students. Also, accidental degeneracies can occur for more complicated systems like 2,4-pentanedione. In these cases, it is helpful to have the students also collect the spectrum of an NMR tube filled with the neat liquid. Once the assignments for the liquid features are established, the remaining features can be assigned as signals arising from the vapor molecules.

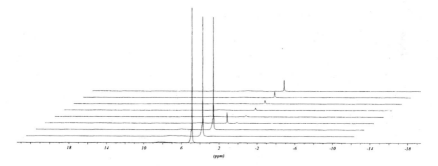

Figure 1: Overlay of ^1H NMR spectra of H_2O at 265K, 275K, 285K, 325K, 335K, 345K, 355K, 365K, and 375K (front-back).

Chemical Shifts and Shielding Constants

One obvious laboratory exercise would be to compare the observed shielding constants between molecules in the vapor phase to those in the neat liquid or in solution. The relationship between the chemical shift (δ), the shielding constant for the nucleus of interest (σ), and the shielding constant for the reference (σ^0) is: *(5)*

$$\delta = (\sigma^0 - \sigma) * 10^6 \qquad (1)$$

An increase in σ corresponds to a decrease in δ. Three terms contribute to σ.

$$\sigma = \sigma_{local} + \sigma_{neighbor} + \sigma_{solvent} \qquad (2)$$

The local contribution (σ_{local}) is from the electrons on the nucleus in question. The neighboring group contribution ($\sigma_{neighbor}$) is from the rest of the atoms in the molecule. The final contribution is from the solvent. Since the contribution

146

from $\sigma_{solvent}$ is expected to be small in the gas phase, the difference between the measured chemical shifts for the liquid and for gas phase provides a reasonable estimate of $\sigma_{solvent}$.

The OH signal in methanol provides a dramatic illustration of the change in the chemical shift for the vapor compared to the neat liquid (see Figure 2). The OH signal shifts from 5.3 ppm in the neat liquid to 2.3 ppm in the vapor. While it is easy to rationalize this change as resulting from the extensive hydrogen bonding in the liquid, changes in the chemical shifts are also observed for non-hydrogen bonded systems. The chemical shift change for acetone, a molecule with a dipole moment, was from 3.0 ppm to 4.7 ppm (see Figure 3). Even non-polar solvents like heptane have observable differences in the liquid- vapor chemical shifts (see Figure 4). In general, the CH₃ and the CH₂ protons of all molecules were deshielded by ~ 2 ppm in the gas phase when compared to the

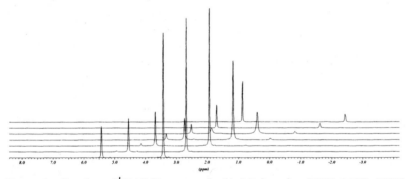

Figure 2: Overlay of ¹H NMR spectra for l/v Methanol at 265K, 275K, 285K, 296K, 315K, and 325K (front to back). The spectrum at 265 K is only from the liquid. The spectrum at 325 K is only from the vapor.

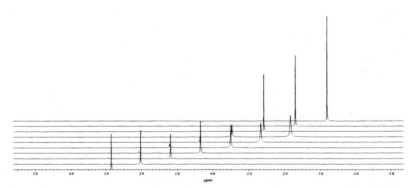

Figure 3: Overlay of gas phase ¹H NMR spectra of acetone at 235K, 245K, 255K, 265K, 275K, 285K, 296K, 305K, 315K, and 325K (front-back).

neat liquid. A comparison between the gas phase spectrum and the spectrum of the molecule in different solutions could also be done to further explore the effects of various solute –solvent interactions.

The chemical shifts observed for the liquid and vapor molecules at 300 K are presented in Table I. The only adjustment made for the values reported in Table I was to add the correction factor needed to bring the liquid CH_3 into agreement with values reported previously *(6,7)* at 300 K to each observed

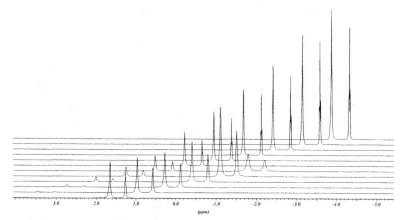

Figure 4: Overlay of ^1H NMR spectra of 0.10 ml of Heptane at 255K, 265K, 275K, 285K, 296K, 305K, 315K, 325K, 335K, 345K, 355K, and 365K

signal. Since the chemical shifts from the liquid and vapor molecules were measured simultaneously, the differences in chemical shifts can be determined even if the spectra are not corrected. The gas phase chemical shifts for ethanol and water relative to methanol can also be compared to the results obtained for theoretical calculations *(4f)* or measurements *(4d)* from the literature.

Table I: Table of Chemical Shifts at 300 K

Molecule	Liquid (corrected) CH_3	CH_2	OH	Vapor (corrected) CH_3	CH_2	OH
Heptane	1.4	1.8	--	3.8	4.2	--
Water	--	--	4.7	--	--	3.4
Methanol	3.8	--	5.3	5.9	--	2.5
Ethanol	1.3	3.7	5.5	3.6	6.0	2.8
Acetone	3.0	--	--	4.7	--	--
2,4-pentanedione						
enol	2.6	6.2	16.2	4.6	8.1	18.3
keto	2.7	4.0	--	4.8	6.1	--

148

Enthalpy of Vaporization

Since the liquid and the vapor were in equilibrium, the intensity of the vapor signals should be proportional to the sample vapor pressure. In principle, the vapor signal intensity could be used as a measurement of the vapor pressure for the compound. However, the signals were not reproducible enough (largely due to the uncertainty in tuning the spectrometer) to use the uncalibrated intensities for these measurements. Two approaches were used to calibrate the signals. The first was to use an internal standard. Any peak that is expected to have constant intensity throughout the temperature range used could serve as the internal standard. The NMR peak arising from the OH bonded to the glass proved to be an acceptable internal standard for the measurement of the enthalpy of vaporization for water. In this procedure, this peak was assigned a value of 1 and the intensity of the vapor peak was determined relative to this value. The intensity of the OH vapor peak was then substituted for pressure in the integrated form of the Clausius – Clapeyron Equation to determine $\Delta_{vap}H$. The graph obtained is shown in Figure 5.

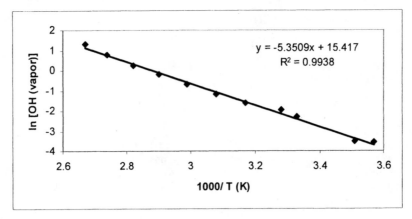

Figure 5: A Typical Clausius – Claperyon Plot Obtained from the Variable Temperature NMR Data for Vapor Water Peak Referenced Against the Water Peak Adsorbed to the Glass

Adding 0.5 ml of liquid water to the sample tube enabled us to use the liquid water peak as an internal standard. Since liquid water was in large excess, its amount was assumed to be constant. Integrating the vapor peak relative to this peak also gave an acceptable value for the enthalpy of vaporization of water. A double walled NMR tube could also be used to contain an internal standard separate from the equilibrium system.

The procedure that gave acceptable results for systems containing small, and hence changing, amounts of liquid was to assume that the sum of the intensities of the liquid and the gas phase NMR peaks were constant for all samples. All peaks were integrated relative to one of the NMR peaks and the ratio I_{vap} / (I_{vap} + I_{liq}) was substituted for the pressure in the integrated Clausius – Clapeyron Equation. An example of a graph obtained for methanol using this procedure is shown in Figure 6.

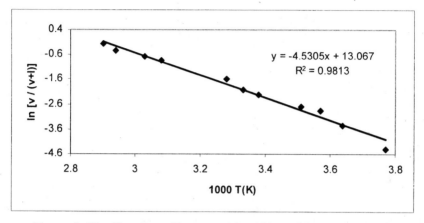

Figure 6: The Clausius – Claperyon Plot Obtained from the Variable Temperature NMR Data for the Vapor Methanol Peak Referenced Against the Total Intensity of the Liquid + Vapor Methanol Peaks

As shown in Table II, the error observed for $\Delta_{vap}H$ was approximately $\pm$ 5 kJ/ mol when samples were allowed to reach temperature for 10 minutes and no in-situ temperature corrections were done. If more precision is desired, the probe temperature can be calibrated and longer delays to allow the temperatures to reach equilibrium could be used. A double walled cell would permit the temperature measurement to be made simultaneously with the vapor pressure measurement.

Keto-Enol Equilibrium for 2,4- Pentanedione

A thorough discussion of the theory and the results obtained from using variable temperature NMR to measure the equilibrium constant for the keto-enol tautomerism of 2,4-pentanedione in various solvent systems was published by Grushow and Zielinski. (10) Measuring the gas phase equilibrium constant would provide a "solvent free" value for this equilibrium. The "solvent free" value obtained can be directly compared to the results of molecular modeling calculations and to the gas phase value obtained by Folkendt et al.(11)

Table II: Comparison of typical measured enthalpies of vaporization to those in the literature. Only water was done using an internal standard. All values are in kJ/ mol

Molecule	$\Delta_{vap}H$ (obs)	$\Delta_{vap}H$ (lit)	Ref
Heptane	33	36.6	11
Water	42	45.054 (273 K)	
		40.657 (373 K)	5 a, b
Methanol	40	38 (298 K)	5 a, b
Ethanol	46	43.5	5 a, b
Acetone	28	31.3	5 b
2,4- Pentanedione	44.1	41.8	11

The NMR spectrum of liquid- vapor 2,4-pentanedione is shown in Figure 7. While it is not trivial to assign each signal, assignments can be made by starting with the spectrum for the neat liquid, assigning the signal from the OH proton in the enol form, and using the expected integration ratios to assign the enol signals. The remaining signals can be assigned to the keto form. Once the liquid assignments are made, the vapor peaks can be assigned using the same procedure. Alternatively, the students can be referred to reference 11 and use their assignments.

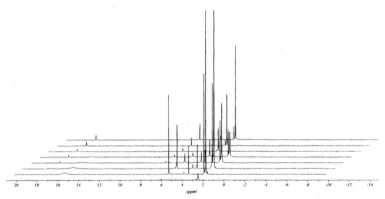

Figure 7: Overlay of ^1H NMR spectra of 2,4-pentanedione at 265K, 285K, 296K, 315K, 325K, 335K, 355K, and 375K (front-back).

As with all spectroscopy, the amount (concentration or partial pressure) of each species is directly proportional to the intensity of its NMR signal. Assuming the proportionality factors are the same for the six CH_3 protons in the keto and the enol forms, the equilibrium constant is given by

$$K = \frac{I_{CH3} \text{ (enol)}}{I_{CH3} \text{ (keto)}} \qquad (3)$$

The van't Hoff plot made using equilibrium constants obtained is presented in Figure 8. The values obtained for the enthalpy and entropy of reaction are given in Table III for averaged student data. Individual student data generally shows more scatter. The most common problem is not allowing the temperature adequate time to stabilize throughout the tube.

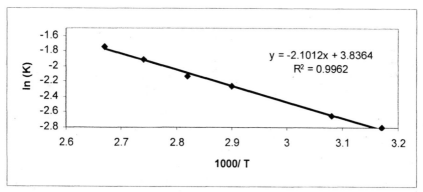

Figure 8: The van't Hoff Plot Observed for the Keto-Enol Equilibrium of Vapor Phase 2,4-Pentanedione. The Values for the Equilibrium Constants Used to Make This Plot Were the Averages from Several Individual Experiments.

Table III: Enthalpies of Reaction for the Keto-enol Tautorism of 2,4-Pentanedione in the neat liquid and the Gas Phase

Solvent	$\Delta_{rxn}H$ / kJ mol^{-1}	$\Delta_{rxn}S$ / J K^{-1} mol^{-1}	T (K)	Ref
Neat Liquid	- 11.5	-28	325	TW
Neat Liquid	- 11.7		311	4i
Vapor	- 17.5	-31	350	TW
Vapor	- 18.0		388	4i, 11

Acknowledgements

TD and TG would like to acknowledge the support of the Virginia Equipment Trust fund for the purchase of the FT-NMR and NSF Award # USE-

9152585 for the purchase of the variable temperature accessory. CD wishes to thank the JMU-NSF-REU program (CHE-0097448) for supporting her during this project.

References

1. Rabi, I.I.; Millman, S.; Kusch, P.; Zacharias, J.R.; *Phys. Rev.* **1939**, 55, 526.
2. Kellog, J.M.B.; Rabi, I.I.; Ramsey, Jr., N.F.; Zacharias, J.R.; *Phys. Rev.* **1939**, 56, 728.
3. Schneider, W.G.; Bernstein, H.J.; Pople, J.A.; *J. Chem. Phys.* **1958**, 28, 601.
4. Some selected examples include:
   a. Ross, B.D.; True, N.S.; Decker, D.L.; *J. Phys. Chem.* **1983**, *87*, 89.
   b. Chauvel, Jr., J.P.; True, N.S.; *J. Phys. Chem.* **1983**, *87*, 1622.
   c. Miyajuma, T.; Hirano, T.; Sato, H.; *J. Molec. Struct.* **1984**, *125*, 97.
   d. Chauvel Jr., J.P.; True, N.S.; *Chem. Phys.* **1985**, *95,* 435.
   e. Moreno, P.O.; True, N.S.; *J. Phys. Chem.* **1991**, *95*, 57.
   f. Fleischer, U.; Kutzelnigg, W.; Bleiber, A.; Sauer, J.; *J. Am. Chem. Soc.* **1993**, *115*, 7833.
   g. Taha, A.N.; Neugebauer Crawford, S.M.; True, N.S.; *J. Am. Chem. Soc.* **1998**, *120*, 1934.
   h. Rittig, F.; Aurentz, D.J.; Coe, C.g.; Kitzhoffer, R.J.; Zielinski, J.M.; *Ind. Chem. Res.* **2002**, *41*, 4430-4434.
   i. Temprado, M.; Roux, M.V.; Umnahanant, P.; Hui Zhao; Chickos, J.S.; *J. Phys. Chem. B* **2005**, *109*, 12590-12595.
   j. Makulski, W.; *J. Molec. Structure* **2005**, *744-747*, 439-446.
   k. Krusic. P.J.; Marchione, A.A.; Roe, C.; *J. Fluorine Chem.* **2005**, *126*, 1510-1516.
5. For recent example see:
   a. Atkins, P.; de Paula, J.; *Physical Chemistry, Seventh Edition*, Freeman, New York, **2002**.
   b. Laidler, K.J.; Meiser, J.H.; Sanctuary, B.C.; *Physical Chemistry, Fourth Edition*, Houghton Mifflin, Boston, **2003**.
   c. Wade, Jr., L. G.; *Organic Chemistry, Fifth Edition*, Prentice Hall, Upper Saddle River, NJ, **2003**.
   d. Carey, F.A.; *Organic Chemistry, Fifth Edition*, McGraw-Hill, Boston, **2003**.
6. Pouchert, C.J.; *The Aldrich Library of NMR Spectra, Ed. II, Volumes 1,2*, Aldrich Chemical Co, **1983**.
7. *http://www.muhlenberg.edu/depts/chemistry/chem201woh/nmrexamples.html*
8. M.A. Natal-Santiago and J.A. Dumesic. *J. Catal.* **1998**, *175*, 252.
9. Hoffmann, M.M.; Conradi, M.S; *J. Supercritical Fluids* **1998**, *14*, 31.
10. Grushow, A.; Zielinski, T.J.; *J. Chem. Ed.* **2002**, *79,* 707.

11. Folkendt, M.M.; Weiss-Lopez, B.E.; Chauvel, Jr., J.P.; True, N.S.; *J. Phys. Chem.* **1985**, *89*, 3347.

12. *CRC Handbook of Chemistry and Physics, 76 th Student Edition*, Lide,D.R., ed. in chief, CRC Press, Boca Raton, FL, **1995-6**.

Appendix I: Derivation of the equation used for the Vapor Pressure Measurements

The relationship between the vapor pressure of any liquid and temperature can be derived from the Clausius-Clapeyron equation. (5) If the enthalpy of vaporization ($\Delta_{vap}H$) is assumed to be constant, the Clausius- Calperyron equation can be easily integrated. The form of this equation derived from this integration that is most useful here is

$$\ln\ p = \quad \Delta_{vap}H\ /\ RT \qquad + \qquad C$$

The NMR signal intensity is related to the vapor pressure p using Beer's Law and the ideal gas law. In this case, Beer's law can be written as

$$I = f\,[C]$$

Where I is the intensity, [C] is the concentration of the sample, and f is the factor that relates the two. The ideal gas law expressed in terms of concentration is

$$p = [C]\ RT$$

Substituting for [C], p becomes

$$p = I\,R\,T\,/\,f$$

Substituting this into the Clausius- Clapyron equation gives

$$\ln\ I = \Delta_{vap}H\ /\ RT\ + \qquad C'$$

where C' = C- ln (R T/ f). The temperature was put in the constant because I is referenced against either an internal standard of the sum of the NMR intensities. In either case, the effect of the temperature will at least partially cancel.

Appendix II: Derivation of the equation used to determine the equilibrium constant for the ket-enol equilibrium.

The equilibrium constant for the reaction in the gas phase

$$\text{Keto} \Leftrightarrow \text{Enol}$$

is

$$K = P_{Enol} / P_{Keto}$$

The NMR signal intensity is related to the vapor pressure p using Beer's Law and the ideal gas law. In this case, Beer's law can be written as

$$I = f[C]$$

Where I is the signal intensity, [C] is the concentration of the sample, and f is the factor that relates the two and the ideal gas law can be given as

$$p = [C] RT$$

Substituting for [C], p becomes

$$p = I R T / f$$

Since there are 6 CH_3 protons in each tautomer, f, R, and T can be assumed to be the same for both and the equilibrium constant becomes

$$K = I_{CH3\ enol} / I_{CH3keto}$$

Chapter 12

Measuring Molecular Motion: Using NMR Spectroscopy to Study Translational Diffusion

Jennifer N. Maki and Nikolaus M. Loening[*]

Department of Chemistry, Lewis & Clark College, Portland, OR 97219

This chapter demonstrates how to use NMR spectroscopy to measure the rate of translational diffusion. Once measured, this rate can provide insight into the size and shape of the molecules in a sample and can even be used to match peaks in the NMR spectrum with the different components of a sample mixture. Of the possible applications of diffusion, the experiment presented focuses on using measurements of the rate of translational diffusion to reveal how the viscosity of a solution varies with the concentration of the solutes.

Introduction

The term *diffusion* describes a variety of physical processes that involve the random motion of particles or the dispersal of energy. In the context of NMR spectroscopy, diffusion can refer to translational, rotational, and spin diffusion. Both translational and rotational diffusion have to do with the physical motion of particles. Translational diffusion refers to the seemingly-random motion of a particle due to its kinetic energy and its interactions with other particles as shown on the left in Figure 1. Rotational diffusion, which also arises due to kinetic energy, refers to the reorientation of a particle relative to the rest of the system as shown on the right in Figure 1. The third process, spin diffusion, refers to the spread of spin polarization via dipolar couplings or, more rarely, scalar couplings.

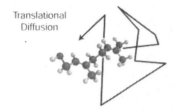
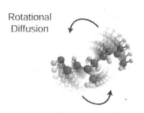

Figure 1. Translational diffusion refers to the seemingly random motion of molecules through space, whereas rotational diffusion describes the reorientation of a molecule relative to the rest of the sample.

The rate at which these processes occur can be measured using NMR spectroscopy. The rate of spin diffusion can be measured using an NOE build-up experiment and the rate of rotational diffusion is usually inferred from measurements of the transverse and longitudinal relaxation rates. The procedure described in this chapter demonstrates how NMR spectroscopy can be used to measure the rate of translational diffusion as well as how this rate can provide insight into the effect of the solute molecules on the bulk properties of the solution.

Background

In a gas or liquid, the individual molecules have enough kinetic energy to overcome the intermolecular forces binding them to other molecules in the sample. Consequently, the molecules move about at a speed determined by their kinetic energy. As a molecule moves through the sample, it changes direction and speed due to collisions with other molecules. Over a period of time, the movement of the molecule forms a zigzag path known as a *random walk* as shown on the left side of Figure 1. This movement occurs even in the absence of external effects (such as concentration gradients or electric fields) and is referred to as *translational diffusion*.

Consider the movement of a molecule along one of the spatial dimensions (the z-axis). In a specified period of time, the molecule will move a certain distance resulting in a displacement Z_1 as shown on the left in Figure 2. For a single experiment on a single molecule, it is hard to specify the value of Z_1 because the random walk of a molecule is (as the name implies) random. In a second experiment, the molecule moves another distance Z_2, which will usually be somewhat different than Z_1. The center part of Figure 2 illustrates the result of repeating such an experiment several times.

Solely based on physical intuition it is possible to deduce several properties of this motion. For example, small values for the displacement should be more likely than large values since the individual steps of the molecule's random walk are more likely to cancel out than to add constructively. Another property is that, on average, the displacement of the molecules will be zero. This is because the molecules are just as likely to move in one direction along the z-axis as the other. Physically, this has to be the case for a stationary sample...if the average displacement of the molecules was non-zero, than the sample would be moving rather than stationary. The results of a large number of experiments (or a large number of molecules in a single experiment) follow a well-defined trend as shown on the right in Figure 2.

Although it is difficult (if not impossible) to predict the distance that a single molecule will move in an experiment, it is possible to statistically define the *probability* that this molecule will move a certain distance. The statistics of the random walks of individual particles on the molecular level result in the bulk property of translational diffusion. The average rate of translational diffusion is described quantitatively by the diffusion coefficient (D). The value of the diffusion coefficient depends not only on the identity of the sample molecules, but also on the solvent and the temperature (and, to a very minor degree, the pressure). Statistically, the probability for a molecule to be displaced a distance Z in time t is described by a Gaussian distribution function (*1*):

$$P(Z) = \frac{1}{\sqrt{4\pi Dt}} \exp\left(-\frac{Z^2}{4Dt}\right) \tag{1}$$

Larger diffusion coefficients correspond to more rapid movement of the molecules; therefore larger displacements are more likely and the distribution function will be broader. Likewise, larger values of t also result in a broader distribution function as the molecules have a longer period of time to diffuse.

As expected, the average displacement calculated using eq 1 is zero:

$$\overline{Z} = \int_{-\infty}^{\infty} Z \times P(Z)\,dZ = \int_{-\infty}^{\infty} \left(\text{odd function}\right) \times \left(\text{even function}\right) = 0$$

The mean-square displacement, on the other hand, is

$$\overline{Z^2} = \int_{-\infty}^{\infty} Z^2 P(Z)\,dZ = 2Dt$$

Therefore, the root-mean-square displacement for a molecule is

$$Z_{rms} = \sqrt{\overline{Z^2}} = \sqrt{2Dt}$$

The diffusion coefficient is on the order of 1×10^{-5} cm^2 s^{-1} for a typical small molecule in solution, and a typical NMR diffusion measurement experiment is 0.1 s long. These values correspond to a root-mean-square displacement along a single axis of 14 μm.

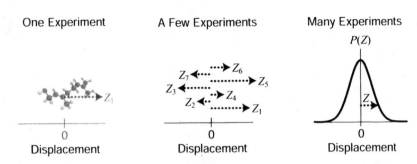

One Experiment   A Few Experiments   Many Experiments

Figure 2. The random walk motion of a single particle during a specific time period will lead to a displacement Z_1, as shown on the left. The center image illustrates the distribution of displacements that result from repeating the experiment several times. After a large number of experiments, the Gaussian distribution shown on the right appears.

Predicting the Diffusion Coefficient. Ideally, a molecule in a gas or liquid will move about at a speed determined by its kinetic energy. In a liquid, however, the distance that the molecule travels is restricted by the "cage" of solvent molecules surrounding it. Only after repeated collisions with the solvent cage will the molecule squeeze into the next cage. This restriction can be modeled as a frictional force, F_f, that is negatively proportional to the velocity, v:

$$F_f = -fv$$

where f is a proportionality constant known as the friction coefficient. For a spherical object moving through a liquid with viscosity η, Stokes' law predicts that the frictional force on the object is

$$F_f = -6\pi\eta rv$$

where r is the radius of the object. Consequently, the frictional coefficient is

$$f = 6\pi\eta r$$

Consideration of the relationship between the kinetic energy of a molecule and the friction due to the surrounding cage of solvent molecules leads to the Stokes-Einstein equation (2):

$$D = \frac{k_B T}{f} = \frac{k_B T}{6\pi\eta r} \tag{2}$$

where k_B is the Boltzmann constant and T is the temperature. This equation was originally derived for spherical colloidal particles and provides an accurate explanation of their motion; it is only an approximation for the behavior of smaller (and/or non-spherical) particles.

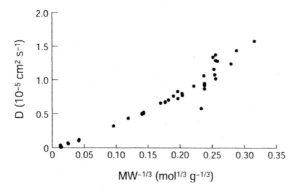

Figure 3. The diffusion coefficient (D) versus the inverse cube root of the molecular weight for a number of dilute solutes in water at 25°C. The lightest molecule included on the graph is methanol (MW = 32 g mol⁻¹) and the heaviest is myosin (MW = 493 kg mol⁻¹). Data are from references 3–4.

Applications of the Diffusion Coefficient. Knowing the value of the diffusion coefficient can be useful for a number of reasons. First, the diffusion coefficient can be used to predict the size of an unknown compound and therefore can be used to estimate the molecule weight. This is possible because eq 2 relates the radius (and therefore the volume) of a spherical molecule to the diffusion coefficient ($V \propto r^3 \propto D^{-3}$). By using this calculated volume and knowing (or assuming) the density, it is possible to estimate the molecular weight. Figure 3 displays the diffusion coefficient versus the inverse cube root of the molecular weight for a number of organic and biological molecules; the trend is apparent although a few outliers exist. As eq 2 is more applicable to large, spherical molecules, it is understandable that the trend shown in Figure 3 is more consistent for large molecules (which appear toward the left side of the graph) than for small ones.

Knowledge of the diffusion coefficient can aid in the determination of the oligomeric state of a compound (*i.e.*, whether a compound is present free in solution or whether it has aggregated to form a dimer, trimer, etc.). The oligomeric state is particularly important in protein NMR where it must be known prior to structure calculations. Although it is possible to determine the oligomeric state using techniques such as dynamic light scattering, these methods are not always applicable at the concentrations used for NMR experiments (*5*). To determine the oligomeric state of a molecule, a comparison is made between the experimentally measured diffusion coefficient and a calculated diffusion coefficient based on the molecule's size and shape (*6*). If the values match, the compound is probably a monomer. If the calculated value is significantly greater than the experimentally measured value, the compound has probably aggregated. This comparison of experimental and calculated values is most useful for detecting dimerization, as the molecular weights of a monomer and a dimer differ by a factor of two. Differentiating higher oligomeric states, such as whether a molecule is a trimer or a tetramer, becomes progressively more difficult as the ratio of the molecular weights (and therefore the variations in the diffusion coefficients) will be smaller. This method is most applicable to large molecules, such as proteins, but also has been used for smaller systems such as non-covalent assemblies of calix[4]arene units (6) and ions in solution (7).

Similarly, diffusion coefficients can be used to estimate binding affinities between substrates and proteins. For example, one screening method used in drug discovery is to observe how the apparent diffusion coefficient for a small molecule changes in the presence of a protein. A decline in the diffusion coefficient of the small molecule in the presence of a protein is indicative of binding, in which case the small molecule is a good candidate for a drug. As will be seen later in this chapter, it is important in such experiments to control for changes in the viscosity upon addition of further solute or other solutes as the solution viscosity, and consequently the diffusion coefficients, is very sensitive to the composition of the sample.

Measurements of the diffusion coefficients can also be used to separate signals from different molecules. In NMR spectroscopy, the spectrum of a mixture is equivalent to the superposition of the spectra of the individual components. Using the experiment described later in this chapter, it is possible to measure a diffusion coefficient for each peak in the spectrum. As the peaks from the non-exchangeable nuclei of a compound will share the same diffusion coefficient, it is possible to use the diffusion coefficient to distinguish signals from different components of the mixture (provided they have appreciably different diffusion coefficients). This method of separating signals is known as diffusion-ordered spectroscopy, or DOSY (8). DOSY is particularly useful in situations where it is not convenient or possible to physically separate a mixture; DOSY is capable of separating the signals from different components without the need to resort to chromatography. Another use for DOSY is in cases where it is the *mixture* of components and their interactions that is of interest.

Magnetic Field Gradients and NMR Spectroscopy

Before running an NMR experiment, it is typical to spend some time "shimming" the magnet. The shimming procedure is carried out in order to make the magnetic field as homogeneous as possible; this is necessary because spatial inhomogeneities in the main magnetic field interfere with the resolution of the NMR experiment. However, observing translational diffusion by NMR spectroscopy requires a spatially-dependent magnetic field. In other words, it is necessary to reintroduce inhomogeneities to the main magnetic field to measure the diffusion coefficient.

Experimentally, these inhomogeneities are generated using gradients. Gradients are coils inside the NMR probe that generate well-defined spatial variations of the magnetic field. Most gradients are designed to generate a linear variation in the magnetic field along the axis of the main magnetic field (B_0); this axis is usually defined as the z-axis and therefore such a gradient is called a z-axis gradient. The symbol G_z represents the strength of such a gradient [$G_z = \partial B(z)/\partial z$]. By changing the magnitude and sign of the electrical current running through the coil of the gradient it is possible to vary the magnitude and the sign of G_z. Likewise, the gradient can be switched off ($G_z = 0$) by switching off the electrical current.

Although z-axis gradients are the most common type of magnetic field gradients, it is possible to generate gradients along other axes, such as the x-axis, the y-axis, or even along the rotor axis of a magic-angle spinning probe. In a typical NMR probe, each gradient coil can generate a magnetic field gradient of several tens of Gauss cm^{-1} for a period of several milliseconds (a "gradient pulse"); weaker gradient pulses can be generated for much longer periods. In all cases, the direction of the magnetic field generated by the gradient is parallel to B_0; the gradient axis is the axis along which the magnitude, not the direction, of the field varies.

Magnetization and Gradients. The Larmor frequency (ω) for the nuclei in a sample depends on the magnetic field and the magnetogyric ratio (γ):

$$\omega = -\gamma B_0$$

When a gradient is applied, the magnetic field varies linearly along the gradient axis and, consequently, the Larmor frequency varies in the same way. As a result, nuclei at different parts of the sample will precess at different frequencies. On a bulk scale, the result is a spatially-dependent phase for the transverse sample magnetization. As shown in Figure 4, the magnetization will begin to vary in phase across the sample after the gradient is switched on, resulting in a helical profile for the transverse magnetization along the gradient axis. As the signal observed in the NMR experiment arises from the sum of all the transverse magnetization in the sample, the spatially-dependent phase results in a diminished signal because the transverse (x and y) components of the magnetization will tend to cancel one another. The degree to which the signal is diminished depends on how tightly the transverse magnetization is "coiled" along the gradient axis. This signal attenuation due to a gradient pulse is referred to as gradient *dephasing*.

Consider an experiment in which two gradient pulses are applied. The first gradient pulse, often referred to as the *phase-labeling* gradient, dephases any transverse magnetization. A second gradient pulse can be used to reverse this

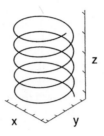

Figure 4. A gradient pulse along the z-axis winds the transverse magnetization into a "helix" (the helix traces out the tips of the magnetization vectors). The pitch of this helix (i.e., how tightly it is wound) depends on the gradient strength and length, and the magnetogyric ratio of the nuclei.

process if the appropriate combination of pulse length and strength are used. Such a gradient is called a *refocusing* gradient because it uncoils the magnetization helix and restores the transverse magnetization to its initial state. For the magnetization to be properly refocused, the gradient values must be chosen such that the spatially dependent phase sums to zero at the end of the experiment. By carefully choosing the strengths of different gradients in an experiment, it is possible to use gradients in place of phase cycling for selecting different components of the NMR signal. However, this is not the focus of this chapter; instead, we will concentrate on how gradients are used to measure translational diffusion.

Gradients and Motion. The measurement of motion by NMR requires at least two gradient pulses in an experiment: the first to dephase the magnetization and the second to refocus it at some later point in the experiment. To completely refocus the magnetization (and completely recover the NMR signal), the molecules of the sample must not move during the time period in between the gradients. If the molecules *do* move, the observed signal will change because the refocusing gradient will not be able to properly restore the transverse magnetization to its initial phase.

In order to visualize how movement affects the result of a gradient-selected experiment it is useful to consider a discrete version of the magnetization helix shown in Figure 4. In the discrete model (shown in Figure 5), the sample is split into thin slices referred to as *isochromats* in which the magnetic field is approximately homogeneous. Before the first gradient pulse the components of the magnetization in the various isochromats are aligned as shown on the left in Figure 5. During the gradient, the nuclei precess at different frequencies in each isochromat. This variation in precession results in the situation shown on the right of Figure 5. In the absence of motion, a suitable refocusing gradient will realign the magnetization. However, if the nuclei move in between the phase-labeling and refocusing gradients, the signal at the end of the experiment will change in phase, amplitude, or both, depending on the nature of the movement.

Coherent Motion. Coherent motions are ones that persist in direction and magnitude over the timescale of an experiment. Examples of coherent motions are laminar flow, plug flow, and convection. The simplest type of coherent motion to deal with is uniform flow of the sample in one direction. As shown in Figure 6, uniform downward motion of the sample in between the phase-labeling and refocusing gradients results in the realignment of the transverse magnetization at the end of the experiment. However, the overall phase of the magnetization is not the same as the initial phase; the magnetization has acquired a phase that depends on the sample velocity. Since the phase of the transverse magnetization is uniform across the sample at the end of the experiment, the spectrum will not change in intensity, but will change in phase depending on the flow velocity. Consequently, the velocity of a sample can be estimated by measuring the change in the phase of the spectrum relative to the spectrum of a stationary sample. Variations of this technique have been used, for example, to measure the flow of blood, the rate of water moving through vascular plants, and the non-Newtonian flow of polymers (*1*).

Incoherent Motion. The effects of incoherent motions, such as diffusion and turbulent flow, on the NMR signal are somewhat different than those from coherent motions. Whereas coherent motions can change the phase and/or the intensity of the signal, incoherent motions only affect the signal intensity. The reason for this is that incoherent motions results in a symmetrical distribution of phases for the individual spins in the sample. The intensity of the NMR signal (*S*) is related to the ensemble average of the phases (ϕ) of the individual spins according to:

$$S \propto \overline{\exp(i\phi)}$$

In the case of a symmetrical distribution of phases about some value, the resulting signal will not change in phase. However, the cancellation of signal components from individual spins will reduce the intensity of the signal.

A mechanism of motion is considered incoherent if the correlation time for movements is short in comparison to the timescale of the experiment. The correlation time refers to the time-scale over which a molecule maintains the same velocity. In molecular diffusion, the velocities of individual molecules change frequently due to collisions, whereas in turbulence the velocities of small regions of the sample change frequently. In NMR spectroscopy, the timescale of the experiment is always much longer than the correlation time for small molecules undergoing unrestricted diffusion, so translational diffusion behaves as an incoherent process.

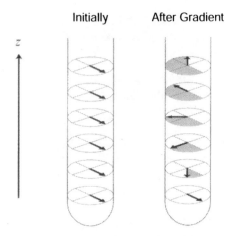

Figure 5. As shown on the left, the components of the sample magnetization in the isochromats are aligned at the start of a gradient pulse. After the gradient pulse, these components will have precessed by different amounts according to their location in the sample as shown on the right. Due to the gradient, the magnetization has a phase (indicated by the shading) that varies with position in the sample.

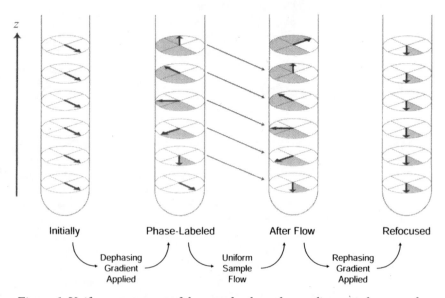

Figure 6. Uniform movement of the sample along the gradient axis between the phase-labeling and refocusing gradient will result in a velocity-dependent phase as shown by the shading.

Measuring Diffusion Coefficients

A Gaussian distribution function (eq 1) describes the distribution of molecular displacements due to translational diffusion. Therefore, in an NMR experiment with dephasing and refocusing gradients there will be a Gaussian distribution of phases due to translational diffusion between these two gradients pulses. As this Gaussian distribution of phases is symmetric the effect of molecular diffusion in the NMR experiment is *only* to attenuate the signal; diffusion does not cause the signal to change in phase. The degree to which diffusion attenuates the signal can be calculated based on a modified version of the Bloch equations (which are a set of differential equations that provide a semi-classical description of the NMR experiment). The result of this derivation (9) is that the attenuation of the signal (*S*) relative to the signal in the absence of diffusion (*S*$_0$) is

$$S = S_0 \exp\left[-D\gamma^2 G^2 \delta^2 s^2 \left(\Delta - \tfrac{1}{3}\delta\right)\right] \qquad (3)$$

where γ is the magnetogyric ratio, G is the gradient strength, δ is the length of the gradient pulses, Δ is the time between the gradients, and s is a shape factor which accounts for how the gradient pulses are ramped on and off. Typically, s will be either 1 (for rectangular-shaped gradient pulses) or $2/\pi$ (for half-sine bell shaped gradient pulses) (10). If we take the natural log of this equation, we find

$$\ln S = \ln S_0 - D\gamma^2 G^2 \delta^2 s^2 \left(\Delta - \tfrac{1}{3}\delta\right) \qquad (4)$$

Consequently, a plot of $-\ln S$ versus $\gamma^2 G^2 \delta^2 s^2 (\Delta - \tfrac{1}{3}\delta)$ should result in a straight line with slope D and y-intercept $-\ln S_0$. One subtle point for using eq 4 is that, when linear regression is used to fit the data, all the points are weighted equally. This can yield inaccurate results, as noise will disproportionately affect some of the data. This problem is not severe if the signal-to-noise ratio for all the data is good; if this is not the case then directly fitting the data to eq 3 using a non-linear fit yields more accurate results (11).

Figure 7 illustrates the diffusion-dependent signal attenuation for a sample of camphene in deuterated methanol in a series of experiments with increasing gradient strength. The spectra were integrated between 1.1 and 0.95 ppm and, as shown on the right of Figure 7, the integrals were plotted versus $\gamma^2 G^2 \delta^2 s^2 (\Delta - \tfrac{1}{3}\delta)$ using a semi-log plot. The slope of this plot corresponds to the diffusion coefficient. For this sample, linear least-squares was used to fit the data, resulting in an estimated diffusion coefficient of 1.339 ($\pm$0.002)$\times 10^{-5}$ cm^2 s^{-1}. The quoted error is the standard deviation of the slope resulting from the linear regression analysis, which usually underestimates the experimental error inherent in the technique. This value for the diffusion coefficient was deduced using the integrals of the methyl peaks; other camphene peaks in the spectrum should provide identical results.

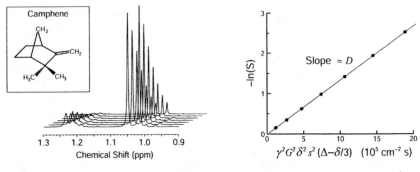

Figure 7. On the left is a stacked plot for a small portion of the camphene ^{1}H NMR spectrum. The spectra differ in the amount of diffusion-dependent signal attenuation; the degree of attenuation was controlled by varying the gradient strength. The semi-logarithmic plot on the right is based on the integrals for the region between 0.95 and 1.1 ppm. The linear least-squares fit to the data is indicated by the solid line. The sample temperature was 300.0 K (26.9 °C).

The Accuracy of Measured Diffusion Coefficients. The value of the diffusion coefficient depends primarily on three physical factors: the identity of the molecule of interest, the solvent, and the temperature. There are also several experimental factors that affect the accuracy of the measurement of the diffusion coefficient, including the uniformity of the gradient (usually only a very minor effect), other sources of motion, distortions of the baseline, and the uniformity of the temperature of the sample. The experiment for measuring diffusion coefficients that is presented later in this chapter was designed to minimize some of these sources of interference.

Solvent Viscosity. According to the Stokes-Einstein equation (eq 2), the diffusion coefficient is inversely proportional to the solvent viscosity. The viscosity depends not only on the temperature of the sample, but also on the *composition* of the sample. For example, changing the concentration of the solute, or adding other solutes to the sample, will alter the solvent viscosity. Whether a solute increases or decreases the viscosity depends on the nature of both the solute and the solvent. For example, geraniol and quinine both increase the viscosity of chloroform, whereas surfactants will often decrease the viscosity of aqueous solutions.

The effect of solvent viscosity is particularly relevant in NMR spectroscopy as, due to the intrinsically low sensitivity of the method, samples with relatively high concentrations are used; concentrations between 0.1 M and 1 M are not unusual. Consequently, it is necessary to know the exact composition of a sample when determining the diffusion coefficient; otherwise the results may not be reproducible.

Restricted Diffusion. The theory used to derive eqs 1 and 3 assumes that the movement of the sample molecules is unbounded. If this is not the case, then the molecules are subject to *restricted diffusion* and the signal attenuation due to diffusion will be less than what eq 3 predicts. Fortunately, a typical solution-state NMR sample tube has a diameter of several millimeters whereas the sample molecules move only a few tens of micrometers during an experiment. Therefore, the effect due to restricted diffusion can be safely ignored and eqs 1, 3, and 4 can be used without difficulty.

Motion. NMR experiments involving gradient pulses are sensitive to both coherent and incoherent motions. This means that any attempt to measure the diffusion coefficient in the presence of another source of movement produces results that reflect *both* movements. Consequently, diffusion measurements should be performed *without* sampling spinning.

In addition, convection can arise due to small temperature gradients in the sample; this is particularly common for low viscosity solvents such as chloroform or acetone (*12*). In diffusion-measurement experiments, sample convection introduces an additional component to the signal decay that results in overestimates of the diffusion coefficient. Fortunately, convection is, at least to a first approximation, a coherent motion and therefore the signal variation due to convection can be removed by using the double-stimulated echo experiment described later in this chapter (*13*).

Gradient Calibration. To make accurate measurements of the diffusion coefficient, an accurate value for the strength of the gradient needs to be known. Typically, this value will be determined when the NMR system is installed and is unlikely to need to be recalibrated. If the value is unknown, it can be determined in two ways. The first method involves making a diffusion measurement using a sample with a known diffusion coefficient. By working backwards from the known diffusion coefficient, it is possible to determine the strength of the gradient.

The second method involves using the spectrometer to acquire an image of a sample with a known geometry; the strength of the gradient can then be deduced by comparing the dimensions of the sample with the frequency separation of the features in the image. For example, calibration of a z-axis gradient can be performed using a vortex plug that had been inserted into an NMR sample tube that is filled with a 1:9 solution of water in deuterated water as shown at the top of Figure 8. The length of the sample (Δz) is related to the frequency separation of the features in the image ($\Delta \nu$) according to

$$\Delta \nu = \frac{1}{2\pi} \gamma G \Delta z$$

where, once again, γ is the magnetogyric ratio, G is the gradient strength.

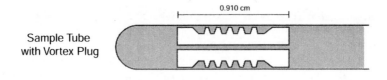

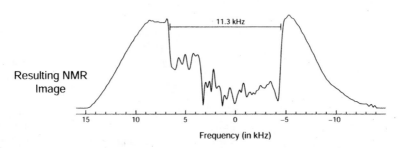

0.910 cm

Sample Tube
with Vortex Plug

11.3 kHz

Resulting NMR
Image

15   10   5   0   −5   −10
Frequency (in kHz)

Figure 8. The strength of the gradient can be calibrated by imaging a sample with known geometry. In this case, the sample consisted of a 5 mm NMR sample tube with a 0.910 cm vortex plug inserted into the active region. The solvent surrounding the plug (indicated by the shading in the figure) was 10% H_2O/90% D_2O (by volume). The bottom part of the figure shows the NMR image of this sample acquired at 300 MHz for ^1H. This image reveals an 11.3 kHz separation when the imaging gradient is set to 5% of its maximum value; this corresponds to a maximum gradient strength of 58.0 G cm^{-1}. The signal goes to zero at the edges of the image due to the limited length of the radiofrequency coil.

The vortex plug shown in Figure 8 is 0.910 cm in length and the resulting image has a frequency separation of 11.3 kHz when the gradient is at 5% of its maximum strength. Therefore, the gradient strength is

$$G = \frac{2\pi \Delta \nu}{\gamma \Delta z} = \frac{2\pi\left(11300 \text{ Hz}\right)}{\left(26750 \text{ rad G}^{\S1} \text{ cm}^{\S1}\right)\left(0.910 \text{ cm}\right)} = 2.92 \text{ G cm}^{-1}$$

and the maximum gradient strength is (2.92 G cm^{-1} / 5%) = 58.0 G cm^{-1}.

Pulse Sequences. Measuring the diffusion coefficient requires the use of an NMR experiment which includes both dephasing and rephasing gradients. The simplest experiment used for measuring diffusion is the gradient spin echo (SE) experiment shown at the top of Figure 9. To achieve a significant amount of diffusion-dependent signal attenuation, the delay time (Δ) must usually be between 0.1 and 0.5 s. This delay is long enough that effects due to transverse relaxation and homonuclear scalar couplings interfere with the diffusion measurement.

Compared to the SE experiment, the stimulated echo (STE) experiment shown in the middle of Figure 9 is much better for measuring diffusion coefficients as the sample magnetization is "stored" as longitudinal magnetization in the middle part of the experiment. During this storage period, the molecules continue to diffuse but the magnetization does not evolve due to couplings and is subject to longitudinal relaxation instead of transverse relaxation.

The preferred pulse sequence for measuring diffusion coefficients is the double stimulated echo (DSTE) experiment shown at the bottom of Figure 9. The advantage of this sequence is that it is only sensitive to incoherent motion (*i.e.*, diffusion). Effects due to coherent motions are refocused by the second stimulated echo in the sequence, which means that effects due to sample convection are compensated for (at least to first-order) in the resulting spectra (*14*). Consequently, the DSTE experiment provides more accurate values for the diffusion coefficient, particularly for samples in low-viscosity solvents such as chloroform or acetone.

Experimental

Sample Preparation. First, prepare a series of five samples of geraniol in deuterated chloroform; the samples should range in concentration from around 0.02 M to 1.00 M. It is useful to make sure that the samples are filled to the same height as this will greatly reduce the amount of shimming necessary when changing samples. Next, prepare a similar series of quinine samples. These samples should cover a similar range of concentrations as the geraniol samples. Studying concentrations above 1 M is unfortunately not possible as the solution approaches saturation at this concentration. Note that quinine is usually only 90% pure; the remainder is usually hydroquinine, which differs from quinine only in that the terminal alkene is saturated.

Data Acquisition. Run a simple one-dimensional (1D) experiment for the first sample to make sure that the sample is shimmed adequately, that the 90° pulse time is calibrated properly, and that the spectral width includes all the peaks in the spectrum. Next, setup and run a DSTE experiment. On Bruker systems, the standard DSTE sequence is called "dstegp3s". This sequence should be used with an eight-step phase cycle. Usually, eight values for the gradient strength are sufficient to observe the attenuation of the signal due to diffusion.

Repeat the 1D and DSTE experiments for each of the geraniol samples and then for each of the quinine samples. For the more concentrated quinine samples the linewidths will be noticeably wider. This effect is due to the increased

170

viscosity of for the concentrated samples. Due to this increase, the quinine molecules tumble more slowly, which leads to faster transverse relaxation rates and, in turn, broader lines in the spectrum.

The variable temperature (VT) control system should be on during the experiments. This is necessary because the diffusion coefficients depend strongly on the sample temperature. For consistent results, it is best to wait a few minutes after inserting each sample to ensure that the sample has had time to equilibrate to the system temperature.

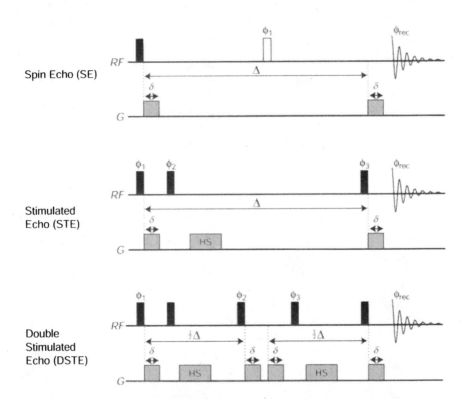

Figure 9. Pulse sequences for the spin echo (SE), stimulated echo (STE) and double stimulated echo (DSTE) experiments. The top line for each experiment indicates the radiofrequency (RF) pulses; filled and open rectangles represent 90° and 180° RF pulses, respectively. The bottom line for each experiment indicates the gradient pulses (G). The gradient pulses labeled "HS" are homospoils; to ensure that the signal is properly refocused all other gradient pulses in a sequence should have the same length, magnitude, sign, and shape. The phase cycles for the experiments are as follows. SE: $\phi_1 = y, -y; \phi_{rec} = x$. STE: $\phi_1 = x, x, x, x, -x, -x, -x, -x; \phi_2 = \phi_3 = y, -y, x, -x; \phi_{rec} = x, x, -x, -x, -x, -x, x, x, x. DSTE: $\phi_1 = x, y, -x, -y; \phi_2 = -x, -y; \phi_3 = -x, -x, -x, -x, x, x, x, x; \phi_{rec} = x, x, -x, -x, -x, -x, x, x. All other RF pulse phases have a phase of x.

Data Processing. Once acquired, each DSTE experiment should be phased and Fourier transformed in the directly-detected dimension. It is usually a good idea to baseline correct the spectra as baseline errors generate systematic errors in the calculated diffusion coefficients. Next, select a region and integrate each of the rows in the data set. Usually, the best region to use is one that has a strong signal and that is well removed from solvent and impurity peaks. For geraniol, the methyl region (1.8 – 1.5 ppm) works well. For quinine, one of the peaks in the aromatic region (8.5 –7 ppm) or the methyl peak at ~3.8 ppm are suggested.

On Bruker spectrometers running XWIN-NMR the analysis of the data can be performed using the relaxation (t1/t2) environment. However, the interface for this function is somewhat Byzantine, so we use a program developed by one of the authors to integrate each row (*xiv*) and then analyze the data using a separate program.

Once measured, the integrals should be fit to eq 4 using linear regression (or to eq 3 using a non-linear fitting routine) to determine the diffusion coefficients. For our data, we fit the data to eq 3 using the solver macro in Excel.

Results and Discussion

The results from measuring the diffusion coefficient for a series of geraniol and quinine samples are shown in Figure 10. In both cases, the diffusion coefficients decrease as the concentrations of the solutes increase, but the variation is much more drastic for the quinine solutions than for the geraniol solutions. The observed variation is primarily due to changes in viscosity. For example, geraniol is a liquid at room temperature and has a higher viscosity than chloroform. Therefore, combinations of the two liquids should have intermediate viscosities. According to the Stokes-Einstein equation (eq 2), increases in viscosity reduce the diffusion coefficient, a conclusion that is in agreement with the observed trend.

The quinine samples show a similar, but much more pronounced, variation, as indicated by the percentage scale on the right side of each graph in Figure 10. This difference can be attributed to the larger size of the quinine molecules as well as the presence of stronger intermolecular forces than those found in the geraniol solutions.

Using the diffusion coefficient for the least concentrated samples and assuming that the viscosity of the solution is the same as the viscosity of chloroform, it is possible to use eq 2 to estimate the molecular radius. Based on our results, the estimated radii for geraniol and quinine are 0.3 nm and 0.5 nm, respectively. These results agree well with average radii determined using molecular modeling programs.

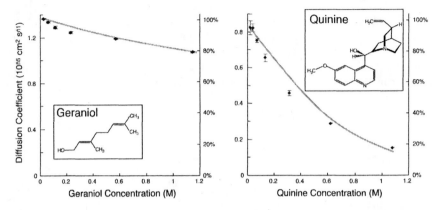

Figure 10. Experimental results for the measurement of the diffusion coefficients for a series of geraniol samples (on left) and a series of quinine samples (on right). The right axis on each graph shows the percent change (100% corresponds to the diffusion coefficient for the least concentrated sample). The data were acquired on a Bruker NMR spectrometer operating at 300 MHz for ^1H. At each concentration, seven (geraniol) or nine (quinine) distinct peaks were integrated and fit to determine the diffusion coefficient. The data represent the average value for each concentration along with the standard deviation. The gray lines indicate "theoretical" values of the diffusion coefficients predicted using eq 2 as explained in the text. The sample temperature was maintained at 300.0 K (26.9 °C) throughout the diffusion measurement experiments.

We measured the viscosity of solutions of geraniol and quinine in chloroform at concentrations ranging from 0 to 1 M using an Ostwald viscometer. Based on these results and our estimated molecular radii, we used eq 2 to calculate the variation of the diffusion coefficient with concentration. These "theoretical" diffusion coefficients are shown as gray lines in Figure 10. The close correspondence between the experimental diffusion coefficients and the calculated values indicate that the sample viscosity is the primary reason for the observed variation in the diffusion coefficients.

Conclusion

NMR spectroscopy is a powerful technique for measuring molecular motion since the technique is non-invasive and does not require physically "tagging" the sample. The rate of translational diffusion can be easily measured in a relatively short NMR experiment, and this information can be used to make inferences about the size of the sample molecules as well as to observe the effect of the

173

solutes on the bulk properties of the solution. Additionally, measurements of the diffusion coefficients can provide insights into the oligomeric state of molecules or be used to monitor protein-ligand binding. However, as the results of such measurements are highly dependent on the sample composition, temperature, and solvent viscosity, these influences need to be included in any interpretation of the diffusion coefficients.

References

1. Callaghan, P. T. *Principles of Nuclear Magnetic Resonance Microscopy*; Oxford Science Publications: Oxford, UK, 1991; Chapter 6, pp 333.
2. Chang, P. T. *Physical Chemistry for the Chemical and Biological Sciences, 2nd ed.*; University Science Books: Sausalito, CA, 2000; Chapter 21, pp 881.
3. Longsworth, L. G., *J. Chem. Phys.* **1954**, *58*, 770.
4. Mortimer, R. G. *Physical Chemistry*; The Benjamin/Cummings Publishing Company: Redwood City, CA, 1993; Appendix A, pp A-24.
5. Krishnan, V. V. *J. Magn. Reson.* **1997**, *124*, 468.
6. Timmerman, P.; Weidmann, J. L.; Jolliffe, K. A.; Prins, L. J.; Reinhoudt, D. N.; Shinkai, S.; Frish, L.; Cohen, Y. *J. Chem. Soc. Perk. T. 2* **2000**, *10*, 2077.
7. Mo, H. P.; Pochapsky, T. C. *J. Phys. Chem. B* **1997**, *101*, 4485.
8. Morris, K. F.; Johnson, C. S., Jr. *J. Am. Chem. Soc.* **1992**, *114*, 3139.
9. Stejskal, E. O.; Tanner, J. E. *J. Chem. Phys.* **1965**, *42*, 288.
10. An additional small correction should be introduced to eqs 3 and 4 to further compensate for the shape factor, but in most cases this is negligible. For more information, see Price, W. S.; Kuchel, P. W. *J. Magn. Reson.* **1990**, *94*, 133.
11. Johnson, C. S., Jr. *Prog. in NMR Spectrosc.* **1999**, *34*, 203.
12. Loening, N. M.; Keeler, J. *J. Magn. Reson.* **1999**, *139*, 334.
13. Jerschow, A.; Müller, N. M. *J. Magn. Reson.* **1997**, *125*, 372.
14. The integration routine (which works from within XWIN-NMR or TOPSPIN) can be downloaded from the author's website at: http://www.lclark.edu/~loening/software/introws

Modern NMR in Laboratory Development

Advanced Organic Chemistry

Chapter 13

NMR Exchange Spectroscopy

Holly C. Gaede

Department of Chemistry, Texas A&M University,
College Station, TX 77843

Two-dimensional exchange spectroscopy (EXSY) is a useful approach for studying dynamic processes with NMR. This method may be used in undergraduate laboratories as an alternative to the established techniques of line shape analysis or saturation transfer. In this chapter, the theory is briefly outlined for all three methods, and their experimental approaches are compared.

Introduction

Dynamic NMR involves the study of samples that undergo chemical or physical changes with time. Because no sample is truly rigid, *all* NMR is dynamic NMR. The timescale of motions that can be observed ranges from nanoseconds to minutes, depending on the sampled experimental observable, such as chemical shift, relaxation rate, or coupling constant. The accessible timescale includes important molecular motions, including cis-trans isomerization and boat-chair cyclohexane intraconversions. Timescales for molecular motions and NMR experiments that can be used to study them are summarized in Figure 1. Besides molecular motions, important dynamic processes that occur in the available time scale include keto-enol tautomerization and the formation of intermolecular complexes. The focus of this chapter will be on the exchange of two species, A and B, between two different environments. Examples of this kind of dynamic equilibrium include internal rotations about a bond, ring puckering, and proton exchange.

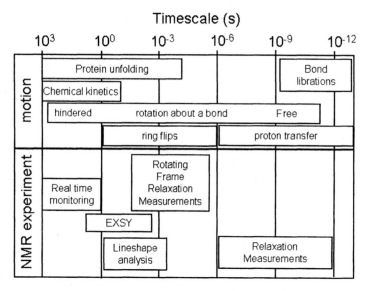

Figure 1. Timescales for molecular motions and NMR experiments that can be used to study them.

The NMR parameters measured are reflective of the environment. If the parameter measured is chemical shift, the exchange between two environments is considered *slow* if $k << |\delta_A - \delta_B|$; *intermediate* if $k \approx |\delta_A - \delta_B|$; and *fast* if $k >> |\delta_A - \delta_B|$, where the chemical shifts of the two environments are δ_A and δ_B. Chemical shift differences in NMR spectrometers available in most well-equipped undergraduate labs (^1H resonance frequencies of 200 – 400 MHz) are typically 10 – 500 Hz, meaning that fast exchange occurs in systems that are exchanging with rate constants $k \gtrsim 10^3$ s^{-1}, corresponding to lifetimes of milliseconds or less. Since chemical shift differences are magnetic field dependent, a system that is in the fast exchange regime on a low field instrument may enter the slow exchange regime when studied at a higher field. Under slow exchange, each chemical shift is observed distinctly. In contrast, under fast exchange, only a single chemical shift is observed at the population-weighted average position.

$$\delta_{obs} = p_A \delta_A + (1 - p_A)\delta_B, \tag{1}$$

where p_A is the fractional population of A. Figure 2 shows the expected appearance of NMR spectra under slow, intermediate, and fast exchange.

Note that different NMR parameters give windows into different time regimes. Averaged ^1H coupling constants are observed for all but the slowest exchange processes, as ^1H coupling constants have differences on the order of only 10 Hz, making 100 ms the upper lifetime limit of fast exchange. Relaxation times,

178

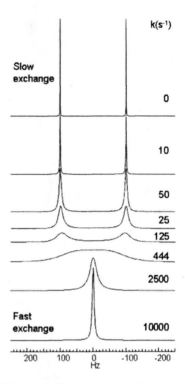

Figure 2. Simulated NMR spectra for a system with two resonances separated by 200 Hz undergoing mutual site exchange at different rates.

with differences on the order of 1 ms in large systems, are averaged only in systems with lifetimes ~0.1 ms or less.

One intuitive method for extracting the rate constant of an exchange process is through two-dimensional NMR exchange spectroscopy (EXSY), where the exchange network is visually apparent in the spectrum. In fact, one of the first published two-dimensional spectra was a study of the exchange of N,N-dimethylacetamide methyl protons.(1) The EXSY experiment serves as a nice introduction to two-dimensional NMR, as its mechanism of coherence transfer is more intuitive than that in COSY or NOESY experiments.

This chapter will compare this 2D approach to two other approaches used to measure exchange in undergraduate laboratories: lineshape fitting (2) and saturation transfer.(3) For ease of comparison, the same prototypical system will be used as an example throughout, the cis-trans isomerization of N,N-dimethylacetamide.

This system is an example of mutual site exchange because the departing methyl group is replaced by an equivalent one. The partial double bond character of the amide bond yields a high internal rotation barrier, and the ^{1}H NMR spectrum is simple, just three well-resolved singlets under slow exchange. Beyond teaching NMR principles, these experiments can be used to illustrate concepts in many sub-disciplines of chemistry. These experiments could be used in physical chemistry to illustrate basic kinetic principles and give the students experience at manipulating the Eyring equation. In organic chemistry these experiments could be used to illustrate the nature of rotation about a double bond.(*4*) Because these amides serve as simple models for peptide bonds, these studies could be used to explain the conformations found in proteins. Ideally, any of these experimental approaches could be combined with a computational study of the rotational barrier.(*5*)

Lineshape Analysis

As outlined above, chemical exchange processes involve characteristic changes in the NMR lineshape. For the simple case of mutual site exchange for two equally populated states, the Bloch equations may be used to derive these lineshapes. More general cases will require a density matrix approach, which is found in an excellent recent review as well as classic texts.(*6-8*)

A spin-1/2 nucleus in a static magnetic field B_0 has a magnetization, M_z, along the same axis. There are two horizontal magnetization components, u and v, that oscillate about this axis at the larmor frequency $\omega_0 = \gamma B_0$, where γ is the gyromagnetic ratio of the nucleus. In the NMR experiment, there is an additional B_1 radiofrequency field. The Bloch equations give the time behavior of the magnetizations.

$$\frac{du}{dt} = \left(\omega_0 - \omega\right)v - \frac{u}{T_2} \tag{2}$$

$$\frac{dv}{dt} = \gamma B_1 M_z - \left(\omega_0 - \omega\right)u - \frac{v}{T_2} \tag{3}$$

With a complex magnetization defined as, $M = u + iv$, the Bloch equation then becomes

$$\frac{dM}{dt} = i\gamma B_1 M_z - i(\omega_0 - \omega)M - \frac{M}{T_2} \qquad (4)$$

In the case of mutual site exchange between two sites A and B at a rate of k, the Bloch equations for A and B become:

$$\frac{dM_A}{dt} = i\gamma B_1 M_{zA} - i(\omega_A - \omega)M_A - \frac{M_A}{T_2} + kM_B - kM_A, \qquad (5)$$

$$\frac{dM_B}{dt} = i\gamma B_1 M_{zB} - i(\omega_B - \omega)M_B - \frac{M_B}{T_2} + kM_A - kM_B \qquad (6)$$

These equations are known as the McConnell equations after H. M. McConnell, who first derived them in 1958.(9) Here it is assumed that $M_{zA} = M_{zB} = M_z$ since the B_1 field is so small as not to disturb the magnetization along z. Note that some derivations define $1/2M_z = M_{zA} = M_{zB}$, and as a consequence the results that follow differ by a factor of 2.

At equilibrium, $dM_A/dt = dM_B/dt = 0$. Neglecting the T_2 terms, the following expression for lineshape may be obtained, recalling that the observable part of the magnetization is the imaginary part of $M_A + M_B$ at steady-state.

$$v = \gamma B_1 M_z \frac{k(\omega_A - \omega_B)^2}{k^2((\omega_A - \omega) + (\omega_B - \omega))^2 + (\omega_A - \omega)^2(\omega_B - \omega)^2}. \qquad (7)$$

In this equation, the units are in radians per second. Using $\omega = 2\pi v$, the absorption lineshape can be given in frequency units.

$$g(v) = \gamma B_1 M_z \frac{4k(v_A - v_B)^2}{\left(\frac{1}{2}(v_A + v_B) - v\right)^2 + \pi^2(v_A - v)^2(v_B - v)^2} \qquad (8)$$

This equation is valid as long as linebroadening from processes other than exchange is negligible. Historically, equation 8 was approximated under certain limiting conditions, and k could be estimated from the spectrum. In particular, coalescence, where the appearance of the spectrum changed from two separate peaks to one flat-topped peak, was sought. At coalescence the first and second derivative of the lineshape is zero, and $k_{coalesence} = \pi(v_A - v_B)/\sqrt{2}$. A laboratory experiment published nearly 30 years ago uses this approach to find k for the cis-trans isomerization of N,N-dimethylacetamide, where $\Delta G^\dagger$ was found to be 71 kJ/mole. (Figure 3)

These approximations are generally severe, and today a better approach is a full-curve fitting of equation 8.(*10*) Optimal values for k, v_A, and v_B can be obtained from simulations in which deviations between the calculated and experimental spectrum are minimized. Computational packages are available for this fitting (e.g. WinDNMR(*11*)). Alternatively, fitting routines can be written with symbolic programs such as MathCad or Mathematica.

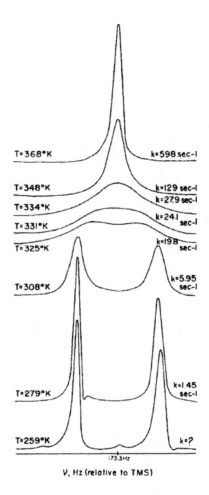

V, Hz (relative to TMS)

Figure 3. Effect of temperature on line shapes and values for k, in the cis-trans isomerization of N,N-dimethylacetamide. Used with permission from the Journal of Chemical Education, Vol. 54, No. 4, 1977, pp. 258-261; copyright © 1977, Division of Chemical Education, Inc.

Saturation Transfer

Another method for evaluating the rate of exchange exploits the fact that irradiation of resonance A will cause changes in the intensity of resonance B in the case of slow exchange. The rate equations for the lower spin state populations [A] and [B] are given by

$$\frac{d[A]}{dt} = -k[A] + k[B] = -\frac{d[B]}{dt}. \tag{9}$$

Similarly, the rate equations for the upper spin-state populations [A*] and [B*] are given by

$$\frac{d[A*]}{dt} = -k[A*] + k[B*] = -\frac{d[B*]}{dt}. \tag{10}$$

Spin-lattice relaxation processes that keep the upper and lower spin state populations in equilibrium in the absence of chemical exchange are given by

$$\frac{dM_A}{dt} = \frac{M_{0A} - M_A}{T_{1A}} \text{ and } \frac{dM_B}{dt} = \frac{M_{0B} - M_B}{T_{1B}} \tag{11a,b}$$

where the net magnetizations are given by $M_A = [A] - [A*]$ and $M_B = [B] - [B*]$, and M_{0A} and M_{0B} are the magnetizations at equilibrium.

Combining equations 9-11 gives the change of net magnetization of A and B due to spin-lattice relaxation and chemical exchange

$$\frac{dM_A}{dt} = -k(M_A - M_B) + \frac{M_{0A} - M_A}{T_{1A}}. \tag{12}$$

$$\frac{dM_B}{dt} = k(M_A - M_B) + \frac{M_{0B} - M_B}{T_{1B}} \tag{13}$$

If a decoupler pulse is selectively applied to resonance B, the population difference between the upper and lower spin states is removed, and $M_B = [B] - [B*] = 0$. The B resonance will not be observed in the spectrum, since the NMR signal is proportional to this population difference. Over time, the saturated spin-population will be passed on to A through chemical exchange, resulting in a reduced signal intensity compared to spectra observed in the absence of saturation. The effect of spin saturation on the N,N- dimethyl-

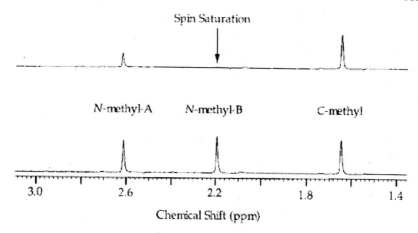

Figure 4. Room temperature ^{1}H-NMR spectrum of DMA in toluene-d_8 with and without decoupler saturation (upper and lower spectrum, respectively). The decrease in the N-methyl-A peak is from the spin-saturation transfer process from the N-methyl-B peak which is saturated and hence not observed. Used with permission from the Journal of Chemical Education, Vol. 74, No. 8, 1997, pp. 978-982; copyright © 1997, Division of Chemical Education, Inc.

acetamide system is shown in Figure 4. The $\Delta G^\dagger$ was found to be 73.7±2.0 kJ/mole.

If the saturation is maintained long enough to reach a steady state, then $dM_A/dt = 0$. Equation 13 can be solved directly for the rate. Experimentally, the spin-lattice relaxation time must be measured independently.

$$k = \frac{1}{T_{1A}}\left(\frac{M_{0A} - M_A}{M_A}\right)$$

(14)

2D Exchange Spectroscopy (EXSY)

A third method for determining the rate constant of slowly exchanging systems uses two-dimensional NMR spectroscopy. The pulse sequence used in NOESY to investigate cross-relaxation due to dipolar coupling may also be used to investigate cross-relaxation due to chemical exchange. When used to investigate chemical exchange, the experiment is usually referred to as EXSY, and results in a visual representation of the exchange process. In this experiment, transverse magnetization is created by a 90° pulse, and all components are "frequency labeled" by precessing at their characteristic frequencies during the evolution period, t_1. The magnetization is stored along z with another 90° pulse for a mixing time, τ_m. During this time, chemical exchange may occur. The magnetization is detected during time t_2 with a final 90° pulse. Nuclei that have not undergone exchange will resonate at the same characteristic frequency during t_1 and t_2, and will lie along the diagonal of the 2D spectrum. Nuclei that have undergone exchange will have different frequencies in t_1 and t_2, and will lie off-diagonal. The magnitude of these cross-peaks is directly related to the exchange rate. The pulse sequence and schematic spectrum in an EXSY experiment is shown in Figure 5.

The cross peak intensities as a function of mixing time, $I_{ij}(\tau_m)$, are given by

$$I_{ij}\left(\tau_m\right)=\left[\delta_{ij}-\tau_m R_{ij}+\frac{1}{2}\tau^2\sum_m\sum_k R_{ik}R_{kj}-\cdots\right]M_j^0,\qquad(15)$$

where M_j^0 is the initial magnetization of j, and the off-diagonal elements of the exchange matrix $\mathbf{R}$ are $R_{ij}=-k_{ji}$.

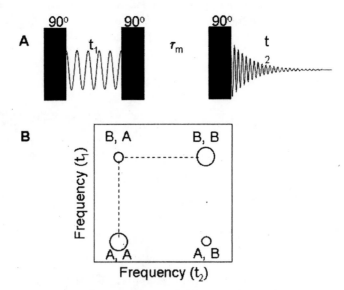

Figure 5. EXSY pulse sequence and schematic 2D exchange spectrum.

At short mixing times, for i ≠ j a linear approximation can be made, and equation 15 simplifies to

$$I_{ij}\left(\tau_m\right)\approx\tau_m R_{ij}M_j^0=k_{ji}\tau_m M_j^0.\qquad(16)$$

For simple two-site exchange between uncoupled A and B spins, assuming equal populations and exchange rate constant $k=k_{AB}+k_{BA}$,

$$k = \frac{1}{\tau_m}\ln\frac{r+1}{r-1}, \qquad (17)$$

where $r = (I_{AA} + I_{BB})/(I_{AB} + I_{BA})$. An example EXSY spectrum of N,N-dimethylacetamide obtained by students is shown in Figure 6.

The choice of mixing time is important. In systems with multiple sites, indirect cross peaks can arise from i to j and then j to k. The use of short mixing times eliminates this possibility, but is inherently inaccurate because the exchange is small at short mixing times, so the crosspeaks are weak. One alternative is to measure I_{ij} as a function of τ_m and determine the rate constant as the slope in a plot of $I_{ij}(\tau_m)$ versus $Mj^0\tau_m$, but this approach can be time consuming.

In EXSY experiments, the cross-peaks and diagonal peaks are all positive, making the phasing of the spectrum straightforward. Moreover, the magnitude of these crosspeaks is generally large compared to NOESY crosspeaks. Furthermore, NOESY crosspeaks are positive in the slow tumbling regime and negative in the extreme narrowing limit. So, for small molecules, cross-peaks due to exchange and NOE are easily distinguished.

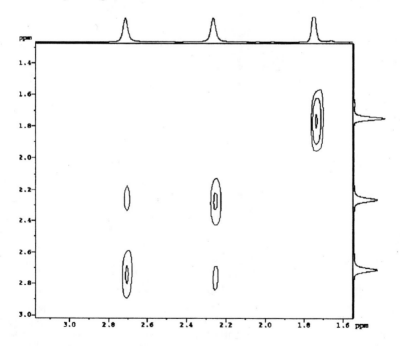

Figure 6. Student-acquired EXSY spectrum of N,N-dimethylacetamide taken at 45 °C at $\tau_{mix}=100$ ms.

Temperature Dependence

A study of the temperature-dependence of k yields thermodynamic parameters of the exchange. Through the Arrhenius equation, $k = A\exp\left(-E_a/RT\right)$, the empirical activation energy, E_a, may be obtained. The Eyring equation may be used to extract the free energy of activation, $\Delta G^\dagger$, and the enthalpy and entropy of activation, $\Delta H^\dagger$ and $\Delta S^\dagger$, respectively.

$$k = \frac{k_b T}{h}\exp\left(-\Delta G^\ddagger/RT\right)$$ (18)

$$k = \frac{k_b T}{h}\exp\left(\Delta S^\ddagger/R - \Delta H^\ddagger/RT\right)$$ (19)

A plot of ln(k/T) versus 1/T will yield both $\Delta S^\dagger$ and $\Delta H^\dagger$. Figure 7 shows this type of plot, for rate constant data obtained from EXSY experiments of N,N-dimethylacetamide in benzene-d_6. This study led to a $\Delta G^\dagger$ value of 72 ± 5 kJ/mole. Because $\Delta S^\dagger$ is obtained from the significant extrapolation from the intercept at 1/T=0, caution should be used when reporting these values.

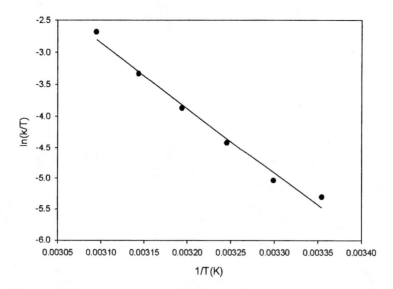

Figure 7. Experimental rate constant data obtained from student-acquired 2D EXSY spectra of N,N-dimethylacetamide plotted as ln(k/T) vs. 1/T.

Experimental Considerations

Sample Preparation

A dilute solution of N,N-dimethylacetamide is prepared in a deuterated solvent, or one that does not have overlapping resonances. Since the activation parameters will be concentration dependent, and the rotation barrier may be influenced by the solvent, it is advisable to use the same, sealed sample throughout. Indeed, a study of rate dependence on solvent would be an interesting variation, and it has been reported that the activation energy increases with increasing dipole moment of the solvent.(*3*) For the lineshape experiments, the sample must be degassed to remove O_2 to eliminate it as a source of indeterminate linebroadening.

Temperature Range

For the lineshape fitting, a broad temperature range should be investigated (25 – 70 °C). Intervals of 10 °C are generally acceptable, but smaller temperature intervals are useful around coalescence. For the spin saturation experiments, a narrower range (0 – 25 °C) may be sufficient. The EXSY experiment's useful temperature range is about 20-60 °C at 300 MHz. In any case, the sample should be allowed to equilibrate inside the NMR probe for 15 minutes at each temperature, and the experiments should be accompanied by a careful temperature calibration of the NMR probe, which may be achieved using ethylene glycol above room temperature and methanol below room temperature.(*12*)

Data Acquisition

Lineshape Analysis
Spectra should be acquired at several temperature intervals before, during, and after coalescence, and the spectra should be processed with minimal or no line broadening. The natural line-width for each spectrum should be determined by measuring the width of a reference peak. In the example of N,N-dimethylacetamide, the non-exchanging methyl C resonance is an ideal internal reference.

Spin Saturation
The decoupler frequency is set on resonance with the N-methyl B resonance for a period of at least $5T_1$ before acquiring the spectrum. The spin-lattice relaxation time will depend on temperature, the solvent, and concentration, but has been measured to be between 4-5 s in toluene-d_8.(*3*) The methyl A and C peaks are each integrated to find the values for M_A and M_{0A}, the latter is possible because the C resonance is not undergoing exchange and should be equal in intensity to the methyl A peak when it is not exchanging. In addition, at each temperature a spin-lattice relaxation measurement must be made.

EXSY

The NMR experiments were acquired on a Bruker DPX spectrometer with the pulse sequence noesytp. A mixing time of 100 ms was chosen so that exchange from NOE would be negligible. In the direct time dimension 1k data points were acquired, and 64 in the indirect dimension. Sixteen scans were averaged for each t_1 point, resulting in an acquisition time of about 3.5 hours. A reference 1D spectrum was also obtained at each temperature. Peak volumes for all three diagonal peaks and two cross-peaks were obtained by using Bruker's 2D integration routine.

Discussion

There are advantages and disadvantages to each of these approaches to measuring chemical exchange with NMR. The lineshape fitting is the least demanding experimentally, but requires a detailed analysis with a computational package in order to extract the rate constants. Furthermore, uncertainty in the natural linewidth can complicate the analysis. In contrast, the spin-saturation experiments require more comfort with the NMR spectrometer, but the analysis is straightforward. The EXSY experiment is not experimentally demanding, but it can be time consuming. However, the use of gradient sequences can dramatically reduce the acquisition time, where only 2 scans is required for each t_1 point.(*13*) Alternatively, students could easily set up an automated multiple temperature sequence of experiments to run overnight. The analysis could then be completed in a regularly scheduled laboratory period. A clear advantage of the 2D approach is the visually apparent exchange network in the spectrum. Because of the difficulties in accurately integrating peak volumes of small cross peaks, the EXSY experiment may not be the best approach for quantitatively determining the rate in all systems. The true value in the 2D approach is in measuring multi-site exchange; however, in these cases a full evaluation of the exchange matrix is needed.

The results obtained from each approach are comparable. The $\Delta G^\dagger$ values obtained for the line shape analysis, saturation transfer, and EXSY experiments were 71, 73.7±2.0, and 72 ± 5 kJ/mole, respectively. However, each of these samples were obtained in different solvents, so the results are expected to differ somewhat.

Students reacted to the EXSY experiment quite favorably. In an end of semester survey, 42% of the students selected the NMR lab as the lab they enjoyed the most. Several students remarked that they learned a lot from the experiment. Specific comments included: "I like the 2D NMR lab because it was clear and straight forward and easy to understand. I also liked learning about 2D NMR, which I had not been exposed to in any other class." Another student remarked, "I enjoyed the kinetics experiment the most because I was able to see what was happening in the experiment visually." Only one negative comment was received: "too many calculations."

Variations or additions to the standard set of EXSY experiments acquired at different temperatures include a study of the mixing time dependence, a study of the solvent dependence, and a study of different molecules. The simplest candidates for study would involve mutual site exchange between uncoupled resonances. Ideal candidates would be analogs of N,N-dimethylacetamide, with different substituents in place of the acyl methyl group. Many other interesting possibilities can be found in the recent literature.(*14-18*)

Conclusion

EXSY experiments are an attractive approach for probing the kinetics of chemical exchange. These experiments may be used to introduce two-dimensional NMR or illustrate principles in biochemistry, organic chemistry, and physical chemistry. The results obtained with EXSY compare favorably to those obtained with other dynamic NMR approaches of lineshape analysis and saturation transfer.

Acknowledgements. The author gratefully acknowledges support from NSF-DUE 9750847 and the students in the physical chemistry laboratories at Ursinus College for help in developing the 2D NMR experiment.

References

1. Jeener, J.; Meier, B. H.; Bachmann, P.; Ernst, R. R. *J. Chem. Phys.* **1979**, *71*, 4546-4553.
2. Gasparro, F. P.; Kolodny, N. H. *J. Chem. Educ.* **1977**, *54*, 258-261.
3. Jarek, R. L.; Flesher, R. J.; Shin, S. K. *J. Chem. Educ.* **1997**, *74*, 978-982.
4. Barrows, S. E.; Eberlein, T. H. *J. Chem. Educ.* **2005**, *82*, 1329-1333.
5. Dwyer, T. J.; Norman, J. E.; Jasien, P. G. *J. Chem. Educ.* **1998**, *75*, 1635-1640.
6. Bain, A. D. *Prog. Nucl. Magn. Reson. Spectrosc.* **2003**, *43*, 63-103.
7. Jackman, L. M.; Cotton, F. A. *Dynamic Nuclear Magnetic Resonance Spectroscopy*; Academic Press: New York, 1975.
8. Sandstrom, J. *Dynamic NMR Spectroscopy*; Academic Press: London, 1982.
9. McConnell, H. M. *J. Chem. Phys.* **1958**, *28*, 430-431.
10. Brown, K. C.; Tyson, R. L.; Weil, J. A. *J. Chem. Educ.* **1998**, *75*, 1632-1635.
11. Reich, H. J. WinDNMR; *JCE Software* **1995**.
12. Vangeet, A. L. *Anal. Chem.* **1968**, *40*, 2227.
13. Parella, T. *Magn. Reson. Chem.* **1998**, *36*, 467-495.
14. Lam, P. C. H.; Carlier, P. R. *J. Org. Chem.* **2005**, *70*, 1530-1538.
15. Leskowitz, G. M.; Ghaderi, N.; Olsen, R. A.; Pederson, K.; Hatcher, M. E.; Mueller, L. J. *Journal of Physical Chemistry A* **2005**, *109*, 1152-1158.
16. Tarkanyi, G.; Jude, H.; Palinkas, G.; Stang, P. J. *Organic Letters* **2005**, *7*, 4971-4973.
17. Contini, A.; Donatella, N.; Trimarco, P. *J. Org. Chem.* **2006**, *71*, 159-166.
18. Wik, B. J.; Lersch, M.; Krivokapic, A.; Tilset, M. *J. Am. Chem. Soc.* **2006**, *128*, 2682-2696.

Chapter 14

Dynamic NMR in an Advanced Laboratory

Fadwa Odeh and Yuzhou Li

Department of Chemistry, Clarkson University, Potsdam, NY 13699

In this experiment, undergraduate students of an advanced laboratory course are introduced to dynamic nuclear magnetic resonance spectrocopy (NMR). The students used pulsed-field gradient methods (PFG-NMR) to measure the diffusion coefficient (D) of components of micelle solutions containing solubilizates (alcohols) with varying alkyl chain length. From the data obtained, the partition coefficient (P) of the alcohols between the aqueous and micelle phases was calculated and the effect of alkyl chain length on P is elucidated. Students used the diffusion coefficients of the micelles to calculate radii and compared the PFG-NMR results with those obtained from dynamic light scattering (DLS) experiments.

Nuclear Magnetic Resonance (NMR) has become a very powerful technique in studying the structural and dynamic aspects of a wide variety of systems at the molecular and supramolecular levels (1). Complex molecular structures can be determined through an array of different NMR experiments. Multidimensional NMR has helped deconvolute complex spectra without the need for assignment through empirical methods. The dynamics of a system may be probed by NMR via a variety of methods such as spin-lattice relaxation, spin-spin relaxation and self diffusion (2). Self diffusion coefficients of complex mixtures can now be routinely measured using pulsed field gradient nuclear spin echo techniques (3). In a typical FT-NMR experiment, the entire spectrum of frequencies of interest is stimulated by a pulse of radio frequency energy (RF) and the response of the system is measured as a function of time. The NMR signal is called the free induction decay (FID). The frequency spectrum is then generated using Fourier Transformation (FT) which converts the time domain data (FID) to the frequency domain spectrum (Figure 1).

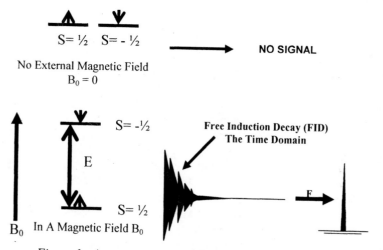

Figure 1. A representation of the basics of NMR signal.

Micelles are small, mainly spherical aggregates of surfactant molecules, where the surfactant concentration is above a certain critical value (critical micellar concentration, or cmc) at a certain temperature. A micelle might contain from few dozens to several thousand molecules that associate due to the amphipathic nature of the surfactants which contain both hydrophobic and hydrophilic ends (4). Figure 2 illustrates the two types of micelles formed from sodium dodecyl sulfate (SDS). In a polar environment (e.g water) the SDS molecules form a spherical association where the polar heads are exposed to the aqueous environment and the hydrophobic tails collectively form a nonpolar core. In a nonpolar environment the polar heads are buried in the center away from the nonpolar environment, called a reverse micelle.

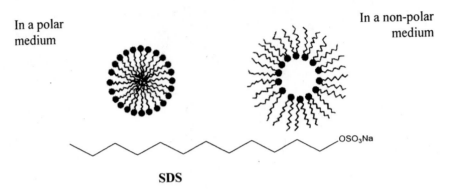

In a polar medium

In a non-polar medium

OSO₃Na

SDS

Figure 2. An illustration of a micelle (in polar medium) and a reverse micelle (in non-polar medium) made from SDS.

The phenomena of solubilization, aggregation and micelle formation are at the core of aqueous surfactant chemistry (5). Below the cmc, molecules are present in solution in their free dissociated form. Upon increasing the concentration above the cmc, micelles form; micellization is accompanied by an abrupt change of many physical properties such as surface tension, viscosity and conductivity. These changes enable the use of a wide variety of analytical techniques to investigate micellization and micellar properties such as solubilization: light scattering (6), fluorescence (7), surface tension (8), viscosity (9), calorimetry (10), sound velocity (10), FT-NMR (11, 12) and many others. Each of these techniques has its own unique advantages and, sometimes, limitations. For example, dyanamic light scattering can effectively measure particle size and particle size change in a surfactant system (13, 14). However, the technique is not chemical specific. Also, dust particles and air bubbles can cause interference. Fluorescence is widely used to study solubilization and microenviromental changes in micelles, but requires that a probe molecule be added to the system and thus increases the number of components; probe molecules might disturb the original eqilibria (15). PFG-NMR, however, can discriminate among particles on the basis of chemical composition. Furthermore, PFG-NMR can provide information on molecular dynamics, which yield valuable information about the surfactant system such as solubilization and the partition coefficient of a solute. In this experiment, students of an advanced undergraduate organic laboratory course investigated a typical surfactant system using pulsed field gradient nuclear magnetic resonance (PFG-NMR).

The experiment was designed for students who have already been exposed to the principles of NMR in other courses such as organic chemistry. Such courses usually cover basic theory and working knowledge behind FT-NMR, including one and two dimensional ^{1}H- and ^{13}C-NMR spectroscopy for structure elucidation of organic compounds. This experiment will expand their NMR knowledge base from a static to a dynamic perspective. More specifically, student examined the dynamic nature of alcohol molecules which are in equilibrium between micelle-bound and solvent environments. The effect of alkyl chain lengths of various alcohols on their differential solubility in water versus micelle is studied in this experiment. Students also compared the particle size data obtained using dynamic NMR and dynamic light scattering (DLS). The students in the advanced labory course are generally familiar with DLS and related techniques for the study of colloidal systems.

NMR Diffusion Measurements

Pulsed field gradient NMR (PFG-NMR) is a powerful, non-destructive technique for measuring self-diffusion coefficients (3). The technique can be used for a variety of systems without the need for any special sample handling or isotopic labeling of the species of interest. PFG-NMR is used to study the aggregation of molecules and the molecular interactions. Molecules associated with the aggregate will diffuse slowly over relatively large distance. The basis of the PFG-NMR method in determining diffusion coefficients is that it is possible to label a nuclear magnetic moment (spin) with respect to its position in the NMR tube. This is accomplished by applying a magnetic field gradient.

There are several PFG sequences that can be used; the simplest is the Stejskal and Tanner PFG-NMR sequence which consists first of a 90° pulse (excitation) followed by a gradient with magnitude g and duration δ (encoding); then a 180° pulse is applied followed by another gradient (decoding). This will take a total time of Δ, which is the diffusion time. Finally, acquisition will take place and the intensity of the signal will be measured at each value of g at constant temperature. In this experiment, both δ and Δ will have fixed values throughout the experiment and g will vary. The PFG-NMR sequence used in this experiment is a longitudinal eddy current delay pulse sequence with bipolar pulse gradients (LED-BPP) which is used to eliminate most of the eddy current effect (shown in Figure 3). In this sequence the gradient pulse is divided into two pulses opposite to each other and with duration $\delta/2$ for each (16, 17).

The diffusion coefficient D can then be calculated from equation 1:

$$I_{(g)} = I_0 e^{(-Dq^2\Delta')} \qquad (1)$$

where $I_{(g)}$ is the intensity of the signal at a gradient g, D is the diffusion coefficient, $q=\gamma g\delta$ (γ is the gyromagnetic ratio of protons, g is the amplitude of the applied gradient and δ is the duration of the applied gradient) and $\Delta' = \Delta - (\delta/3) - \tau_g/2)$ (where Δ is the diffusion time, τ_g is the time between the negative $\delta/2$ gradient pulse and the π RF pulse). A plot of $I_{(g)}$ vs. $q^2\Delta'$ will give an exponential decay curve, which is fitted to obtain the diffusion coefficient D. An alternative is to plot $ln(I_{(g)}/I_0)$ vs. $q^2\Delta'$; the result will be a straight line with a slope of $-D$.

The diffusion coefficients of the micelles can also be used to determine the size of these aggregates. Assuming that students already have basic knowledge of DLS for particle size determination, both the dynamic NMR and light scattering techniques can be compared and contrasted for their advantages and disadvantages. The radius of an aggregate can be calculated from diffusion coefficient data according to the well known Stoke-Einstein equation:

$$D = kT / f \qquad (2)$$

where k is the Boltzman constant, T is the temperature in Kelvin and f is the friction coefficient and is defined as

$$f = 6\pi\eta r_s \qquad (3)$$

where r_s is the radius of the aggregate (assuming it is spherical) and η is the viscosity.

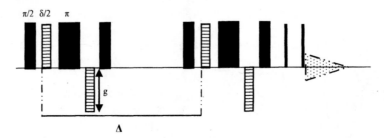

Figure 3. The LED pulse sequence with bipolar gradient pulses.

Experimental Section

Materials

Four alcohols were used in this experiment without further purification: methanol (Fisher Scientific), propanol (J.T Baker), heptanol (ACROS) and hexanol (Aldrich). Sodium Dodecyl Sulfonate (SDS) (AVANTI) was purified by crystallization twice in methanol. Deuterated water (Aldrich) was used (99% D) as purchased. Tetramethyl silane (TMS) - (Aldrich).

Instrumentation

The NMR spectrometer used in this experiment was a Bruker DMX 400 AVANCE equipped with a 5 mm BBO probe with z-gradients and a temperature control unit. The dynamic light scattering experiments were performed on a non-intrusive back scattering system (NIBS) manufactured by ALV (Langen, Germany).

Procedure

Four reference samples were prepared by dissolving each alcohol in D_2O to yield a uniform concentration (6.0 µL/1.0 mL). Micelle samples were prepared by dissolving 70.0 mg SDS, 6.0 µL alcohol and 6.0 µL TMS in 1.0 mL of D_2O. The typical volume of each micellar sample was about 20 mL per student to have sufficient amount for NMR and DLS measurements. 1.0 mL of each reference or micelle sample was transfered to a clean and dry NMR tube labeled with its contents. A ^{1}H-NMR spectrum was first obtained in order to optimize the parameters for the PFG-NMR experiment such as spectral width, receiver gain and number of scans needed. From the ^{1}H-NMR experiment file a new file was opened which will automatically copy all the optimized parameters for the PFG-NMR experiment. PFG-NMR experiments can be conveniently performed in a two dimensional fashion using the DOSY (Diffusion Ordered SpectroscopY) experiment. In the new file, one changes the pulse program to ledbpgp2s, the experiment environment to 2D, and the number of scans to 8 or greater. An adjustment to the diffusion parameters should be done. A Δ value of 100-200ms, δ value of 2-4ms and d1 of at least $5T_1$ were used depending on the type of sample. In DOSY, gradients from 10-90% of the maximum field were used.

The diffusion coefficient D of each alcohol in D_2O (D_{free}) and micellar solutions (D_{app}) along with the diffusion coefficients of TMS in SDS ($D_{micelle}$) were measured at a constant temperature of 300K using the parameters described above. For DOSY data fitting, the xwin-nmr (t1/t2) protocol was used after processing (xf2) and phase correcting the data. In (t1/t2) menu, parameters were adjusted such as δ, Δ and γ.

Results and discussion

The diffusion coefficients of all key components in a micellar solution (the micelle and the solubilizate in its free and aggregated states) were measured using PFG-NMR. The diffusion coefficients were then used to determine the solubility of alcohols with various alkyl chain lengths in water (free and associated with micelles). When a small hydrophobic molecule (insoluble in water) is associated with a micelle it assumes the same diffusion coefficient of the micelle. For alcohols, the molecules are likely distributed between free (in water) and associated (in micelles) states. Due to fast equilibrium, the observed diffusion coefficient measured by NMR is an average of the two states. The weight by which each state will contribute to the observed diffusion coefficient is the mole fraction of the solubilizate in each state.

The present FT-PFG NMR experiment monitors the Brownian motion of individual molecules during a time of typically 100-200 ms, during which a micelle diffuses nearly 1000 times its own diameter (micelle's diameter (d) is ~ 50 A° and diffusion coefficient D is ~ 5 x 10^{-11} m^2sec^{-1}). Solubilizate diffusion within the micelle is therefore unimportant and since there is fast exchange on the NMR time scale, the solubilized molecule (A) acquires an apparent self-diffusion coefficient D_A^{app} will equal to:

$$D_A^{app} = p.D_A^{micelle} + (1-p)D_A^{free} \qquad (4)$$

By rearranging this equation we get:

$$p = (D_A^{free} - D_A^{app})/(D_A^{free} - D_A^{micelle}) \qquad (5)$$

where p represents the fraction of molecules A that are associated or solubilized in the micelle ($0 \leq p \leq 1$). According to the argument above, $D_A^{micelle}$ can be taken as the diffusion coefficient of the micelle itself which can be estimated by monitoring D^{app} of a very hydrophobic solute (tetramethyl silane, TMS, was used in this experiment and co-solubilized in trace amounts). Alternatively, $D^{app}_{surfactant}$ can be utilized, provided that the surfactant has a low critical micelle concentration (cmc) and that the total concentration of surfactant is much larger than the cmc. It should be noted that the measurement approach is independent of any assumptions concerning the micelle's size or shape; $D^{micelle}$ is measured in the actual solution and accounted for. D_A^{free} can be measured in the solutions without surfactant.

Representative data generated by the students at Clarkson University are shown in Figures 4 and 5. In Figure 4, the ^{1}H-NMR spectrum and the PFG-NMR data of methanol in D_2O (signal at 3.3 ppm) are shown. Figure 5 shows the ^{1}H-NMR spectrum for methanol in the presence of SDS micelles and the PFG-NMR data for both methanol and SDS micelles (at 3.4 and 4.1 ppm respectively).

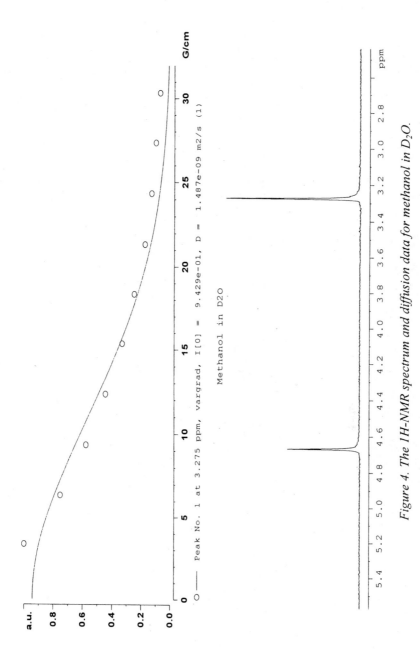

G/cm

Peak No. 1 at 3.275 ppm, vargrad, I[0] = 9.429e-01, D = 1.487e-09 m2/s (1)

Methanol in D2O

Figure 4. The 1H-NMR spectrum and diffusion data for methanol in D₂O.

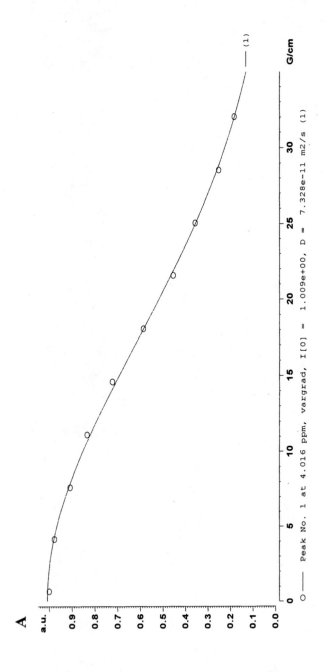

O —— Peak No. 1 at 4.016 ppm, vargrad, I[0] = 1.009e+00, D = 7.328e-11 m2/s (1)

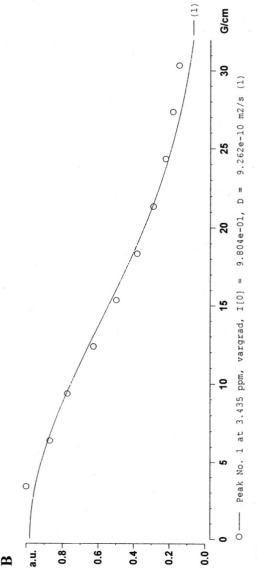

Figure 5. The diffusion data for A) SDS B) methanol and the 1H-NMR spectrum of the mixture of both at 300K.
Continued on next page.

B

Methanol in SDS micelle solution

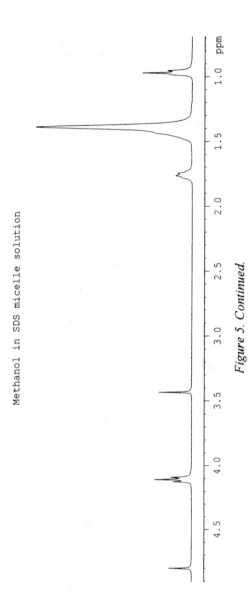

Figure 5. Continued.

Table 1 summarizes the diffusion coefficients for the free alcohols, the alcohols in presence of SDS micelles and the diffusion coefficient of the micelles (per SDS proton signals). The partition coefficients, P, were then calculated based on the diffusion coefficient data and listed in Table 1 as well. It is clear that the fraction of the alcohol associated with the micelle is increasing with increasing number of carbons in the alkyl chain of the alcohol. This is expected since increasing the carbon atoms will increase the contribution of the hydrophobic part in the alcohol and hence, decrease the alcohols solubility in water until it becomes insoluble in water and P approaches 100% (total solubility in the micelle system) (Figure 6).

Table 1. The distribution coefficient of alcohols with various hydrocarbon chain lengths between the micelles and water at 300K

Solubilizate	D_{Free}	$D_{Apparent}$	$D_{Micelle}$	$P\%$
Methanol	1.49E-09	9.26E-10	7.46.E-11	39
Propanol	8.12E-10	3.78E-10	7.81E-11	59
Hexanol	2.12E-10	5.40E-11	2.32E-11	87
Heptanol	6.61E-10	5.30E-11	2.69E-11	93

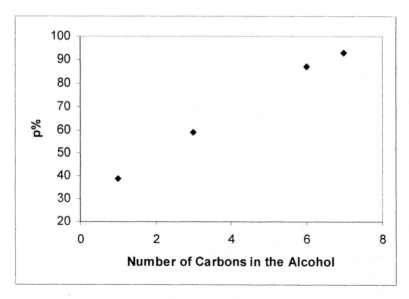

Figure 6. The distribution coefficient of alcohols with various hydrocarbon chain lengths between the micelles and water at 300K.

Table 2 lists the particle sizes for micelle systems measured by DLS and PFG-NMR. Other than propanol, PFG-NMR data show a clean trend of increasing particle size with the increase in alcohol hydrophobicity or partition coefficient inside the micelle; however, it was emphasized to students that the particle size increase trend should not be taken as an absolute trend since the parameters used were not as good for the hexanol and heptanol system, as they were for methanol and propanol. In hexanol and heptanol, the attenuation of the signal intensity did not exceed 50% but due to lack of enough time devoted to this experiment the results were taken as is. This might result from micelle swelling. The data obtained using DLS, however, consistently show an artifact at ca. 1 nm. Furthermore, the particle size result for the methanol system is extremely suspicious. The difference between the two techniques can be explained by the fact that NMR monitors the micelles with chemical specificity. In other words, only the entity that gives the right chemical shift can contribute to the diffusion coefficient measurement. Therefore, PFG-NMR has its intrinsic advantage over DLS. More specifically, the apparent particle size obtained by DLS is often affected by small amount of impurities, such as air bubbles or dust particles in the sample. Furthermore, the presence of particles with high radius values, even with small population sizes overshadows the presence of the more minute particles. The typical micelle size measured using NMR was in the range of 3-9 nm. The reported micelle size using various techniques such as DLS give a radius of 4-5 nm for SDS micelles. Our results are close to reported data. One of the possible sources for the discrepancy between measured data and reported values is the fact that a constant viscosity term was used during the calculation of particle size. In order to improve the quality of these data, students could measure the viscosity of the micelle solution directly and use it in the calculations if time is available.

Table 2. Micelle's radius obtained using PFG-NMR and DLS at 300K.

Micelle	R using DOSY (nm)	R using DLS (nm)
Methanol/D_2O/SDS	3.3	0.44
Propanol/D_2O/SDS	3.2	1.11, 10.99
Hexanol/D_2O/SDS	7.8	1.06, 17.33
Heptanol/D_2O/SDS	9.2	1.04, 14.34

Conclusion

In an attempt to expose third-year chemistry students to an effective analytical technique, an aqueous micellar system was invstigated using PFG-NMR and DLS techniques. The advantages of DLS include the ability to measure particles over a large size range (3-3000 nm). The instrumentation could be relatively simple and inexpensive for undergraduate laboratory courses. Furthermore, DLS can measure systems with extremely dilute particle concentration. For PFG-NMR, the dynamic range in particle size is smaller (3-300 nm). The technique is also difficult to use for a chemical system that is slow in motion and/or low in concentration. The availability of the instrument to the class is also an issue under certain circumustances.

The designed experiment spanned a two-week period, including one lecture hour covering the background and appliocations of PFG-NMR. The data obtained by the students were self-consistent and similar to the reported values, even though this was their first experience with dynamic NMR in concept and practice. Due to the prominence of Clarkson University research programs in colloid and surface science and the high percentage of undergraduate students participating in research, many students in the lab course have already been exposed to some concepts and issues related to colloids such as particle size and phase behavior of surfactants. Over 80% of the student who participated in this experiment recommended the continuation of this experiment in future advanced laboratory courses at Clarkson University. More than 75% of the students surveyed reported that they felt that the experiment provided a valuable experience. Some of them even introduced PFG-NMR into their own research projects. One such student is Mr. Norman Marshal, a chemistry senior who enrolled in this class during the Fall 2004 semester. He was working on a project concerned with size distribution of liposome systems using DLS, and this experiment shifted his interest to include PFG-NMR measurements. In addition, various students became interested in other types of non-classical NMR techniques for their research projects, such as relaxation times, to study molecular interactions. Overall, this laboratory experiment successfully introduced another dimension of NMR to the students and broadened their view in NMR spectroscopy. More importantly, the experiment equipped the students with a powerful non-invasive analytical technique for their own independent research projects.

Acknowledgments

We would like to thank Nicole Heldt for her valuable suggestions and proof reading of the manuscript. Also we would like to thank the undergraduate students in Dr Li's research group for their wonderful help in planning and testing this experiment. Our thanks also go to Nuha Salem and Maxim Orlov, the teaching assistants for the advanced laboratory course at Clarkson University.

References

1. R. Ernst, G. Bodenhausen and A. Wokaun, Principles of Nuclear Magnetic resonance in One and Two Dimensions, Oxford University Press, 1990.
2. Dynamic NMR Spectroscopy," J. Sandström, Academic Press, New York, 1982.
3. W. Price, Concepts Magn. Reson., 1997, 9, 299-336.
4. R. Laughlin, The Aqueous Phase Behavior of Surfactants, Academic Press, San Diego, 1994.
5. A. Priev, S. Zalipsky, R. Cohen and Y. Barenholz, Langmuir, 2002, 18(3), 612-617.
6. B. De Meulenaer, P. Van Der Meeren, M. De Cuyper, J. Vanderdeelen and L. Baert, J. Colloid Inter. Sci., 1997, 189, 254-258.
7. W. Caetano and M. Tabak, J. Colloid Inter., 2000, 225(1), 69-81.
8. Y. An, E. Carraway and M. Schlautman, Water Res., 2002, 36(1), 300-308.
9. Y. Maylonas and G. Staikos, Langmuir, 1999, 15(21), 7172-7175.
10. G. Basu Ray, I. Chakraborty, S. Ghosh, S. Moulik and R. Palepu, Langmuir, 2005, 21(24), 10958-10967.
11. J. Carlfors and P. Stilbs, J. Colloid Inter. Sci., 1985, 104(2), 489-499.
12. G. Kossena, W. Charman, B. Boyd, D. Dunstan and C. Porter, J. Pharm. Sci., 2004, 93(2), 332-348.
13. F.Odeh, N. Heldt, M. Gauger, G. Slack and Y. Li, PFG-NMR Investigation of Liposome Systems Containing Hydrotrope, J. Disp. Sci. Tech., 2006 in press.
14. N. Heldt, J. Zhao, S. Friberg, Z. Zhang, G. Slack, and Y. Li, Tetrahedron, 2000, 56(36), 6985-6990.
15. S. Matzinger, D. Hussey and M. Fayer, J. Phys. Chem. B, 1998, 102, 7216-7224.
16. D. Wu, A. Chen and C. Johnson, Jr., J. Magn. Reson., 1995, A115, 260-264.
17. W. Price, Concepts Magn. Reson., 1998, 10, 197-237.

Modern NMR in Laboratory Development

Biochemistry and Biophysics

Chapter 15

Angiotensin and Oxytocin: 1D and 2D NMR of Biologically Active Peptides in a First-Semester Biochemistry Course

Toni A. Trumbo

Department of Chemistry, Bloomsburg University of Pennsylvania,
Bloomsburg, PA 17815

The peptide hormones, angiotensin and oxytocin, were used to
teach fundamental NMR theory and application to
undergraduates that have no background in physical chemistry
and limited background in mathematics and physics. The
exercises presented here helped students taking Biochemistry 1
understand the basic principles of NMR, gave students a taste
of the applications of NMR, and taught students the meaning
of and how to assign 2-D NMR spectra. A vector approach,
enhanced by a human-based reconstruction of experiments,
was employed. Concepts included 1-D proton NMR, scalar
coupling, and 2-D techniques. Students assigned TOCSY
spectra and 1-D ^{1}H spectra for angiotensin and oxytocin, and
wrote short reports discussing aspects of peptide NMR.
Students were also held responsible for concepts in
fundamental NMR theory on lecture exams.

Introduction

While theory, mechanics, and application, of most instruments employed in the chemistry laboratory is easily discussed, NMR is much more complex and very difficult to summarize. A purist may say that it is impossible or even blasphemous to teach any NMR theory without the mathematics. However, most students are first introduced to NMR in organic chemistry courses, where the majority of students are not chemistry majors. These students will never have the mathematics necessary to approach NMR from a purist standpoint; thus, the black box approach is the standard. With the black box approach, students are taught little or no theory. At minimum, they learn to confirm molecular connectivity, with only a few ambitious instructors showing how NMR can follow a reaction and give yield ratios. The students go on to the next stage of their careers, graduate school or a job, without knowing the almost limitless capabilities of NMR. To use such a costly instrument only to verify molecular connectivity, does not illustrate the value NMR provides. For a graduate to never think of NMR spectroscopy again when confronted with a chemical challenge limits potentially valuable research options.

Two biologically active peptides were used to pique the interest of the students and to present NMR spectroscopy as a useful tool for studying biomolecules. The exercises presented here were developed to further three goals:

1. to help students taking Biochemistry 1 understand the basic principles of NMR
2. to give students a taste of the many applications of NMR and
3. to teach students to understand the meaning of and to assign 2-D NMR spectra.

Classroom time constraints prevent an extensive treatment of NMR; however, devotion of two lab periods towards discussion and application of structural methods to biochemical molecules is time well invested. Extensive treatment of NMR theory is more appropriate for an advanced spectroscopy course for chemistry majors. The angiotensin peptide was used to teach students to assign TOCSY spectra. The angiotensin is a hormone that helps to regulate blood pressure in humans (1). The fragment employed in this experiment is under investigation for agonist activity against angiotensin converting enzyme (ACE), the only enzyme for which hypertension drugs have been approved (1). The heptapeptide (^{1}DRVYIHP^7) provides a terrific initiation into the world of biomolecular NMR due to its non-repeating and varied sequence, and the presence of proline. Since proline is the only amino acid without an amide hydrogen, it presents an interesting dilemma in NMR. Oxytocin (^{1}CYIQNCPLG^9) is nonapeptide hormone that is responsible for many biological activities, most famously, uterine contraction during childbirth. Oxytocin also facilitates mother-baby bonding, and research has shown evidence of increased levels of oxytocin after sexual intercourse in both men and women (*2*). The amino acids are non-repeating, with the exception of two cysteines, and the sequence also contains proline. Most students are familiar with oxytocin and find it interesting.

Many types of NMR experiments lend themselves well to studying biomolecules, particularly the two-or-more-dimensional varieties. *It is not possible to fully discuss the techniques and pulse sequences presented here. For a thorough review, the reader is referred to Roberts (3) and Hornak (4).* Many organic chemists are familiar with the benefits of COSY (Correlated SpectroscopY). With the COSY experiment, a correlation between coupled nuclei allows energy exchange during a cycling pulse sequence. The resulting spectrum has off-diagonal peaks (crosspeaks) at the intersection of signals for two nuclei experiencing scalar coupling (J-coupling). When the crosspeaks are traced horizontally and vertically to the diagonal, the signals for two protons that are coupled are found. If one knows the assignment for one of the signals, assignment of the others follows in sequence by "walking around" the spectrum. After assignment of the signals on the diagonal in the COSY, it is a simple matter to match the pattern of peaks on the diagonal with the pattern of peaks in the 1-D spectrum. The crosspeaks also facilitate measurement of scalar coupling constants for overlapping signals, providing yet more valuable information about interactions among the protons. Unfortunately, COSY quickly loses its effectiveness when trying to assign proton spectra for peptides of more than three or four amino acids, when even the crosspeaks become overlapped. At this point, TOCSY NMR becomes very useful.

The TOCSY (TOtal Correlation SpectroscopY) NMR pulse sequence replaces the second 90° pulse of the COSY with a spin-locking pulse. As a result of spin-locking, coupled nuclei transfer energy over unusually long distances, forming an extended network of correlated nuclei. TOCSY shows correlation that are separated by a number of bonds that is limited by the spin-spin relaxation time between protons. For example, it is not unusual to see a correlation between the amide and delta protons in leucine. With a typical organic compound, the TOCSY produces an overly complicated spectrum; however, for biomolecules that have small repeating units the technique is invaluable. Proteins consist of only twenty or so amino acids in different sequences. Each amino acid is separated from the others by a carbonyl, effectively isolating the amino acid from scalar coupling with other amino acid. Thus, each side chain forms its own set of coupled nuclei that give a characteristic set of signals in NMR spectra. The result is twenty different and discernable patterns that appear in the TOCSY spectrum. Carbohydrates also exhibit this phenomenon. Although the spectrum appears more complicated, learning to recognize the patterns is much easier than learning to use reverse logic to determine connectivity in the 1-D proton spectrum of a small organic compound. Biochemistry students who have never heard of NMR can learn to assign a TOCSY spectrum in a couple of hours. From there, assignment of other two or more-dimensional NMR spectra is an easy leap.

TOCSY NMR may also be thought of as a gateway technique for other NMR experiments. The chemical shift of a nucleus does not change when comparing TOCSY, COSY, NOESY, and ROESY spectra. With an assigned TOCSY spectrum, one may easily assign a COSY by matching the crosspeak assignments and regain ability to measure coupling constants. NOESY (Nuclear Overhauser and Exchange SpectroscopY) and ROESY (Rotating frame Overhauser Enhancement SpectroscopY) crosspeaks indicate where two nuclei share dipole-dipole interactions through-space at distances of up to 5Å. The intensity of the crosspeak is proportional to the negative sixth power of distance between the nuclei, thus calibration of the peak intensities may be used to generate distance constraints for structure computation. A TOCSY allows assignment of the NOESY and ROESY crosspeaks and enables identification of the nuclei involved in through-space correlations. NOESY and ROESY spectra show through-space connectivities that allow assignment of a TOCSY spectrum for a peptide when amino acids appear more than once in the sequence. In this manner, for example, one may differentiate between valine number two and valine number eight in the sequence. The TOCSY-enabled assignment of 1-D proton spectra for a peptide or nucleic acid ligand under variable conditions may be used to track changes in proton chemical shifts due to intermolecular interactions. Two or more spectra are acquired for a ligand under different conditions or in the presence of other molecules. For example, in a technique employed by Ni, *et al.* (*5*), two spectra were acquired for a peptide ligand, one with only peptide, and one of peptide and an enzyme to which it binds. The differences in the spectra revealed nuclei in the peptide that came in contact with the enzyme surface and were important for enzyme-ligand interactions. There are many more types of experiments than what may be summarized here. The information gained through combining NMR experiments in this way is only limited by the inventiveness of the scientist.

Materials and Methods

Materials

Angiotensin (1-7) and oxytocin peptides were purchased from Sigma (St. Louis, MO). Deuterium oxide was purchased from Cambridge Isotope Laboratories (Landover, MA). Water magnetic susceptibility-matched NMR sample tubes were purchased from Shigemi, Inc. (Allison Park, PA). The specialized NMR sample tubes allow minimal amounts of the somewhat expensive peptides to be used. If peptide expense is not an issue, standard NMR tubes may be used.

Methods

Sample Preparation

The samples consisted of 5 mg of angiotensin or oxytocin dissolved in 400 μl of 90% deionized H_2O/10% D_2O. The samples were loaded into low-volume water magnetic susceptibility-matched Shigemi NMR sample tubes.

Acquisition of Spectra

The spectra were obtained at the University of Louisville on an 11.7 T (500 MHz) Varian INOVA spectrometer with a SUN workstation and Varian software. Each 1-D spectrum had 32 scans to increase the signal-to-noise ratio. Each TOCSY had 128 increments of t_1 with 8 scans per increment. Water was suppressed employing the WET pulse sequence and a gradient probe (*6*). The TOCSY acquisition time was approximately 2 hours at 22°C. The spectra were Fourier-transformed, phased, referenced, and the water further suppressed mathematically using Felix software (Accelyrs, Inc., San Diego, CA) on a Silicon Graphics workstation (Silicon Graphics, Inc., Mountain View, CA). Please note that solvent suppression will result in loss of signal for any nucleus that has a resonance frequency near that of the solvent due to signal overlap. Water has a proton chemical shift of 4.8 ppm and it is not unusual to lose the signal for one or two alpha protons that have the same or nearby chemical shifts. (Electronic copies of the spectra are available at no charge from the author. Please e-mail requests to tbel2@bloomu.edu.)

The spectra may also be obtained on instruments with lower field strength and without a gradient probe. In this case, water proton signal may be suppressed using presaturation. Please note that presaturation may result in reduction of amide signals, but careful processing will help to mitigate the effect.

Time constraints and instrumentation limits have prevented students in this course from acquiring their own TOCSY spectra. It is important that the students use or at least see it in operation to demonstrate the theory presented in this lesson. A one-dimensional spectrum of the peptides was easily acquired for this purpose in a minimal amount of time.

Course Set-Up

The Biochemistry 1 course had one four hour laboratory period each week. There was a maximum of eighteen students per lab section with a single instructor teaching the lessons. Students were required to purchase a laboratory manual (*7*) that included a summary of NMR theory, the NMR laboratory exercise, chemical shift tables that were adapted from Wisart, *et al.* (*8*), and copies of the TOCSY spectra for both peptides. Protein folding, protein structure, and amino acid chemistry, including the acidic nature of the amide hydrogen, preceded these lab exercises in the lecture portion of the class. Everything was presented in a discussion-type format with as much student participation as possible. Two lab periods were used to discuss and complete exercises and experiments in two methods for determining or investigating the structure of proteins: x-ray crystallography and NMR. A comparison of techniques was made, including the complementary nature of the techniques and the limitations of each. Approximately 80% of the time allotted was spent in NMR. The discussion was facilitated by PowerPoint slides illustrating the concepts under consideration.

Introduction of NMR Theory to Students

Many students in the class had been exposed to 1-D NMR assignment in organic chemistry courses, thus students were asked to recount aloud what they recalled about NMR. Students that were new to NMR were reassured that they had an advantage other students in that they had no pre-conceived fears about NMR. Then, NMR theory was discussed from a vector approach that combined and adapted concepts from many sources (*3-4, 9-11*). In brief, the vector approach allows a visualization of what occurs during each step of a pulse sequence. A Cartesian coordinate system is set up with a ray that shows the direction and strength of the forces that evolve as the experiment progresses. Thorough treatment of the vector approach may be found in the references stated above. A fictitious molecule with fourteen nuclei was followed as a pulse sequence progressed was used to demonstrate what is accomplished by each step in the sequence and how the signal came into being. Spin-spin and spin-lattice relaxation mechanisms were discussed in relative detail. NMR was compared to exciting the nuclei by playing them music on a radio, then allowing the nuclei to sing back during relaxation. The scientist listens to the songs and compares the songs in order to assign each signal to a type of nucleus.

For students with difficulties following vectors, the class was lead through a human-based reconstruction "Dance of the Nucleus." Each student represented a nuclear magnetic moment, the arm pointing spin up or spin down. The student-nuclei were excited by music, their magnetic moments aligned, one student was assigned to spin-flip, and then they were allowed to relax and sing (hum, whistle) back. The human-based reconstruction method was also employed to discuss scalar coupling, as well as the statistics that result in the familiar peak-splitting phenomenon that student relied upon heavily for assignment in organic chemistry classes. The reasons why a Fourier transform is performed on the FID were discussed, followed by how this process results in the spectrum with which they are familiar. Finally, the reason for and meaning of the delta scale was explained.

The parts of the 1-D spectrum were illustrated, including the amide region, the aromatic region, and the aliphatic region. A remnant of the suppressed water signal is usually visible in the 1-D proton spectrum. This was shown to the students to open up discussion about the affects of having water in the sample at ten thousand times the concentration of the peptide and the need for solvent suppression.

Unfortunately, time does not allow for every student to operate the current instrument. Instead, the students were taken to see the physical operation of the NMR spectrometer during acquisition of a 1-D spectrum. In past organic chemistry courses, the instructor and perhaps a lab technician obtained spectra for the students, therefore only the chemistry majors had ever seen the instrument. A simple sample was run to illustrate the pulse sequence, with special care to remind them of what each step in the sequence accomplishes. At this time, concepts of the signal lock, use of deuterated solvents, and shimming were introduced. The parts of the instrument were shown to demonstrate the radio-like nature of the instrument and to further tie-in the music concept.

After discussion of how the simple 90° pulse sequence results in the 1-D proton spectra with which they were familiar, pulse sequences resulting in 2-D spectra were discussed. The first lesson was a COSY by simplified vector approach, as illustrated in Figure 1.

The first 90° pulse aligns the nuclear magnetic moments in the y-direction such that the net magnetization vector points along the y-axis, just as it does with the 1-D experiment. Relaxation is allowed to occur for a short period of time, t_1, on the order of milliseconds. During the relaxation time nuclei that are coupled exchange energy and magnetization is passed from one nucleus to another, much like a baton in a relay race. Full relaxation is not reached during this short time. A second 90° pulse shifts the nuclear magnetic vectors such that they have a -z component. The net magnetization vector is in the x-z plane. Then, the nuclei are allowed to fully relax during a second relaxation period, t_2, during which the signal is recorded. The t_1 relaxation period is changed a small increment and the sequence is repeated. Again, the t_1 relaxation period is changed a small increment again and the sequence is repeated, and so forth until a complete data set is collected. For each t_1 relaxation time, the magnetization transfer occurs to a different extent before being subjected to the second pulse. Each t_1 relaxation time will have a different FID that contains information about the nuclei that were exchanging energy due to coupling at the given t_1. The t_1 provides one time domain, while the FIDs provide a second time domain. Both are Fourier-transformed into frequencies and a plot of one with respect to the other produces the COSY spectrum. To help illustrate the experiment, the student-nuclei were once again employed to walk through the experiment.

The TOCSY pulse sequence was then introduced by comparison with the COSY. It was explained that the spin-locking pulse allowed energy transfer to be extended much farther. NOESY and ROESY were discussed in brief. The

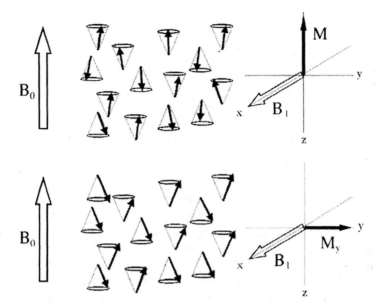

Figure 1- Fourteen nuclei in a fictitious molecule are used to illustrate what occurs during the first 90° pulse in the COSY experiment. B_0 is the external magnetic field. The RF signal induces a magnetic field (B_1). A force is produced along the y-axis and all of the nuclear magnetic moments align in the y-direction. The net magnetization vector (M) points along the y-axis.

applications of TOCSY were presented much as they are discussed previously, with figures included for illustration.

Assignment of Angiotensin TOCSY Spectra

Students were asked to bring a mechanical pencil, a non-marking white eraser, and a straight-edge. Each student received a print of the full TOCSY for angiotensin, along with an enlarged print of each of the fingerprint region, the aliphatic region, and the amide-aromatic region, including the diagonals. The fingerprint region contains the crosspeaks that occur between the backbone amide protons and the side chain protons. The aliphatic region encompasses the diagonal and all associated crosspeaks from 0 ppm to about 5 ppm.

The students were first asked to draw and label the peptide sequence DRVYIHP, including all hydrogens with corresponding Greek designations. The spectrum and enlargements were projected onto a screen by an overhead projector or white board by a projector linked to a computer. Students were shown how to identify and label each enlargement as one quadrant of the full spectrum.

Teaching assignment began with learning to recognize patterns in the fingerprint region. Vertical lines were drawn on the board with a meter stick and on paper by the students with a straight edge, to connect the crosspeaks for each amino acid that has a pattern in the fingerprint region (Figure 2). Care was taken to show that the line extends all the way to an amide signal on the diagonal. Some students were more comfortable working on the full spectrum so they could see the amide proton signal, others were able to just extend the line mentally and do the work on the enlarged fingerprint region. Students, working together in groups of two or three, were asked to compare the patterns from the chemical shift chart with those in the fingerprint region. Due to its distinct pattern, it was typical for the first resonance pattern recognized in the fingerprint region to be isoleucine, usually within a few minutes. If a group needed prodding, they were asked to look for a pattern to match that for isoleucine. Then the students worked their way through the other resonance patterns. Some students were puzzled by a lack of signals for proline in the fingerprint region. They were encouraged to voice reasons for this phenomenon, and could usually recall that proline has no hydrogen on the amide and it would not be possible for it to have correlations with the aliphatic side chain.

Students labeled every signal in the fingerprint region and the corresponding amide proton signals on the diagonal, as demonstrated in Figure 2. The assignment had to include the residue number and identification of the proton; for example, "I3 NH-α," indicating that this resonance is generated by the correlation of the amide and alpha protons of isoleucine #5. The signals were traced back to the diagonal and labeled there also. Then, the students were shown how to carry through in assignment of the signals in the aromatic and aliphatic regions of the spectrum, including identification of signals for proline.

Assignment of Angiotensin 1-D Proton Spectra

The correlation between the diagonal in the TOCSY spectrum and the one-dimensional spectrum was demonstrated to the students individually, as each finished assigning the TOCSY during the first lab period. The topographical map-like qualities of the TOCSY spectrum allowed students to assure themselves that the match was correct. Energy exchange between two nuclei that are close together results in a large crosspeak with concentric circles that are close together, indicating an intense peak (a "broad and tall mountain"). Students were required to write complete assignments were written in pencil on the spectrum.

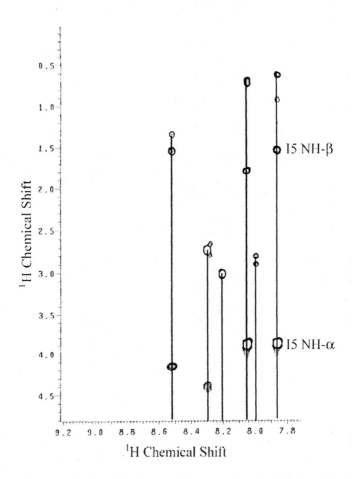

Figure 2- An example of students' results. Drawing lines helps identify the amide-to-aliphatic correlations for each amino acid in the sequence. Note that the fingerprint region for angiotensin (1-7) only has six lines. Proline does not have an amide proton and will not have crosspeaks in this region. An example of labeling is shown.

Assignment of Oxytocin TOCSY and 1-D Proton Spectra

At the beginning of the second lab period, students received the spectra for the oxytocin peptide (found in the Appendix) with the sequence CYIQNCPLG. They were allowed to confer with each other as much as they needed but interference by the instructor was limited. Instructions were given to follow the same technique as the week before to assign as many peaks as possible on the spectra.

Student Report

The following points for discussion in a short report were given in the lab manual and were due the following week.

"For which amino acid(s) are there no signals in the fingerprint region? Why is this so? Considering the sequence of oxytocin, what interesting feature could form that would induce a secondary structure in such a small peptide? (Hint: Oxytocin DOES have this feature.) The sample used to acquire these spectra was made with a solvent mixture of 90% H_2O and 10% D_2O. What would be the result in the spectra if 100% of the solvent were D_2O? In organic chemistry, we use deuterated organic solvents such as $CDCl_3$ to make the NMR samples. If we intended to use these spectra for studies on how this peptide folds, what would be the problem with using $CDCl_3$ or any other deuterated organic solvent?"

Evaluation of Student Performance

Assignment of the TOCSY was worth 47% of the grade, assignment of the 1-D proton spectrum was worth 20%, and the short report was worth 33% (points for discussion at 6.7% each). The TOSCY and 1-D spectra with the most correct assignments were given full credit, with all other spectra being graded on a curve with respect to the full-credit spectra. Students were also held responsible for all discussed NMR concepts on subsequent lecture quizzes and exams. Questions included variations of the following:

"Use a diagram to explain why one sees a triplet for a proton in 1-D ^1H NMR due to scalar coupling with a pair of neighboring methylene protons."
"What would result in the TOCSY NMR spectrum if a peptide sample were made with 100% D_2O?"
"How many amide proton signals would one expect to see in the NMR spectrum of a peptide with sequence CYIQNCPLG?"

Results and Discussion

Angiotensin (1-7) and oxytocin are interesting biologically active peptides. Students recognized oxytocin immediately and angiotensin with some prodding. Once the students understood that a crosspeak signifies that "this proton is connected to this proton by bonds" and nothing more complicated, the assignment went very well. Pitfalls of which and instructor should be aware include: students that make it more difficult on themselves by trying to project more meaning into the crosspeaks, and students that begin with the "I never understood NMR in organic chem so this is beyond me" disposition. These challenges are averted by repetition of the "this proton is connected to this proton by bonds" mantra and gentle reassurance, respectively. Working in small groups enabled students to discuss assist each other with assignments. It is also challenging to avoid overwhelming the students with too much theory in on lab period. The material should be split over the two-week period in a manner that is logical. For example, week one could include basic NMR theory and 2-D assignment, while saving discussion of 2-D techniques for the second week.

Of twenty-seven groups, more than half assigned more than 70% of the peaks. Credit subtracted was almost never due to resonances that were assigned incorrectly but instead for resonances not assigned at all. Due to the assignment being a relatively small percentage of the course grade, some groups were reluctant to invest the time to fully assign the spectrum. Students were usually able to find and discuss the answers to the questions posed for the short report. Some drew mechanisms to illustrate their report. If a difficulty was encountered, students were helped by breaking the question into a series of smaller questions to assist them in discovering a logical answer.

Students indicated that they did not find the exercise difficult, however they did find it tedious. It is important for students to know that, in science, many tasks are tedious but the rewards are worth the effort. Furthermore, once one understands how to do assignments, many labs that use multi-dimensional NMR techniques in research have computer software that will automatically assign the spectra. The software, however, is not fail-proof and the experimenter needs to understand the process.

Use of angiotensin and oxytocin has had a positive effect on the ability of students to answer theory-related questions on lecture exams. As the instructor has improved in delivery, student performance has improved such that now approximately 80% of students are able to correctly answer all questions concerning NMR theory. Of the remaining 20%, 10% are able to answer most questions, with the most difficult being, "Use a diagram to explain why one sees a triplet for a proton in 1-D ^{1}H NMR due to scalar coupling with a pair of neighboring methylene protons."

Future directions include NMR-based molecular modeling of the oxytocin peptide when an adequate ROESY spectrum is acquired. Short peptides, such as angiotensin and oxytocin, do not usually have secondary structure. There is not enough length of polymer to adequately fold. Oxytocin's cysteines, however, are involved in a disulfide bridge that induces a turn in one end of the peptide. After assigning the spectra, students would be shown how the inter-proton distances indicated by the intensity of the off-diagonal resonances could be calibrated and then entered into Spartan (Wavefuction, Inc., Irvine, CA) as constraints. Students would perform a simple geometry optimization experiment to illustrate some concepts in molecular modeling. A newer modern instrument has recently been acquired that will allow the students to acquire their own spectra. Additionally, to gain further insight into the effectiveness of this experiment, a pre- and post-experiment survey about NMR misconceptions is planned.

Acknowledgments

The author thanks Muriel C. Maurer for allowing a graduate student to try to teach NMR to undergraduates, and John P. Morgan for providing helpful criticisms of this manuscript.

References

1. Santos, R. A. S.; Frezard, F.; Ferreirs, A. J. *Curr. Med. Chem. Cardiovasc. Hematol. Agents*, **2005**, *3*, 383-391.
2. Roberts, J. D. *ABCs of FT-NMR;* University Science Books: Sausalito, CA, 2000.
3. Hornak, J. P. *Basics of NMR*; http://www.cis.rit.edu/htbooks/nmr/nmr-main.htm, 1997-1999.
4. Lippert, T. H.; Mueck, A. O.; Seeger, H.; Pfaff, A. *Horm. Res.,* **2003**, *60*, 262-271.
5. Ni, F.; Konishi, Y.; Bullock, L.D.; Rivetna, M.N.; Scheraga, H.A. *biochemistry* **1989**, *28*, 3106-3119.
6. Smallcombe, S. H.; Patt, S. L.; Keifer, P, A. *J. Mag. Res. Ser. A*, **1995**, *117*, 295-303.
7. Pugh, M. E.; Schultz, E.; Bell, T. T. *Concepts and Techniques in the Biochemistry Laboratory;* Bloomsburg University: Bloomsburg, PA, 2005.
8. Wishart, D. S.; Sykes, B. D.; Richards, F. M. *J. Mol. Biol,.* **1991**, *222*, 311-333.
9. Kemp, W. *Organic Spectroscopy,* 3rd ed.; W. H. Freeman and Company: New York, NY, 1995.
10. Lambert, J.B;, Shurvell, H. F.; Lightner, D. A.; Cooks, R. G. *Organic Structural Spectroscopy;* Prentice Hall: Upper Saddle River, NJ.
11. Skoog, D. A.; Holler, F. J.; Nieman, T. A. *Principles of Instrumental Analysis, 5th ed.;* Saunders College Publishing: Philadelphia, PA, 1998.

Chapter 16

Experimentation with [15]N Enriched Ubiquitin

David Rovnyak, Laura E. Thompson, and Katherine J. Selzler

Department of Chemistry, Bucknell University, Moore Avenue,
Lewisburg, PA 17837

Nuclear magnetic resonance (NMR) spectroscopy is of
fundamental importance in the advancement of the life
sciences. Students obtaining chemistry degrees are
increasingly likely to enter the life sciences and to intersect
with the application of NMR to the study of biological
systems. Increased undergraduate laboratory instruction in
biological NMR is needed to help bridge the gap to the
sophisticated uses of NMR in the life sciences in graduate and
industry settings. We present several biomolecular NMR
experiments using human ubiquitin for upper level chemistry
coursework. We review prior work and describe extensions,
such as temperature variation, to illustrate protein dynamics in
more detail. We exploit direct ^{15}N detection of the NMR
signal to achieve very high quality spectra on limited
instrumentation and using a commercially available sample.

Introduction

The National Institute of Health's "Roadmap for Medical Research"
promotes the discovery of the detailed structural and functional relationships
among all parts of cellular machinery in humans (1). NMR has a critical role in
meeting the goals of the Roadmap since it is uniquely suited for studying
membrane proteins, intermolecular interactions, dynamic processes, and
macromolecules in native or near-native environments.

Enhancing the NMR training of the next generation of scientists will help to meet a national priority and will reflect the wide scope of NMR utilization in the life sciences. There is also a dangerously widening chasm between the level of NMR instruction in undergraduate education and the complexity of NMR experimentation at the graduate level. To address these concerns, we have been developing and testing biomolecular NMR laboratories in upper level course work for majors in our 'Biochemistry and Cell Biology' program. We have set out to create robust, student-tested experiments that represent *authentic* training in biomolecular NMR and are accessible to the widest possible range of undergraduate educators.

We decided upon directly observing the ^{15}N NMR signal for ^{15}N enriched proteins. Although the typical practice is to perform ^{1}H observation, our approach provides compelling benefits to faculty and students.

Goals for Protein NMR Laboratory Experiments

Practical Considerations

NMR instrumentation for performing structural biology research has additional components beyond those found in standard double-channel NMR instruments, and often requires additional levels of maintenance. Biomolecular NMR spectroscopy uses four independent RF channels, specialized probes, and proton (^{1}H) signal detection. Unfortunately such instrumentation is not yet widely available to principally undergraduate institutions (PUIs). Moderate field double-channel instruments are commonly operated by PUIs, and are optimized for X channel signal detection, so-called 'direct detection' (^{15}N through ^{31}P). Operating at 200-400 MHz frequencies for ^{1}H, these instruments are not generally suitable for performing conventional protein NMR studies.

We performed protein NMR via such "direct detection". Detecting a ^{13}C or ^{15}N NMR signal eliminates the need to suppress the water signal, which greatly simplifies experiments. Also, at the end of the pulse sequence, it is no longer necessary to return a coherence from an X nucleus back to ^{1}H for observation. This so-called "back" step of "out and back" experiments involves several additional pulses and delay times. Eliminating this step further simplifies experiments and also saves time in the pulse sequence, which reduces some of the deleterious T_1 relaxation of the protein signals. Also, the transverse relaxation time constants (T_2) of heteronuclei such as ^{13}C or ^{15}N proteins are larger for moderate fields (200-400 MHz for ^{1}H) as shown in Figure 1. Larger values of T_2 translate into narrower line shapes and higher signal to noise ratios. The effect is modest, but not insignificant.

As the frequency of the signal decreases, so does the sensitivity of detecting that signal (3). In principle, detecting ^{13}C nuclei is favored over detecting ^{15}N

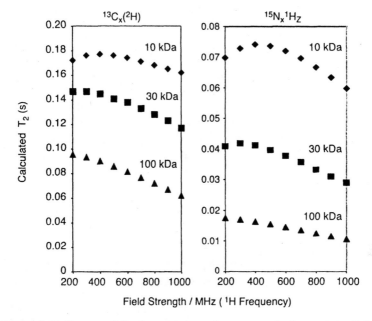

Figure 1. Estimates of T_2 time constants for prototypical proteins. Values depend on numerous parameters and are approximate; rather the trend towards larger T_2's as field decreases should be noted. Left: ^{13}C T_2's for a deuterated protein. Right: antiphase $^{15}N\{^1H\}$ T_2's for a protonated amide nitrogen.

nuclei since the ^{13}C Larmor frequency is over twice as large as that for ^{15}N. It is extremely expensive to purchase ^{13}Cenriched proteins at this time, while ^{15}Nenriched proteins are significantly more accessible for purchase (see notes in Experimental Details). The use of ^{15}N enriched proteins leads also to a number of pedagogical advantages over ^{13}C enriched samples

Pedagogical Considerations

The starting point of a protein NMR investigation is to acquire an experiment that correlates in a two-dimensional (2D) representation all directly bonded ^1H-^{15}N spin pairs of an ^{15}N enriched sample. This experiment involves a heteronuclear single quantum coherence (HSQC) transfer among the nuclei (4). This 2D-HSQC experiment will confirm that the protein is folded, and is used to identify the first peak lists which help to complete the protein assigbnments. The HSQC is a critical building block for nearly all other protein NMR experiments and has value as a gateway experience into protein NMR, allowing students to experience how a NMR structural study begins.

Furthermore, the relationships between relaxation properties of ^{15}N nuclei and the global and local dynamics of proteins are very well known. It is common in an NMR structural study to characterize the global reorientation and backbone dynamics by ^{15}N relaxation. Similar to the usage of the HSQC, investigators often begin a structural study with the determination of the protein rotational correlation time by the use of ^{15}N relaxation studies. Broadly, it is our impression that most undergraduate curricula do not include experimentation with hydrodynamics or protein dynamics. It is highly desirable to find ways of allowing students to experimentally measure protein hydrodynamic properties and to directly observe that proteins are not at all rigid, but instead have substantial motional freedom with functional significance.

Experimental Details

A sample of approx. 11 mg uniformly ^{15}N-enriched human ubiquitin was purchased from VLI Research (Malvern, PA, http://www.vli-research.com) at 1 mM concentration in 50 mM potassium phosphate buffer at pH 5.8. Additional evaporative concentration was performed and the sample transferred to a Shigemi (Allison Park, PA) sample tube. Additional vendors also providing isotopically enriched ubiquitin are Spectra Stable Isotopes (Columbia, MD, 21046, http://www.spectrastableisotopes.com) and Cambridge Isotope Laboratories (Andover, MA, http://www.isotope.com).

All pulse sequences were implemented on a Bruker ARX spectrometer operating at 300 MHz for ^{1}H (Figure 2), and may be downloaded at (http://facstaff.bucknell.edu/drovnyak/). The T_1 and especially T_2 measurements should be acquired with sample spinning off.

^{15}N Detected 2D-HSQC

This first laboratory has been used for three years in an upper level chemistry course entitled 'Biological Physical Chemistry' involving mainly senior students in a Biochemistry and Cell Biology major who are intending to proceed in the life sciences in medicine or graduate research.

The NMR instrument is initially demonstrated to groups of 3-5 students who then carry out a series of short exercises that facilitate a guided inquiry approach to some fundamentals of protein NMR. The instructor should be present to help facilitate discussion, but all experiments can be independently acquired and processed by the students.

Students first acquire a ^{1}H NMR spectrum with no water suppression and are asked to discuss why they do not observe signal from the protein (e.g. role of receiver gain and dynamic range). Students enable a flag that implements water suppression via presaturation and acquire another 1D proton spectrum, then

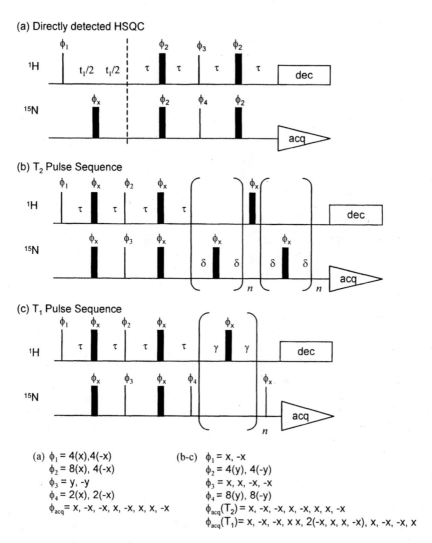

(a) $\phi_1 = 4(x), 4(-x)$ (b-c) $\phi_1 = x, -x$
$\phi_2 = 8(x), 4(-x)$ $\phi_2 = 4(y), 4(-y)$
$\phi_3 = y, -y$ $\phi_3 = x, x, -x, -x$
$\phi_4 = 2(x), 2(-x)$ $\phi_4 = 8(y), 8(-y)$
$\phi_{acq} = x, -x, -x, x, -x, x, x, -x$ $\phi_{acq}(T_2) = x, -x, -x, x, -x, x, x, -x$
$\phi_{acq}(T_1) = x, -x, -x, x\ x, 2(-x, x, x, -x), x, -x, -x, x$

Figure 2. ^{15}N detected pulse sequences for proteins. Thin lines are 90° pulses and thick bars are 180° pulses. Pulses labeled with ϕ_x are applied with x phase. Sequences are (a) a 2D-HSQC experiment, and (b-c) experiments for measuring T_1 and T_2 time constants. The coherence transfer uses $\tau = 2.7$ ms $\sim 1/4J_{NH}$, (^{1}H-^{15}N J $\sim$ 92 Hz). In (b) we use values of δ between 0.45 – 0.60 ms. In (c) we use value of γ between 5-7 ms. Adapted with permission from (2).

discuss the changes in the spectrum. Presaturation avoids the use of gradients or shaped pulses, and gives more than sufficient water suppression to appreciate the proton NMR spectrum of a protein. Students should then be given the typical signal regimes for proteins (5), which we summarize in Table 1. The signal regions for aqueous proteins are generally consistent with the qualitative rules that students have previously learned for organic compounds in chloroform. These initial ^{1}H NMR exercises help to build confidence in handling protein NMR spectra.

Table 1. Typical Signal Regions for Protein NMR Spectra[a]

Protein Moiety	Chemical Shift Range (ppm)
backbone amide	10.0 - 7.0
aromatic	8.0 – 6.5
backbone alpha	6.0 – 3.5
aliphatic side chain	3.5 – 1.0
methyl	< 1.5

(a) Data obtained from reference (5), Ch. 8.

A ^{1}H spectrum is then acquired with ^{15}N decoupling; increased sensitivity and spectral simplification are observed in the amide signal region due to the collapsing of doublets to single lines. Important questions to discuss with students include how dispersion in the proton NMR spectrum is consistent with a folded protein, and whether useful studies of proteins can be carried out solely with ^{1}H 1D-NMR spectra. A 1D-NMR ^{15}N spectrum of the ^{15}N enriched ubiquitin sample is acquired by setting $t_1 = 0$ in Figure 2a. This experiment may be performed without and with ^{1}H decoupling in order to demonstrate the collapse of ^{15}N doublets upon ^{1}H decoupling. We ask students to discuss signal resolution in the ^{15}N spectra, and to compare it with the ^{1}H spectra. (example of the ^{15}N 1D-NMR spectrum in Figure 5).

The goal of this discovery exercise is to illustrate the severe limitation on signal resolution in one-dimensional spectra. To this end, we ask students to estimate the number of ^{1}H signals and ^{15}N signals they expect from ubiquitin. Next, we present the principle of 2D-NMR in the context of achieving resolution of ^{1}H-^{15}N spin pairs. A graphic we use in lecture notes is shown in Figure 3.

Students then configure and acquire a ^{15}N{^{1}H} HSQC, consuming about two hours, and perform interpretation directly from the spectra. Configuring the experiment (Figure 2a) requires primarily specifying 90° pulse lengths on the ^{1}H and ^{15}N channels. This is similar to the use of standard biomolecular pulse sequences that are supplied with commercial NMR systems; such experiments are often automated to such a degree that most need only be configured by

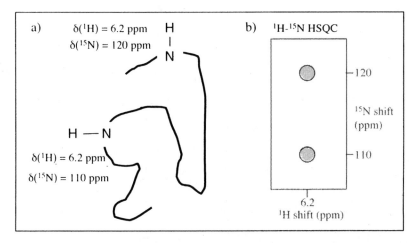

Figure 3. Graphic used in lecture notes to illustrate the principle of spreading signals to achieve higher resolution in 2D-NMR. Panel (a) indicates two backbone amide spin pairs in which the the protons have identical shifts; the ^{15}N shifts of the attached nuclei are not equal. Since it is possible to generate correlated signals in 2D-NMR (b) the degeneracy of the ^1H signals is broken.

specifying calibrated RF powers on every channel. One caveat is that optimizing water suppression when doing proton detection can be more involved and require additional experience.

Students are expected to count the number of signals, to think about how they distinguish between signals and noise, and consider if their count makes sense for human ubiquitin. We again ask students to discuss why the well-dispersed appearance of the spectrum is consistent with a folded protein.

Protein Hydrodynamics/Dynamics

We have developed a follow-up laboratory in the same course, Biological Physical Chemistry, for performing hydrodynamic studies of proteins. There is a resurgence in the utilization of methods such as dynamic light scattering and analytical ultracentrifugation in biophysics, so we have increased instruction on sedimentation and diffusion in this class. Specialized instrumentation for performing hydrodynamic measurements is expensive and rarely found in principally undergraduate institutions. Macromolecular dynamics measurements are commonly performed on biomolecular NMR instruments, so we adapted these approaches to be used on lower field, double-channel instruments (2b).

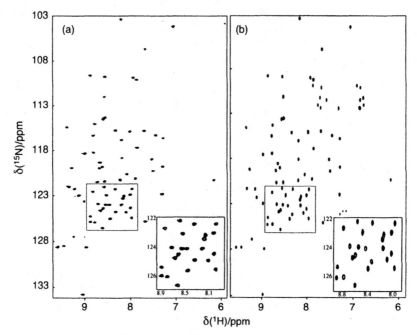

Figure 4. Panel (a): ^{15}N detected $^{15}N\{^1H\}$ HSQC obtained on a spectrometer operating at 300 MHz for ^1H in 6 hr to show the high fidelity of the method; a good HSQC can be acquired in 2 hr. Panel (b) : ^1H detected $^1H\{^{15}N\}$ HSQC obtained on a 500 MHz spectrometer (courtesy Dr. Richard Harris, University College London, Bloomsbury Center for Structural Biology) in approximately 25 minutes. Adapted and reproduced with permission from (2a) .

We devised a method for measuring average T_1 and T_2 relaxation time constants for the backbone amide ^{15}N nuclei of proteins using series of 1D ^{15}N NMR spectra of ^{15}N-ubiquitin. The ratio of T_1/T_2 is then analyzed with standard relaxation equations to yield the rotational correlation time (τ_c), a sensitive reporter on the timescale of rotation for the macromolecule. This experiment complements and extends the study of translational diffusion. Einstein related a macroscopic property, the diffusion constant D, to a microscopic property, the frictional coefficient f, (7)

$$D = \frac{k_B T}{f} ,$$

(1)

where k_B is Boltsmann's constant and T is the temperature in Kelvin. Stokes developed several analytic expressions for f. For translation of a sphere

$$f_{trans} = 6\pi\eta r \ , \tag{2}$$

where η is the viscosity of the solution and r is the particle radius. The frictional coefficient for rotation of a sphere in solution is

$$f_{rot} = 6\eta V \ , \tag{3}$$

where V is the particle volume. Equations 1 and 2 yield a Stokes Einstein equation (Eq. 4) that correlates translational diffusion to particle radius. Equations 1 and 3 yield a Stokes-Einstein equation (Eq. 5) that correlates rotational diffusion to particle volume.

$$D_{trans} = \frac{k_B T}{6\pi\eta r} \tag{4}$$

$$D_{rot} = \frac{k_B T}{6\eta V} \tag{5}$$

The NMR measurement of translational diffusion requires gradient methods and is not considered here (6). Rotation governs nuclear spin relaxation by modulating small local magnetic fields at the nuclei (5,8,9). Rotational motion follows in analogy to translational motion and is a natural extension in coursework, but the undergraduate biophysical texts we have encountered to date neglect rotational motion. Since D is closely related to the rotational correlation time, $D_{rot} = 1/6\tau_C$, then

$$\tau_C = \frac{\eta V}{k_B T} \ . \tag{6}$$

Using Eq. 6 the molecular radius can be estimated in the spherical approximation, which is a reasonable assumption for ubiquitin.

The rotational correlation time, τ_c, is an exponential time constant for the decay lifetime of an autocorrelation function that describes molecular motion. Two qualitative pictures help to give τ_c a physical interpretation. A widely cited view is that τ_c is the approximate time for the particle to rotate by one radian. A second view is that τ_c is the average duration of a mode of rotation. Both are acceptable physical pictures that can be represented to students.

Figure 2b-c shows pulse sequences for measuring T_1 and T_2. The experiments are repeated for different values of the delay periods (δ or γ in Figure 2b-c respectively). The T_1 and T_2 experiments are designed to yield a series of decaying spectra that can be integrated and fit to an exponential function to discover the time constant of decay. As little as five spectra can be acquired for each of T_1 and T_2. The experiments are commonly performed in an hour (30 minutes for each of T_1 and T_2), but increasing this time can yield more detailed results. As a rough guidline one should use several delay periods both less than and greater than $1*T_1$ or $1*T_2$.

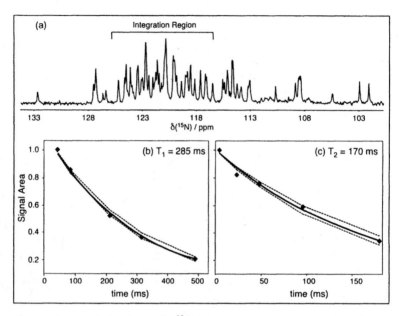

Figure 5. Panel (a) shows the ^{15}N 1D-NMR spectrum and a typical region for integration. Sample student data and analysis of T_1 and T_2 relaxation times shown in panels (b-c). (adapted and reproduced with permission from (2b)). Students estimated errors by qualitative use of bounding lines, (dashed lines).

Several approximations are involved in this approach to measuring T_1 and T_2 (2b), but the principle assumption is to integrate a broad region of the ^{15}N 1D-NMR spectrum to observe average protein dynamics. Example data and two-parameter exponential fits performed by students in the class and using Microsoft Excel are shown in Figure 5. Three parameter fits will be of added value if a systematic offset in the data is suspected, however a two-parameter fit forces the exponential to 0, a legitimate physical assumption. Bruker Xwinnmr software can automate fitting relaxation data, but we do not permit this.

We have implemented relaxation equations into an Excel spreadsheet (http://www.facstaff.bucknell.edu/drovnyak/ubiq_hsqc.html) for determining the rotational correlation time from the ratio T_1/T_2. We use the Lipari-Szabo formalism (10, 11) in the assumption of a rigid protein. Students use the spreadsheet by a trial and error approach in which they guess a value of the rotational correlation time and are given a computed T_1/T_2 ratio for their guess. They refine their guess for τ_c until a computed T_1/T_2 agrees with their experimental T_1/T_2 ratio. For the data in Figure 5, the students found $\tau_c = 5.1 \pm 0.7$ ns, whereas the accepted value is 4.1 ns (at T ~ 298 K) (12). We obtained $\tau_c = 4.1 \pm 0.4$ ns in control experiments with higher sensitivity. There is more uncertainty in the measurement of T_2 (2). Student results commonly fell between 4-5 ns. In the spherical approximation, the radius estimated from $\tau_c = 5.1$ ns is 17.2×10^{-10} m, while the radius obtained from $\tau_c = 4.1$ ns is 16.1×10^{-10} m. The radius determined by size exclusion FPLC is 15.7×10^{-10} m (13).

Extensions of the Motional Studies

We have noticed several areas where pedagogically useful variations on the dynamics laboratory could be achieved with modest additional laboratory time, and by better use of the experiment itself. These extensions have been tested and will now be incorporated in the upcoming year's laboratories.

In devising the original dynamics experiment we had to deal with the need to obtain the net rotational correlation time for the whole protein. The solution for averaging backbone dynamics was to integrate a broad range of signals, which did not allow students to see evidence of local dynamics, contradicting our goal to illustrate protein flexibility and the existence of fast local motions. All too often students do not have the opportunity to witness proteins as dynamic objects or to consider that dynamics correlate with function.

Although the T_1, T_2 experiments require measuring and integrating the individual spectra, we steered away from having students analyze the spectra further since our goal was to focus on global dynamics. This was a missed opportunity and Figure 6 illustrates an eight point series of ^{15}N spectra where it is evident that there are signal regions that do not relax with the same time constant. There are variations in T_2 across the protein backbone (which are easily seen with fewer than eight spectra). Fast local motions vary significantly from residue to residue and tend to correlate with regions of functional significance. Active sites of proteins require greater local mobility in order to be able to sample large conformational spaces and to accommodate substrates. Therefore spectra (T_1 or T_2) such as in Figure 6 are a compelling *direct* demonstration of the existence of local motions in proteins.

Students will then discuss stack plots such as Figure 6 to consider the source of T_2 variation and its significance in characterizing protein function. This experiment is unique for allowing students to easily notice the existence of variation in motion across the protein backbone. We also investigated the possibility of quantifying the dynamics of individual residues since there are several distinctly resolved signals in the ^{15}N spectrum; however we found the sensitivity to be poor and did not obtain good agreement with literature values. In principle this is feasible with improved sensitivity and baseline stability.

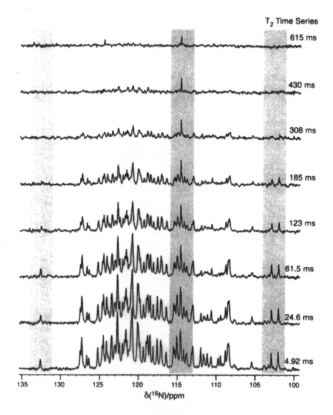

Figure 6. Spectra for measuring the T_2 relaxation time constant using the sequence of Figure 2b with $\delta = 0.60$ ms and $n = 2, 10, 25, 50, 75, 125, 175$. Shaded bars highlight a few areas where variation in T_2 relaxation is easily seen. These data were acquired with high sensitivity (1400 scans per F1D) for publication but significantly fewer transients still provide high quality data.

When performing this lab in small groups we noticed that variation in the measured τ_c data correlated with slightly different temperature settings used by the different groups. Although this is certainly predicted by Eq. 6, we did not expect that the methodology would be sensitive to differences of a few Kelvin, given that the student data yielded errors on the order of 0.5 ns. We performed a temperature series, using neat methanol for external temperature calibration (5), and found a strong relationship between T_1/T_2 and temperature by increasing the measurements to six time increments (Figure 7). The results confirm that increasing the temperature decreases the rotational correlation time.

Ubiquitin is ideal for this extension since it can reversibly be taken to 310 K or potentially higher. Each group of students will be assigned a different

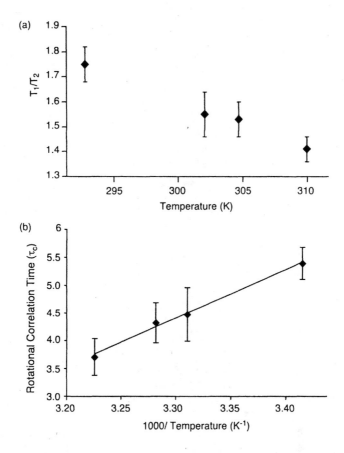

Figure 7. Temperature dependence of T_1/T_2 in (a) and τ_c in (b). A linear relationship between τ_c and the inverse of temperature is supported in (b) with a linear fit ($R^2=0.9934$). Increasing temperature increases molecular motion and reduces the solution viscosity. Errors determined in SigmaPlot v9.0. For T_2, $\delta = 0.6$ ms and $n=2,10,30,60,90,150$; for T_1, $\gamma = 3.5$ ms and $n=6,12,20,40,60,130$.

temperature and will share the data among the class. We will ask the quintessential physical chemistry laboratory question, "Suggest a relationship between τ_c and T that yields a linear relationship and test your prediction"!

Assessment

End of year surveys have shown strong student support to increase instruction in hydrodynamics, which we infer supports enhanced experimentation. It is axiomatic that hands-on experimentation is an invaluable learning opportunity for students. Although we do not have assessment of many aspects of the intended areas of learning, students demonstrated to the instructor better comprehension of and ability to use Stokes-Einstein relations following completion of the laboratory. We intend the relaxation experiment to support instruction in hydrodynamics: no special prerequisites are needed, and no mathematics beyond the use of the Stokes-Einstein relation is required. We used these experiments, however, in a senior course, and decided to give a comprehensive lecture on how molecular reorientation drives nuclear spin relaxation. Students had a great deal of difficulty explaining the qualitative basis for nuclear relaxation in their laboratory reports. In analyzing the student writing it appears that weakness in understanding core concepts in electromagnetism was a hindrance (2b). The extensions presented here were tested with students, but have not been assessed in courses yet.

Broader Impacts

There is a significant startup cost to acquire the sample (ca. $1k), but human ubiquitin has a very long shelf life in solution at 4 °C. We are beginning the fourth year of service for this sample, which has served 60+ students to date. We expect the sample to be used by hundreds of students for many more years. This sample and some of these experiments can be used in multiple courses; some of this work is planned for an upcoming spectroscopy course. The experiments here can be used in a biochemistry course or in advanced seminars; instructors can tailor the work for other purposes. Connections to the molecular biology of ubiquitin could be explored for future maturation of the experiments.

The extended experiments (e.g Figure 6) may be unique among undergraduate experimentation in allowing students to directly witness the flexibility of the backbone in a stably folded protein.

When beginning this work we were candidly surprised at the high spectral quality and ease of data acquisition for performing ^{15}N direct detection on proteins at low fields. This work may have significance for emerging research methods. This work suggests that ^{15}N could be a promising NMR reporter near to paramagnetic centers. Also, emerging dynamic nuclear polarization (DNP) techniques favor the observation of low gamma nuclei in order to minimize relaxation losses (14). The methods here may be of utility in extending DNP experiments to biomolecules.

Finally, we also tested an extension of the HSQC experiment that replaces the conventional practice of uniform incrementation of the indirect proton evolution period with nonuniform sampling (15,16). This approach yielded high quality HSQC spectra with time savings of factors of 3 or better, but requires specialized processing and substantial additional comprehension of NMR principles. We do not see a pedagogical justification for doing this work in classes at this time, but the use of alternative data sampling and processing is expanding and maturing and we may revisit this experiment at a later time.

Acknowledgements

We thank the Bucknell Department of Chemistry for purchasing the ubiquitin sample, and Bucknell University for a summer assistantship (LET, 2005).

References

1. National Institutes of Health, "National Institutes of Health Roadmap for Medical Research". http://nihroadmap.nih.gov/ (accessed 2006).
2. (a) D. Rovnyak, L.E. Thompson, Biochem. *Biochem. Mol. Biol. Educ.* **2005**, 33, 117-122. (b) L. E. Thompson, D. Rovnyak *Biochem. Mol. Biol. Educ.* **2006**, In Press.
3. D. I. Hoult and R. E. Richards *J. Magn. Reson.* **1976**, 24, 71-85.
4. G. Bodenhausen, D. J. Ruben *Chem. Phys. Lett.* **1980**, 69, 185-189.
5. J. Cavanagh, W. J. Fairbrother, A. G. Palmer and N. J. Skelton **1996**,
6. K. F. Morris, C. S. Johnson, Jr. *J. Am. Chem. Soc.* **1992**, 114, 3139-3141.
7. C. R. Cantor, P. R. Schimmel **1980**,
8. M. H. Levitt **2001**,
9. F. F. J. M. van de Ven **1995**,
10. G. Lipari, A. Szabo *J. Am. Chem. Soc.* **1982**, 104, 4546-4559.
11. G. Lipari, A. Szabo *J. Am. Chem. Soc.* **1982**, 104, 4559-4570.
12. A. L. Lee, A. J. Wand *J. Biomol. NMR* **1999**, 13, 101-112.
13. V. Uversky *Biochemistry* **1993**, 32, 13288-13298.
14. C.-J. Joo, K.-N. Hu, J. A. Bryant, R. G. Griffin *J. Am. Chem. Soc.* **2006**, 128, 9428-9432.
15. D. Rovnyak, D. Frueh, M. Sastry, A. S. Stern, J. C. Hoch, G. Wagner *J. Magn. Reson.* **2004**, 170, 15-21.
16. D. Rovnyak, J.C. Hoch, A.S. Stern, G. Wagner *J. Biomol. NMR* **2004**, 30, 1-10.

Chapter 17

Probing the Phase Behavior of Membrane Bilayers Using ^{31}P NMR Spectroscopy

Barry Zee and Kathleen Howard[*]

Department of Chemistry and Biochemistry, Swarthmore College,
Swarthmore, PA 19081

^{31}P nuclear magnetic resonance (NMR) spectroscopy is a
powerful method to characterize the phase preferences of
aqueous dispersion of phospholipids. Here we show ^{31}P
spectra from both a bilayer-forming and a hexagonal-forming
synthetic lipid dispersion and describe why the two line shapes
differ. We also show ^{31}P spectra of a natural mixed lipid
extract that can access both the bilayer and hexagonal phase
depending on temperature.

We have recently introduced a new biophysical experiment into our upper-
level laboratory curriculum at Swarthmore College. The goal of this experiment
is to use ^{31}P NMR to examine phospholipid organization in biomembranes.
Membranes serve as selective permeable barriers between cells and their
environment and as a matrix for proteins (*1*). One of the remarkable properties
of membrane lipids is that they can form a variety of different aggregate
structures depending on the environment when dispersed in water; this is known
as polymorphism (*1,2*). Two examples of such aggregates are the bilayer phase
and the inverted hexagonal phase, shown in Figure 1. It is well accepted that
the primary organization of lipids in biomembranes is the bilayer phase which is
necessary for the maintenance of the membrane barrier. However, biomembranes also contain large amounts of lipids that on their own prefer non-
bilayer structures such as the inverted hexagonal phase. One suggestion for the
existence of non-bilayer lipids within a biological membrane is to allow the

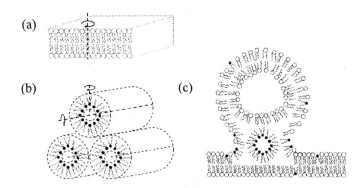

Figure 1. Schematic drawings of (a) a bilayer phase (b) an inverted hexagonal phase and (c) a cellular budding event where both phases are simultaneously populated. Axes of rotation are shown in (a) and (b).

membrane to rearrange and readily form local, transient non-bilayer structures necessary for fusion, division and budding events (2). Shown in Figure 1c is a cartoon of a membrane budding event where a predominantly bilayered membrane adopts a non-bilayer intermediate as part of an essential cellular process.

Small-angle x-ray and neutron diffraction allow for definitive identification of lipid phases. However, these techniques require specialized equipment and are not well suited for obtaining the relative proportions of different phases in multiphase systems. ^{31}P NMR has become suitable alternative to diffraction techniques in many cases and is a convenient, sensitive and widely used technique for lipid aggregate structure determination. The agreement between phase determination by ^{31}P NMR and small-angle x-ray diffraction has been carefully documented (3).

There are several reasons why ^{31}P NMR spectroscopy is a powerful method for studying biological membranes. The majority of lipids found in biomembranes are phospholipids. The phosphate nucleus located in a lipid head group is a built-in non-perturbing probe of the membrane environment. [31]P is a spin-1/2 nucleus and is 100% naturally abundant so there is no need for isotopic enrichment. The lipid phosphorous head group also exhibits a large chemical shift anisotropy (CSA, discussed below) that for large (radius >2000Å) lipid aggregates is only partially averaged by the restricted modes of motion available. Bilayer and hexagonal phases involve different patterns of motional averaging and thus result in distinctly different ^{31}P line shapes. In this paper we show ^{31}P spectra from both a bilayer-forming and a hexagonal-forming synthetic lipid dispersion and describe why the two line shapes differ. We also show ^{31}P spectra of a mixed lipid extract from porcine brain that can access both the bilayer and hexagonal phase depending on temperature.

Experimental

Preparation of Membrane Lipid Samples

1,2-dimyristoyl-*sn*-glycero-3-phosphocholine (DMPC), 1,2-dioleoyl-*sn*-glycero-3-phosphoethanolamine (DOPE) and total lipid extract of porcine brain (Catalogue # 131101) were purchased in powdered form from Avanti Polar Lipids (Alabaster AL). The DMPC, DOPE and brain lipid extract samples that were hydrated to 20 weight % D_2O were all prepared using the same protocol. 50 mgs of the lipid was weighed directly into a standard 5mm ID, 175 mm long NMR tube. 12.5 μL of D_2O was then added to the tube. Samples were extensively vortexed and then subject to three cycles of freeze-thaw to fully homogenize. NMR tubes were spun slowly (500 rpm) in a benchtop centrifuge with a swinging bucket rotor to spin the viscous lipid dispersion down to the bottom of the tube. The DMPC sample used to collect the isotropic spectra was made by preparing a ~1 mg/ml solution of DMPC in D_2O and sonicating the sample in a bath sonicator until the sample became translucent.

Collection of ^{31}P NMR Spectra

[31]P one-dimensional spectra with ^{1}H decoupling were collected at 161.98 MHz on a Bruker DRX NMR spectrometer using a standard solution 5 mm broadband tunable probe. No special "solids" hardware was required. The 90° pulse was 8.6 μs, the sweep width was 200 ppm, and the relaxation delay was 2 s. Spectra were collected using 256 scans and processed using 100-Hz line broadening. During temperature dependence measurements, the samples were allowed to equilibrate for 30 minutes at each temperature before recording the spectra.

Results and Discussion

Powder Patterns from a Bilayered and Hexagonal Phase

We first collected spectra of two different synthetic lipids, DMPC and DOPE, whose structures are shown in Figure 2. DMPC consists of a phosphatidylcholine headgroup, glycerol backbone and two saturated acyl chains fourteen carbons in length. DOPE has a phosphoethanolomine headgroup, glycerol backbone and two acyl chains eighteen carbons in length, with one double bond in each of the acyl chains.

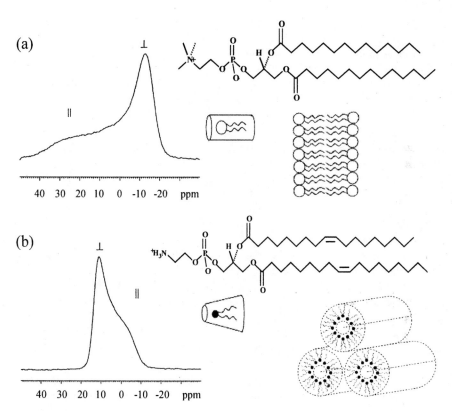

Figure 2. *P NMR spectra of (a) DMPC hydrated to 20 % D₂O, and (b) DOPE hydrated to 20 % D₂O. DMPC forms a bilayer phase . DOPE forms a hexagonal phase.*

The phase a particular lipid adopts when dispersed in water is often explained in terms of a "shape-structure" concept, which posits that the shape of an individual lipid molecule relates to the intrinsic curvature of the aggregate (2). DMPC is a cylindrically shaped lipid and packs together to form bilayers. The van der Waals radius for the head group of DOPE is small relative to the area taken up by its acyl chains and the lipid is thus considered cone-shaped. DOPE forms an inverted hexagonal phase at room temperature.

Figure 2 shows the ^{31}P spectra collected for DMPC and DOPE hydrated with 20 weight % D_2O. As opposed to the sharp line solution NMR spectra most undergraduates are familiar with, the broad spectra observed for these two lipid samples are powder patterns. To explain powder patterns, and why the powder patterns differ for these two lipids, an essential part of this experiment for the students is a review of the theory of chemical shift. There are several texts to describe this (4), and a brief overview is described below.

The phenomenon of chemical shift arises from the electron clouds surrounding atomic nuclei. Electrons within a molecule can generate small magnetic fields that add to or subtract from the external magnetic field experienced by the nuclei. The precession frequency of a particular nucleus is proportional to the local value of the magnetic field. Since the induced field generated by electrons depends on the bonding pattern within a molecule, the chemical shift is different along the various molecular directions. This effect is known as chemical shift anisotropy (CSA).

The orientation dependence or anisotropy of chemical shifts can be described using the mathematics of tensors (4,11). Single crystal studies are used to determine the orientation of the chemical shift tensor within the molecular frame of a molecule. Since lipids do not readily crystallize, the directions of the three principal axes for phospholipids are usually based on model phosphate compounds (5). The orientation of the ^{31}P chemical shift tensor in the molecular frame of a phosphate segment is shown in Figure 3. The σ_{11} element is approximately perpendicular to the O3-P-O4 plane, and the σ_{22} element approximately bisects the O3-P-O4 angle. The principal elements of the shielding tensor vary only slightly among the various phospholipids found in biomembranes and are approximately $\sigma_{11} = -80$ ppm, $\sigma_{22} = -20$ ppm and $\sigma_{33} = 110$ ppm (5).

For large aggregates of hydrated phospholipids, there is diffusion of individual lipids within the aggregate, but not sufficient motion to completely average the chemical shift so that a solution-like sharp line spectrum is observed. Instead, as shown in Figure 2 powder patterns are observed. A powder pattern is a broad line with a characteristic shape derived from the superposition of chemical shifts from different orientations. In the case of phospholipids, there is rapid molecular motion around the long axis of the lipids that partially averages the tensor to axial symmetry. In axial symmetry, two principal values of the chemical shift tensor are degenerate. The two distinguishable principal values may be determined by examination of an axially symmetric spectrum. They are denoted $\sigma_\parallel$ and $\sigma_\perp$ and correspond to B_0 being oriented parallel and perpendicular to the symmetry axis respectively. Since

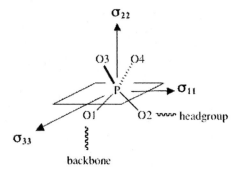

Figure 3. Orientation of the ^{31}P chemical shift tensor with respect to the molecular frame of a phosphate segment. The principal values for the $\sigma_{11},\ \sigma_{22},\ \sigma_{33}$ elements are –80, -20, and 110 ppm respectively.

there are relatively more ways for the static field to be oriented perpendicular, rather than parallel, to the axial tensor, $\sigma_\perp$ corresponds to the greater intensity edge of the axially symmetric lineshape.

The lineshapes for bilayered DMPC and hexagonal DOPE both indicate axial symmetry and the parallel and perpendicular edges of the spectra are indicated in Figure 2. Note, however, that the line shape observed for DOPE has reversed asymmetry compared to the DMPC spectrum and is narrower by about a factor of two. These differences in the line shapes reflect the fact that different patterns of motional averaging are allowed in the bilayer phase than in the hexagonal phase. The principal values of the chemical shift tensor, and the orientation of the chemical shift tensor in the phosphate molecular frame are very similar between these two lipids. However, the geometry of the aggregates leads to the chemical shift being averaged differently.

Motion leads to averaging of the molecular anisotropies. In both bilayer and hexagonal phases lipid molecules rotate around their long axis as shown in Figure 1. Bilayer lipids also move in the plane of the bilayer. For lipids in hexagonal phases there is an additional significant source of motion not present in the bilayer samples that additionally decrease the breadth of the lineshape and reverses the ordering of the parallel and perpendicular components (5). This motion is lateral diffusion around the small (~20 Å diameter) aqueous channels. Lipid diffusion around hexagonal pores occurs about an axis of motional averaging that differs from the long axis of the lipids. For a more mathematical treatment of the bilayer and hexagonal lineshapes see (5).

An Isotropic Phospholipid Dispersion

The connection between powder pattern shapes and the isotropic chemical shift is a useful pedagogical connection to make. The chemical shift measured for istropically tumbling molecules in solution is simply the average of the three principal values. In Figure 4 the spectrum of a sonicated dilute aqueous sample of DMPC is shown. Under dilute conditions, sonicated DMPC forms small vesicles that tumble rapidly on the NMR timescale. The chemical shift anisotropy is thus averaged and a single peak is observed. A simple calculation ($\sigma_i = \frac{1}{3}(\sigma_{11} + \sigma_{22} + \sigma_{33})$) with the principal values of the phosphate tensor listed above indicates the observed shift of ~0 ppm is appropriate.

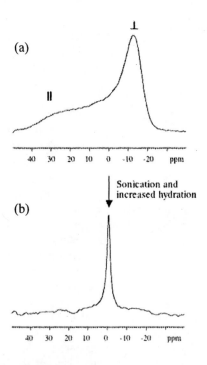

Figure 4. ^{31}P NMR spectra of DMPC (a) hydrated to 20% D$_2$O, showing a bilayer phase and (b) hydrated to 99% D$_2$O and sonicated, leading to rapid tumbling and resulting in an isotropic spectra.

Temperature Dependence of ^{31}P spectra of Brain Lipid Extract

^{31}P NMR was used to investigate the phase behavior of an aqueous dispersion of total brain lipid extract (porcine). The brain lipid extract is a mixture of lipids with the following composition: 16.7% phosphatidylthanolamine, 10.6% phosphatidylserine, 9.6% phosphatidylcholine, 2.8 % phosphatidic acid, 1.6% phosphatidylinositol and 58.7 % other. Although this is a complex mixture of phospholipids, the similarity of the principal values and orientation of the chemical shift tensor among phospholipids leads to overlap of the phosphate signal from all the component lipids into an observed line shape that is diagnostic of the overall phase of the mixture.

Shown in Figure 5 are ^{31}P spectra of the brain lipid extract collected at 10° increments between 30°C and 80°C. The spectrum obtained at 30°C has a low field shoulder and a high field peak separated by approximately 60 ppm. This lineshape is clearly indicative of the bilayer phase expected for data collected near the physiologically relevant porcine growth temperature. Upon increasing the temperature to 40°C, a new spectral component becomes visible. This position corresponds with that of the perpendicular edge of an inverted hexagonal lipid phase. This component increases in intensity as the temperature

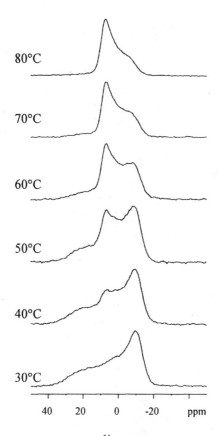

Figure 5. Temperature dependent ^{31}P NMR spectra of porcine brain lipid extract hydrated to 20% in D$_2$O. Samples were equilibrated for 30 minutes after each temperature change.

is raised, until the spectral shape characteristic of a pure hexagonal phase with a breadth of approximately 30 ppm is observed at 70°C. Multiphase systems cause a superposition of the line shapes characteristic of each phase present and the temperature course shown in Figure 5 clearly shows the lipid extract transitions from a bilayer phase to a hexagonal phase as the temperature increases.

Temperature dependent ^{31}P spectra very similar to what is shown in Figure 5 have been demonstrated with lipid extracts from *Escherichia coli* bacterial membranes (*6,7*). Previous work has shown that bacteria grown under different temperature conditions adapt their membrane lipid composition to exist in a "window" between bilayer and hexagonal phases (*6,7*), presumably to facilitate access to essential cellular processes such as membrane fusion and budding.

Educational Goals

At Swarthmore College solution NMR spectroscopy is introduced to students in the first semester organic chemistry courses. Students are familiar with the concepts of ^{1}H and ^{13}C isotropic chemical shifts, scalar coupling patterns, and the identification of small molecules using solution NMR. They are not usually aware, however, that there are spin active nuclei available beyond ^{1}H and ^{13}C, and that solid state NMR (SSNMR) techniques can provide valuable information on structure and dynamics.

The experiment described here has been used in two of our junior/senior level courses, Instrumental Methods and second semester Physical Chemistry. For most students completing the lab, it is their first opportunity to tune a NMR probe and learn some details about the electronics of a NMR spectrometer. The concept of a tensor is also unfamiliar to some of the students. Before the experiment began, the instructor provided some basic background on the mathematics of tensors and tensor properties in NMR *(11)*.

We believe other colleges might find this experiment useful in one of their advanced laboratory courses. The lipids used in this experiment are all commercially available and of reasonable cost. The three lipid samples described in this manuscript (DMPC, DOPE and Brain lipid Extract) cost us a total of ~$150 to purchase in 2005. Each of the lipid samples were used by several different student lab groups throughout the semester and stored in the freezer between use. Two recently published Journal of Chemical Education articles treat material related to the experiment described here; one discusses ^{31}P SSNMR line shapes of glasses (*8*) and the other magic angle spinning of a phospholipid *(9)*.

Finally, the work described here has particular relevance for students with biochemical interests. Swarthmore College has a special major in Biochemistry that requires students to complete two semesters of Physical Chemistry. Implementation of this experiment has fulfilled a need for the introduction of biophysical experiments into our traditional Physical Chemistry laboratory curriculum. Although protein and nucleic acids are routinely treated samples in our Biochemistry laboratory course, this lab is a unique opportunity for students to work with lipid samples. Occasionally we have combined the ^{31}P NMR experiments described in this lab with a complementary differential scanning calorimetry (DSC) experiment on membranes phase transitions (*10*).

Summary

In this paper we show ^{31}P spectra from bilayer-forming, hexagonal-forming and istropically tumbling membrane lipid samples and discuss the relationship between line shapes and aggregate structure. The lipid samples are commercially available and the NMR experiments are straightforward to collect on a solution NMR spectrometer equipped with a tunable broadband probe. Whereas solution NMR is often covered extensively in the introductory chemistry curriculum, solid-state NMR methods are not. This laboratory experiment demonstrates how the examination of a ^{31}P powder pattern spectrum can reveal information about the geometry and symmetry of a lipid aggregate.

Acknowledgements

We would like to thank Swarthmore students from the Instrumental Methods and Physical Chemistry courses who participated in these labs between 2005-2006 for their enthusiasm and feedback. Barry Zee is the recipient of an undergraduate summer fellowship from the Merck Institute of Science Education received as part of a Merck/AAAS Undergraduate Science Research Program grant to Swarthmore College. Kathleen Howard is supported by an NSF CAREER grant (0092940) and a Henry Dreyfus Teacher-Scholar Award.

References

1. Gennis, R. B. *Biomembranes: Molecular Structure and Function*, Springer-Verlag, NY, 1989.
2. de Kruijff, B. *Curr. Opin. Chem. Biol.,* **1997**, *1*, 564-569.
3. Tilcock, C. P. S., Cullis, P.R., Gruner, S.M. (1986) *Chem. Phys. Lipids* **1986**, *40*, 47-56.
4. Macomber, R. S. *A complete introduction to modern NMR spectroscopy*, Wiley-Interscience, NY 1998.
5. Seelig, J. *Biochim. Biophys. Acta* **1978**, *515*, 105-140.
6. Rietveld, A. G., Killian, J.A., Dowhan, W., de Kruijff, B. *J. Biol. Chem.* **1993**, *268(17)*, 12427-12433.

244

7. Morein, S., Andersson, A., Rilfors L., Lindblom, G. *J. Biol. Chem.* **1996**, *271(12)*, 6801-6809.
8. Anderson, A., Saiki, D., Eckert, H., and Meise-Gresch, K. *J. Chem. Ed.* **2004**, 81(7), 1034-1037.
9. Gaede, H., and Stark, R. *J. Chem. Ed.* **2001**, *78(9)*, 1248-1037.
10. Ohline, S., Campbell, M., Turnbull, M., and Kohler, S. *J. Chem. Ed.* **2001**, *78(9)*, 1251-1256.
11. Harris, R. K. *Nuclear Magnetic Resonance Spectroscopy.* pp. 245-248. John Wiley & Sons, NY 1986.

Modern NMR in Laboratory Development

Inorganic Chemistry

Chapter 18

When Nuclei Cannot Give 100%

Chip Nataro, William R. McNamara, and Annalese F. Maddox

Department of Chemistry, Lafayette College, Easton, PA 18042

The presence of NMR active nuclei that are not 100% naturally abundant and/or spin ½ can complicate NMR spectra. Representative examples are presented and the splitting patterns discussed.

The introduction to the fundamental ideas of NMR spectroscopy typically occurs in the first semester of organic chemistry. Most organic texts briefly describe the technique and then focus on ^1H NMR, with some mention of ^{13}C NMR and 2-D techniques such as COSY and HETCOR. The interpretation of ^1H NMR spectra requires gleaning information from splitting patterns, chemical shifts and peak integration. While chemical shifts and peak integration are relatively straightforward, the concept of splitting patterns typically takes some time to digest. To interpret splitting patterns, the 2nI+1 rule (although many texts have simplified that to the n+1 rule which works for ^1H and ^{13}C, both of which have I = ½) (*1*) is presented for determining the splitting pattern that should be observed for a given proton. Unfortunately, this can lead to the incorrect conclusion that only the protons need to be considered to determine splitting patterns in any given molecule.

The possibility of coupling to or coupling between other nuclei is briefly mentioned when ^{13}C NMR is introduced. ^{13}C-^{13}C coupling is typically not observed because "the probability that any molecule contains more than one ^{13}C atom is quite small." (*1*) The possibility of ^{13}C-^1H coupling is dismissed "because a ^{13}C signal can be split not only by the protons to which it is directly attached, but also by protons separated from it by two, three or even more bonds, the amount of splitting might be so great as to make the spectrum too complicated to interpret. Thus, the spectrum is measured under conditions, called broadband decoupling, that suppress such splitting." (*1*) Typically only decoupled spectra, which are represented as $^{13}C\{^1H\}$, are presented, so ^{13}C-^1H coupling in ^{13}C NMR is not seen.

However, ^{13}C-^1H coupling can occasionally be seen in ^1H NMR spectra. The ^1H NMR spectrum of CH_2Cl_2 in $CDCl_3$ is shown below (Figure 1). To understand the experimental spectrum, the two extremes should first be considered. If the sample of CH_2Cl_2 contained only NMR inactive carbon-12 nuclei, the ^1H NMR would be a singlet centered at 5.29 ppm. However, the other extreme would be if the sample only contained NMR active ^{13}C nuclei, in which case the ^1H NMR spectrum for $^{13}CH_2Cl_2$ would be a doublet centered at 5.29 ppm. The experimental spectrum is a weighted average of these two extremes and the weighted average is based on the natural abundance of the carbon isotopes. Carbon-12 has a natural abundance of 98.93% while carbon-13 is 1.07% naturally abundant. In a sample of CH_2Cl_2, 98.93% of the molecules will have a carbon atom that is not NMR active, therefore the ^1H NMR signal will be a singlet. The remaining 1.07% of the molecules will contain an NMR active carbon atom which will give rise to a doublet. The two peaks of the doublet are of equal intensity, so the relative integration of the peaks will be approximately 0.535 to 98.6 to 0.535. The peaks due to the presence of a spin active nucleus that is not 100% abundant are often called satellites. There are many nuclei that have $I = \frac{1}{2}$ and are not 100% abundant, some of the more commonly encountered examples are listed in Table 1.

A second reason one might dismiss ^{13}C-^1H coupling is the magnitude of the $^1J_{C-H}$ coupling constant. In their discussion of coupling constants, most text books suggest that proton-proton coupling constants greater than 20 Hz are rare in ^1H NMR. The largest coupling constants likely to be encounter are $^2J_{H-H}$ couplings in systems with diastereotopic protons. As seen in Figure 1a, the coupling constant ($^1J_{C-H}$) is 177.6 Hz, which is significantly larger than the

Table 1. NMR properties of some nuclei

Isotope	Spin	Natural Abundance	Isotope	Spin	Natural Abundance
^1H	1/2	99.985%	^{29}Si	1/2	4.7%
^2H	1	0.015%	^{31}P	1/2	100%
^6Li	1	7.4%	^{59}Co	7/2	100%
^7Li	3/2	92.6%	^{73}Ge	9/2	7.8%
^{10}B	3	19.6%	^{77}Se	1/2	7.6%
^{11}B	3/2	80.4%	^{95}Mo	5/2	15.7%
^{13}C	1/2	1.11%	^{103}Rh	1/2	100%
^{14}N	1	99.6%	^{183}W	1/2	14.4%
^{19}F	1/2	100%	^{195}Pt	1/2	33.8%
^{27}Al	5/2	100%	^{199}Hg	1/2	16.8%

coupling constant for diastereotopic protons. An additional lesson can be learned from this example. The typical introduction to NMR will teach a student that if two protons are coupled, they will have the same coupling constant. This also holds true to coupling between two different nuclei; the $^1J_{C-H}$ coupling constant of 177.6 Hz is seen in both the ^1H and the ^{13}C spectra (Figure 1b).

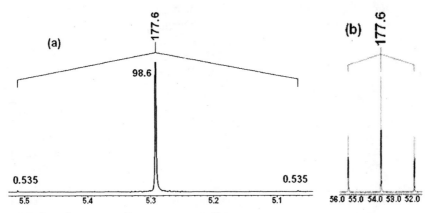

Figure 1. ^1H (a) and ^{13}C (b) NMR spectra of CH_2Cl_2 in $CDCl_3$ at 400 MHz and 100.6 MHz respectively.

The typical introduction to NMR provides one additional opportunity to examine nuclei that are not spin ½ and/or 100% abundant. One of the most common solvents for NMR is $CDCl_3$. Deuterium is NMR active but, unlike ^1H, has a spin of 1. Therefore, based on the equation $2nI + 1$, coupling to one deuterium will result in three peaks. This can be seen in the ^{13}C NMR of $CDCl_3$ (Figure 2). The signal for the carbon is split into three peaks of equal intensity due to presence of the deuterium. For spin ½ nuclei, the intensity of the peaks is based on Pascal's triangle which is built on powers of two; the sums of the first four rows are 2^0, 2^1, 2^2 and 2^3 respec-

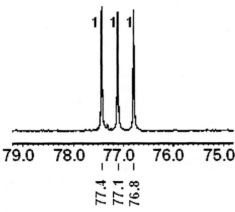

Figure 2. ^{13}C NMR of $CDCl_3$ at 100.6 MHz.

tively. A doublet is two equal intensity peaks, a triplet is three peaks in a ratio of 1:2:1, a quartet is four peaks in a 1:3:3:1 ratio, etc. For nuclei that are not spin ½, a different form of Pascal's triangle must be considered. To determine the

Pascal's triangle for nuclei that are not spin ½ the solution to $2nI + 1$ when n =1 gives the number that, when raised to any particular power, will give the sum of the corresponding rows. For example, the Pascal's triangle for deuterium (and other spin 1 nuclei) will be based on powers of three. For the first four rows, the sum of the rows will be 3^0, 3^1, 3^2 and 3^3 giving ratios of 1, 1:1:1, 1:2:3:2:1 and 1:3:6:7:6:3:1. To determine which row to use, the value of n is used as the exponent. Again, deuterated solvents provide a wonderful example for observing these patterns. In the ^{13}C NMR of CD_3CN (Figure 3), the three deuterium atoms (n = 3 and I = 1) give rise to a seven line pattern for the carbon that the deuterium atoms are bonded to, in a 1:3:6:7:6:3:1 ratio which sums to 3^3. Developing the intensity patterns for all the other spin values is beyond the scope of this book. Suffice it to say that if there is coupling to one and only one nucleus of any spin, no matter what the splitting, the peaks will be of equal intensity.

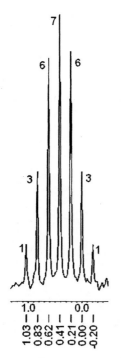

Figure 3. ^{13}C *NMR of* $\underline{C}D_3CN$ *at 100.6 MHz.*

Before moving on to spectra involving nuclei other than ^1H and ^{13}C, it is worth briefly noting that when spectra of these other nuclei are obtained, there are often nearby carbon atoms and protons. While the carbon atoms typically do not complicate the spectra of other nuclei (*previously discussed*), the presence of the protons can be quite troublesome. Consider the spectra of trimethylphosphine, $P(CH_3)_3$. The $^{31}P\{^1H\}$ spectrum of $P(CH_3)_3$ is a singlet at –61.1 ppm while the ^1H spectrum contains a doublet at 0.78 ppm (Figure 4). It is very common for someone inexperienced with multinuclear NMR to be confused by the presence of this doublet. This is coupling between the nine equivalent protons of the methyl groups and the phosphorus. The phosphorus-proton coupling will also be seen in the ^{31}P spectrum of $P(CH_3)_3$, giving rise to a ten line pattern. In addition, the $^2J_{H-P}$ is 2 Hz in both spectra, indicating that the protons are coupled to the phosphorus. While this can provide useful information, it can also make interpreting spectra difficult, and therefore, when protons are present, decoupled spectra are most often obtained.

In progressing from standard ^1H and ^{13}C NMR to interpreting spectra of molecules with other NMR active nuclei, a number of factors must be

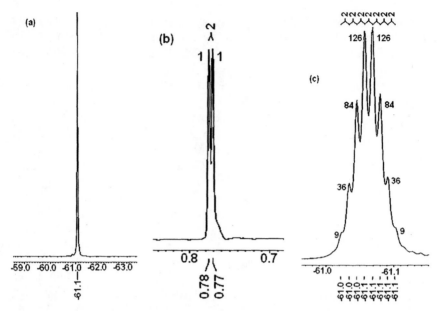

Figure 4. NMR spectra of P(CH₃)₃ in CDCl₃: (a) $^{31}P\{^1H\}$ at 161.9 MHz, (b) ^1H at 400 MHz and (c) ^{31}P at 161.9 MHz.

considered. The first factors to consider are the range of chemical shifts, the magnitude of the coupling constants and the possibility of coupling between different nuclei. An example where all of these considerations must be accounted for is in the ^{31}P and ^{19}F spectra of [N(C₄H₉)₄][PF₆] (Figure 5). The ^{31}P signal will be split into a seven line pattern by the six equivalent fluorine atoms; the intensity of the peaks based on Pascal's Triangle will be 1:6:15:20:15:6:1 the sum of which is 2^6. The ^{31}P chemical shift for the PF₆⁻ is -143.8 ppm and the coupling constant ($^2J_{P-F}$) is 714 Hz. In the ^{19}F spectrum there is a doublet at -72 ppm. The $^2J_{P-F}$ coupling constant of 714 Hz that was observed in the ^{31}P spectrum is identical in the ^{19}F NMR.

Another factor to consider in interpreting NMR spectra is the natural abundance of the NMR active nuclei. In the previous example, coupling between two spin ½ nuclei (neither of which were carbon or hydrogen) that are 100% abundant was addressed. To begin examining spectra in which one of the nuclei is not 100% abundant, consider the $^{31}P\{^1H\}$ spectrum of (1,1'-bis(di-isopropylphosphino)ferrocene)platinum dichloride (Figure 6). (*2*) In this complex, the two phosphorus atoms of the ligand are bound equivalently to the square planar platinum center. If platinum had no NMR active nuclei, the signal in the ^{31}P NMR would be a singlet. However, ^{195}Pt is spin ½ and has a natural abundance of 33.8%. Therefore, ^{195}Pt satellites are observed as a doublet ($^1J_{P-Pt}$

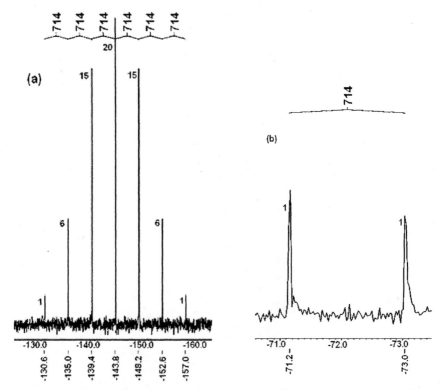

Figure 5. (a) ^{31}P of $[N(C_4H_9)_4][PF_6]$ at 161.9 MHz. (b) ^{19}F of $[N(C_4H_9)_4][PF_6]$ at 376.3 MHz.

= 3800 Hz) that is symmetric about the central peak. The central peak accounts for 62.2% of the signal, while the satellites, which are a 1:1 doublet, are 33.8% of the total or 16.9% each.

When examining spectra in which there is coupling to nuclei that are less than 100% abundant, it is important to remember that coupling to other nuclei is still possible. In the previous example, the two phosphorus atoms are equivalent, and only show coupling to the ^{195}Pt nuclei. This gives satellites that are essentially reflections of the central peak, in this case a singlet. If there were coupling to another nucleus, the central peak as well as the satellites would show this. In the closely related compound ((R)-(-)-1-[(S)-2-(diphenylphos-phino)ferrocenyl]ethyldi-t-butylphosphine)platinum dichloride, (3) the two phosphorus atoms are not equivalent, and therefore, couple (Figure 7). If platinum had no spin active isotopes, the ^{31}P signal for each phosphorus atom would be a doublet. Since ^{195}Pt is active, satellites are observed and they have the same $^2J_{P-P}$ (20.0 Hz) as the central peak. The satellites can be considered to be a doublet of doublets with each peak representing 8.5% of the signal.

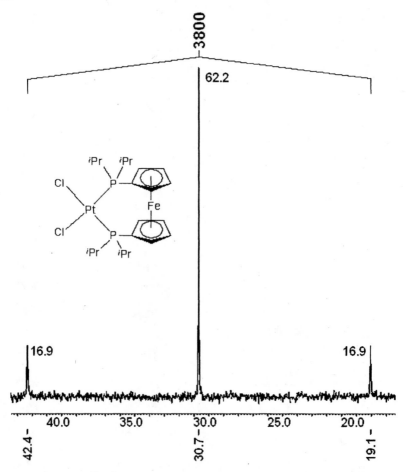

Figure 6. $^{31}P\{^{1}H\}$ spectrum of (1,1'-bis(di-isopropylphosphino)-ferrocene)platinum dichloride in CDCl$_3$ at 161.9 MHz.

The presence, or lack thereof, of satellites can be useful in a variety of different ways. The formation of the two previously mentioned platinum complexes can be followed by monitoring the ^{31}P NMR of the reaction mixture, in particular, noting the growth of the product peaks with the accompanying satellites. In the asymmetric compound, 1-diphenylphosphineoxide-1'-diphenylphosphineselenide-ferrocene, the two peaks in the ^{31}P NMR occur at very similar chemical shifts, making assigning the peaks somewhat problematic. (4) However, the presence of ^{77}Se (7.58%) satellites around the peak at 31.5 ppm allows for the unambiguous assignment of the spectrum (Figure 8).

In all of the examples thus far, coupling between two and only two different NMR active nuclei has been considered (e.g. ^{13}C-^{2}H, ^{31}P-^{19}F and ^{31}P-^{77}Se). If

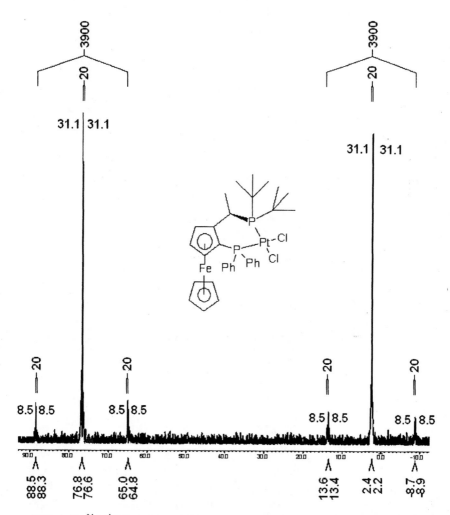

Figure 7. $^{31}P\{^1H\}$ *spectrum of ((R)-(-)-1-[(S)-2-(Diphenylphosphino)ferro-cenyl]ethyldi-t-butylphosphine)platinum dichloride in CDCl$_3$ at 161.9 MHz.*

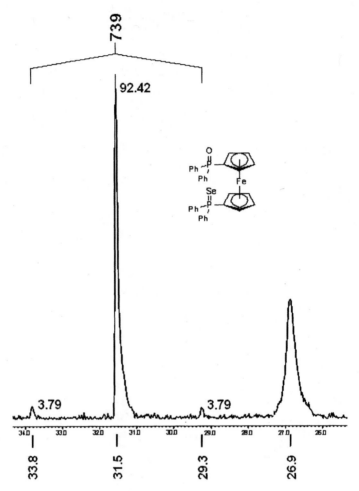

Figure 8. $^{31}P\{^{1}H\}$ *spectrum of 1-diphenylphosphineoxide-1'-diphenylphosphineselenide-ferrocene in CDCl₃. at 161.9 MHz.*

multiple NMR active nuclei are present in a molecule, coupling to each nucleus is possible. While this may seem straight forward, this idea can prove to be very confusing. Consider the molecule *cis*-bis(triphenylphosphine)platinum(II) ethylene-1,2-dithiolate (Figure 9) in which there are four different NMR active nuclei, ^{1}H, ^{13}C, ^{31}P and ^{195}Pt. (5) The $^{31}P\{^{1}H\}$ spectrum of this molecule looks very similar to the spectrum presented in Figure 6; it is a singlet at 18.64 ppm with ^{195}Pt satellites. There are four peaks in the ^{1}H NMR; three peaks are in the range of 7.15-7.60 ppm for the aromatic protons and one peak at 6.62 ppm for the ethylene protons. The $^{13}C\{^1H\}$ spectrum has five peaks, four for the aromatic carbon atoms and one for the ethylene carbons. The last spectrum to consider is the ^{195}Pt spectrum which is most relevant to this discussion. If the platinum nucleus were only coupled to the two equivalent phosphorus atoms, the signal would be a 1:2:1 triplet. However, the platinum nucleus is also coupled to the ethylene protons, thus splitting the triplet further into a triplet of triplets with peak intensities of 1:2:1:2:4:2:1:2:1. The larger splitting is attributed to coupling to phosphorus ($^1J_{P-Pt}$ = 2876 Hz) and the smaller to proton ($^3J_{H-Pt}$ = 83 Hz). These coupling constants would be the same in the spectra of the other nuclei; the $^{31}P\{^1H\}$ spectrum would be a singlet with ^{195}Pt satellites ($^1J_{P-Pt}$ = 2876 Hz) while the ^1H signal would have ^{195}Pt satellites with $^3J_{H-Pt}$ = 83 Hz.

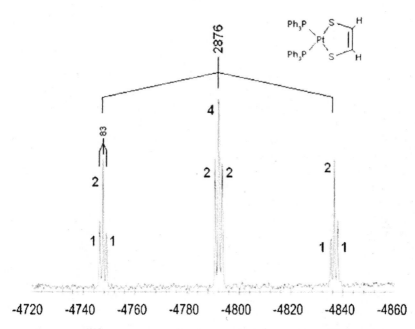

Figure 9. ^{195}Pt spectrum of cis-bis(triphenylphosphine)platinum(II) ethylene-1,2-dithiolate at 64.5 MHz. (Reproduced from reference 5. Copyright 1985 American Chemical Society.)

In the ^{195}Pt spectrum of cis-bis(triphenylphosphine)platinum(II) ethylene-1,2-dithiolate, the platinum is coupled to two different nuclei (^1H and ^{31}P), both of which are spin ½ and 100% natural abundant. If the discussion focuses on spin ½ systems, there are two additional possibilities where coupling to two different nuclei is observed. The first is when one nucleus is 100% abundant and

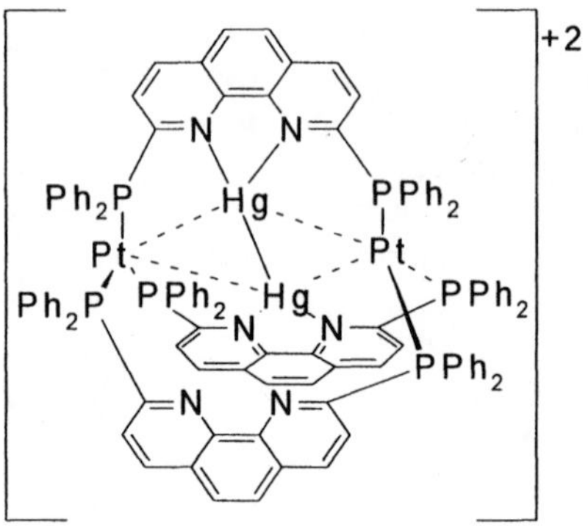

Figure 10. $[Pt_2Hg_2(P_2phen)_3]^{2+}.(6)$

the second nucleus is not. This type of situation was addressed with the $^{31}P\{^1H\}$ spectrum of ((R)-(-)-1-[(S)-2-(Diphenylphosphino)ferrocenyl]ethyl-Di-t-butylphosphine)platinum dichloride (Figure 7) in which a given phosphorus atom couples to the other phosphorus atom (100%) and the platinum (33.8%). The final possible situation is when the nucleus of interest is coupled to two different nuclei, both of which have natural abundances less than 100%. One example of this occurs in the $^{31}P\{^1H\}$ spectrum of $[Pt_2Hg_2(P_2phen)_3][PF_6]_2$ (Figure 10). (6)

While this complex ion appears asymmetric, there is an exchange process in which Hg-N bonds are broken and formed so that each mercury atom spends time bonded to each P_2phen ligand. This process is fast on the NMR time scale, making all three of the P_2phen ligands, and therefore all six of the phosphorus atoms in those ligands, equivalent. Due to this equivalency, the $^{31}P\{^1H\}$ spectrum contains two signals, one for the complex ion and the second for the PF_6^-. The signal for the PF_6^- looks similar to the spectrum shown in Figure 5a while the signal for the complex ion is far more complex. The phosphorus atoms of the P_2phen ligands can couple to ^{195}Pt (33.8%) and/or ^{199}Hg (16.8%). To rationalize the observed pattern the isotopic composition of $[Pt_2Hg_2(P_2phen)_3]^{2+}$ must be analyzed. In 33.8% of the complex ions there will be a ^{195}Pt nucleus; this will give rise to the anticipated ^{195}Pt satellites. Normally, it would be expected that 16.8% of the complex ions would contain a ^{199}Hg nucleus. However, in this case there is actually a 33.6% chance that a given ion contains a ^{199}Hg nucleus. This is due to the fast exchange process involving the two equivalent mercury atoms. Each mercury atom has an equal, but independent, chance of being ^{199}Hg and therefore the possibility of the ion containing one ^{199}Hg nucleus is 33.6%. Of the 33.8% of the complex ions that contain a ^{195}Pt nucleus, 33.6% (or 11.4% of the total signal) will also contain a ^{199}Hg nucleus. This leaves 32.6% of the complex ions in which there is neither a ^{195}Pt nor ^{199}Hg. The spectrum (Figure 11) reflects all four of these possibilities. (6) The central peak (44.2 ppm) shows no coupling to any active nuclei and accounts for 32.6% of the total signal. The ^{195}Pt satellites account for 33.8% of the total signal. Of that 33.8%, 66.4% occurs in the peaks at 45.0 and 41.4 ppm which are ions that do not have a ^{199}Hg nucleus; the remaining 33.6% is contained in the peaks at 45.2, 44.7, 41.6 and 41.1 ppm. The ^{199}Hg satellites comprise the remaining 33.4% of the total signal. A total of 66.2% of that 33.4% is found in the peaks at 43.4 and 42.9 ppm in which there is not a [195]Pt nucleus. The remaining 33.8% is found in the peaks at 45.2, 44.7, 41.6 and 41.1 ppm.

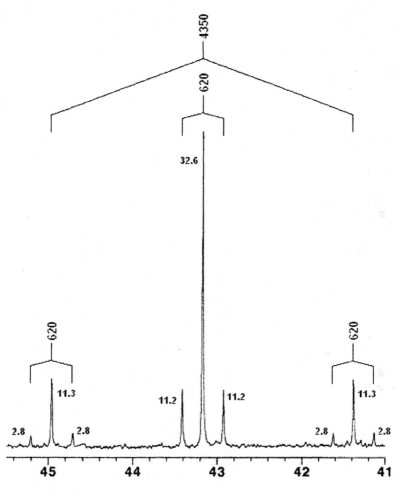

Figure 11. ^{31}P{^{1}H} NMR (P$_2$phen region) of [Pt$_2$Hg$_2$(P$_2$phen)$_3$][PF$_6$]$_2$ in CD$_3$CN at 121.7 MHz. (Reproduced from reference 6. Copyright 2002 American Chemical Society.)

The [195]Pt and [199]Hg spectra (Figure 12) (6) of [Pt$_2$Hg$_2$(P$_2$phen)$_3$][PF$_6$]$_2$ provide additional examples of coupling between different NMR active nuclei where one nucleus is 100% abundant and the other is not. The [195]Pt spectrum is a quartet, due to coupling to three equivalent phosphorus atoms ([1]$J_{P\text{-}Pt}$ = 4350

258

Hz), with ^{199}Hg satellites ($^1J_{Hg-Pt}$ = 1602 Hz). The ^{199}Hg spectrum is a heptet, due to coupling to six equivalent phosphorus atoms ($^2J_{P-Hg}$ = 620 Hz), with platinum satellites having the same $^1J_{Hg-Pt}$ as seen in the ^{195}Pt spectrum. The peaks in the septet are indicated by the splitting on top of Figure 12b. The platinum satellites are a doublet of heptets (the outer most peaks are not intense enough to distinguish from the baseline) and are indicated by the lower splitting pattern in Figure 12b.

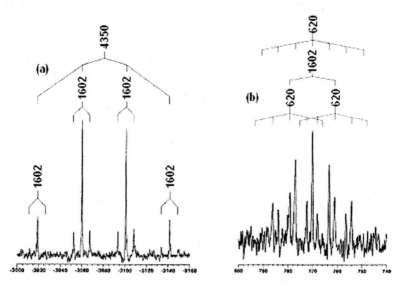

Figure 12. NMR spectra of [Pt$_2$Hg$_2$(P$_2$phen)$_3$][PF$_6$]$_2$: ^{195}*Pt (a) at 106.9 MHz and* ^{199}*Hg (b) at 89.4 MHz. (Reproduced from reference 6. Copyright 2002 American Chemical Society.)*

To this point, the primary focus has been on NMR active nuclei that are not 100% abundant and have a spin of ½. An example of coupling to a nucleus that is not spin ½ has also been presented. Although somewhat rare, there are nuclei that are less than 100% abundant and have spins greater than ½. In these cases, the lessons learned from the previous examples can be combined to explain the spectrum. An example of these combined ideas can be found in the NMR, in particular the ^{14}N and ^{95}Mo spectra (Figure 13), (7) of (η^5-C$_5$H$_5$)Mo(CO)$_2$(NO). Both nuclei are quadrupolar; ^{14}N is spin 1 and 99.6% abundant while ^{95}Mo is 15.7% abundant and has a spin of 5/2. The ^{95}Mo spectrum of (η^5-C$_5$H$_5$)Mo(CO)$_2$(NO) consists of a triplet due to coupling to ^{14}N (Figure 13). (7) The triplet is anticipated to be in a 1:1:1 ratio, however, the authors state that the ideal 1:1:1 triplet could not be observed due to solvent limitations and decomposition of the compound. (7) In the ^{14}N spectrum, 84.3% of the molecules do not contain ^{95}Mo, and therefore, 84.3% of the signal is a singlet which is cut off for scaling purposes. The remaining 16.7% of the signal is split into a 1:1:1:1:1:1 sextet by the ^{95}Mo nucleus. The $^1J_{N\text{-}Mo}$ coupling constant of 46 Hz is seen in both spectra. Due to the width of the central peak and the magnitude of the coupling constant, the middle two peaks of this sextet are obscured by the singlet due to molecules that do not contain a ^{95}Mo nucleus.

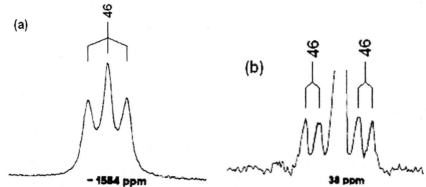

Figure 13. NMR spectra of (η^5-C$_5$H$_5$)Mo(CO)$_2$(NO): ^{95}Mo (a) at 16.3 MHz and ^{14}N (b) at 18.1 MHz. (Reproduced from reference 7. Copyright 1983 American Chemical Society.)

While students are often introduced to NMR in organic chemistry, the usefulness and complexity of NMR spans all chemical disciplines. The literature is full of addition examples of NMR involving nuclei other than ^{1}H and ^{13}C. While many of these nuclei may not give 100%, they present additional means of characterizing compounds and intriguing challenges in interpreting spectra.

Acknowledgement

The authors wish to thank the Kresge Foundation for the purchase of the Jeol Eclipse 400 MHz NMR.

Literature Cited

1. Carey, F. A. *Organic Chemistry, 6th Ed.* McGraw Hill: Boston, MA, 2006; pp 543-578.
2. Ong, J. H. L.; Nataro, C.; Golen, J. A.; Rheingold, A. L. *Organometallics* **2003**, *22*, 5027.

3. Ghent, B. L.; Martinak, S. L.; Sites, L. A.; Rheingold, A. L.; Nataro, C. *manuscript in preparation.*
4. Swartz, B. D.; Nataro, C. *Organometallics* **2005**, *24*, 2447.
5. Keefer, C. E.; Bereman, R. D.; Purrington, S. T.; Knight, B. W.; Boyle, P. D. *Inorg. Chem.* **1999**, *38*, 2294.
6. Catalano, V. J.; Malwitz, M. A.; Noll, B. C. *Inorg. Chem.* **2002**, *41*, 6553.
7. Minelli, M.; Hubbard, J. L.; Christensen, K. A.; Enemark, J. H. *Inorg. Chem.* **1983**, *22*, 2652.

Appendix

Experimental

The following is a general synthetic procedure that we have used to prepare platinum complexes of bidentate phosphines with metallocene backbones. This can be adapted to other bidentate phosphines and used to prepare compounds that can be studied by $^{31}P\{^1H\}$ NMR. The spectra provide examples of coupling to ^{195}Pt which is not 100% naturally abundant. If an asymmetric phosphine, such as Josiphos, Walphos or 1-diphenylphosphino-1'-di-tert-butylphosphinoferrocene is used, the phosphorus atoms will be inequivalent and $^2J_{P-P}$ coupling will be seen in the $^{31}P\{^1H\}$ spectrum. The coupling will be observed in both the main peak and the ^{195}Pt satellites.

Under argon, a 5.0 mL solution of benzene containing approximately 50 mg of bidentate phosphine was added *via cannula* to a 5.0 mL solution of benzene containing 1 molar equivalent of $PtCl_2(MeCN)_2$. The resulting solution was stirred vigorously overnight. Solvent was then removed *in vacuo* typically leaving an orange-yellow product. If the dried product was a solid, it can be used without additional purification. If the resulting product is an oil, the product was dissolved in minimal diethyl ether followed by addition of hexanes which precipitated out the product. The solution was then filtered and the precipitate was dried *in vacuo.*

Problems

1. A series of complexes of the general form (triphos)PtX (triphos = $H_3CC(CH_2PPh_2)_3$; X = CO, PPh_3 or PF_3) have been prepared and characterized by $^{31}P\{^1H\}$ NMR spectroscopy. (*1*) The structure of the molecule is shown in Figure 1 (*note:* all of the phosphorus atoms of the triphos ligand are equivalent. Based on the information provided, sketch the $^{31}P\{^1H\}$ and ^{19}F spectra of these three complexes.

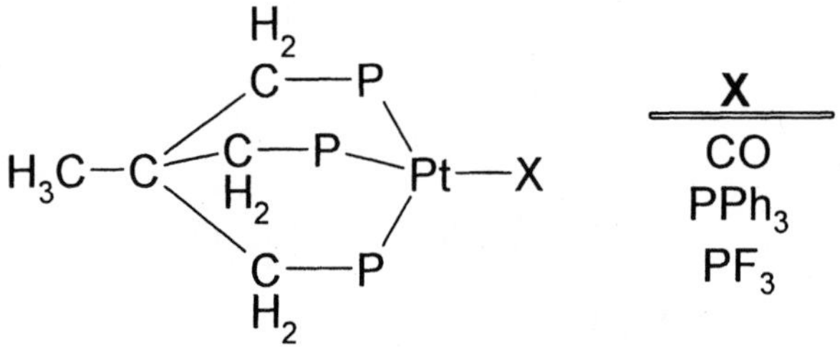

Figure 1. Structure of (triphos)PtX complexes.

Complex	$^1J_{P(tiphos)-Pt}$	$^1J_{P-Pt}$	$^1J_{F-P}$	$^2J_{P-P}$	$^2J_{F-Pt}$	$^3J_{F-P}$
		Coupling constants (in Hz)				
(triphos)PtCO	2837					
(triphos)Pt(PPh$_3$)	3096	5400		51		
(triphos)Pt(PF$_3$)	2867	9500	1111.3	95	895	46.4

2. The molecule shown below (Figure 2) was prepared and characterized by NMR. (2) Based on the information given, sketch the ^{19}F, $^{31}P\{^1H\}$ and $^{103}Rh\{^1H\}$ NMR spectra.

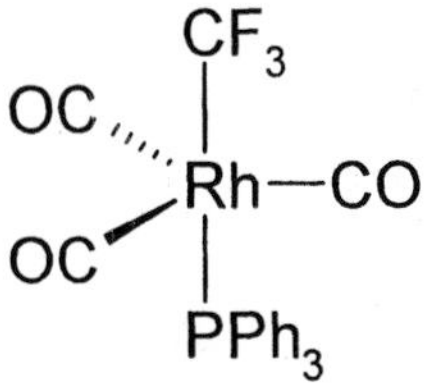

Figure 2. Structure of Rh(CO)$_3$(PPh$_3$)CF$_3$

$^1J_{P-Rh}$ (Hz)	$^2J_{F-Rh}$ (Hz)	$^3J_{P-F}$ (Hz)
	Coupling constants (in Hz)	
69.3	Not observed	62.0

^{19}F	^{31}P	^{103}Rh
	Chemical Shift, δ, (in ppm)	
7.7	32.0	Not reported

3. When the authors initially prepared the complex in problem 2, they were uncertain of the structure of the complex. (2) To fully elucidate the structure, they prepared the ^{13}CO labeled complex and obtained the $^{13}C\{^1H\}$, ^{19}F, $^{31}P\{^1H\}$ and $^{103}Rh\{^1H\}$ NMR spectra at -70 °C. Based on the information provided below, sketch these spectra. For the $^{13}C\{^1H\}$ spectrum, only sketch the signal for the labeled (carbonyl) carbon atoms.

Coupling Constants (in Hz)

$^1J_{C-Rh}$	$^1J_{P-Rh}$	$^2J_{C-P}$	$^2J_{F-Rh}$	$^3J_{C-F}$	$^3J_{P-F}$
71.1	68.9	14.0	8.0	10.3	60.9

Chemical Shift, δ, (in ppm)

^{13}C δ	^{19}F δ	^{31}P δ	^{103}Rh δ
184.3	7.7	31.7	Not reported

4. Shown below (Figure 3) is the structure of *trans*-Pt(D)$_2$(PMe$_3$)$_2$ which was prepared by D.L. Packett and W.C. Trogler. (*3*) Sketch the $^{31}P\{^1H\}$ spectrum for this complex. The coupling constants for this complex are as follow: $^1J_{P-Pt}$ = 2594 Hz and $^2J_{P-D}$ = 20 Hz.

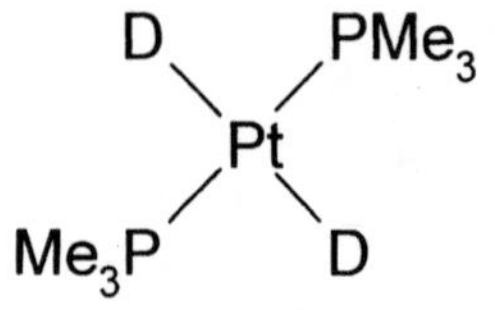

$$\begin{array}{ccc} D & & PMe_3 \\ & Pt & \\ Me_3P & & D \end{array}$$

Figure 3.

Answers

1. The chemical shifts for the complexes were not reported in the original manuscript. However, the patterns and intensities can all be determined.

(triphos)PtCO

The $^{31}P\{^1H\}$ spectrum (Figure 4) is a singlet with platinum satellites. Based on the natural abundance of ^{195}Pt, 33.8% of the phosphorus atoms are coupled to platinum while 66.2% of the phosphorus is not.

(triphos)Pt(PPh$_3$)

In this complex, there are two different environments for the phosphorus atoms, and therefore two phosphorus signals (triphos – Figure 5 and PPh$_3$ – Figure 6). The phosphorus atoms of the triphos ligand are equivalent so the signal is a doublet, due to coupling to the PPh$_3$ phosphorus, with platinum satellites. The satellites also display coupling to the PPh$_3$ phosphorus, so the satellites appear as doublets. The PPh$_3$ signal is a quartet, due to coupling to the three equivalent triphos phosphorus atoms, with platinum

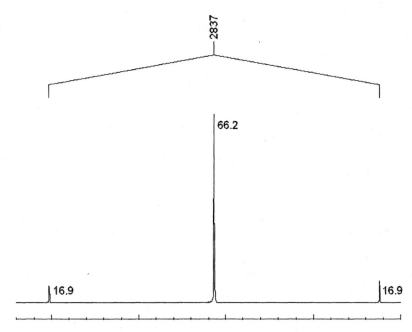

Figure 4. Simulated $^{31}P\{^{1}H\}$ spectrum of (triphos)PtCO.

satellites. In the case of the PPh$_3$ signal, the satellites also exist as quartets due to coupling to the three equivalent phosphorus atoms of the triphos ligand. In each spectrum, the satellites account for 33.8% of the signal while 66.2% of the signal is accounted for in the central doublet or quartet.

(triphos)Pt(PF$_3$)

The signal for the triphos ligand is an overlapping doublet of quartets with platinum satellites (Figure 7). Coupling of the triphos phosphorus atoms to the phosphorus of the PF$_3$ group, splits the signal into a doublet. Each peak of the doublet is then split into a quartet by the three fluorine atoms. The $^{2}J_{P\text{-}P}$ is approximately twice as large as the $^{3}J_{F\text{-}P}$ which results in the direct overlap of two peaks. This give the appearance of a sextet, however the intensities of the peaks are not in the 1:5:10:10:5:1 ratio expected for a sextet. The same pattern is seen for the platinum satellites. As seen for the other complexes in this series, the central set of peaks account for 66.2% of the signal and the satellites account for the remaining 33.8%.

The signal for the phosphorus of the PF$_3$ group is a quartet of quartets with platinum satellites (Figure 8). In this case, the $^{1}J_{F\text{-}P}$ is much larger than the

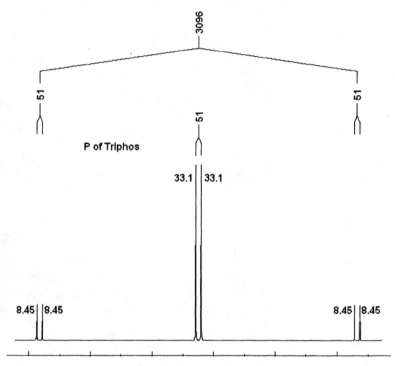

Figure 5. Simulated $^{31}P\{^{1}H\}$ spectrum of the tiphos phosphorus in (triphos)Pt(PPh₃).

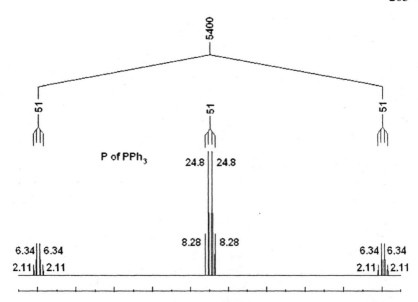

Figure 6. Simulated $^{31}P\{^{1}H\}$ spectrum of the PPh$_3$ phosphorus in (triphos)Pt(PPh$_3$).

^{2}J$_{P-P}$, so there are no overlapping peaks. The larger quartet of quartets accounts for approximately 66.2% of the signal.

The ^{19}F$\{^{1}$H$\}$ spectrum is a doublet of quartets with platinum satellites (Figure 9). The two large quartets account for 66.2% of the signal while the smaller quartets make up the remaining 33.8%.

2. The ^{31}P$\{^{1}$H$\}$ spectrum of this rhodium complex is an overlapping doublet of quartets (Figure 10). In this case, none of the peaks directly overlap. The signal is split into a 1:1 doublet, each peak of which is then split into a 1:3:3:1 quartet.

The ^{19}F$\{^{1}$H$\}$ spectrum is a doublet due to coupling to the phosphorus (Figure 11a). There is no resolved coupling to the ^{103}Rh, but the authors report the ^{19}F peak as a broad doublet. (2) Although the ^{103}Rh spectrum is not reported, the pattern and intensity of the signal can be deduced (Figure 11b). It is assumed that the coupling to fluorine would not be resolved.

The addition of the three ^{13}C carbonyl ligands complicates all of the spectra. The ^{31}P$\{^{1}$H$\}$ spectrum is an overlapping doublet of quartets of quartets (Figure 12). The simulation shows that all of the peaks should be resolved. The intensity of the peaks is based on the signal being split into a 1:1 doublet, which is then split into a 1:3:3:1 quartet, each peak of which is then into a 1:3:3:1 quartet. Overall, this gives eight small peaks (integration of

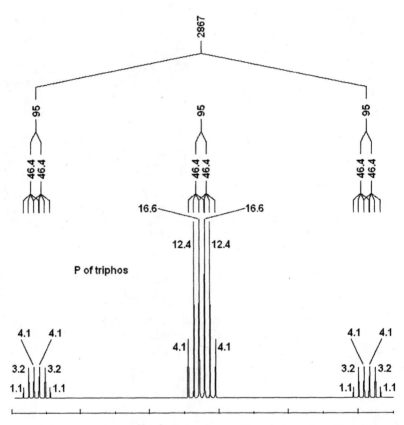

Figure 7. Simulated $^{31}P\{^{1}H\}$ spectrum of the triphos phosphorus in (triphos)Pt(PF$_3$).

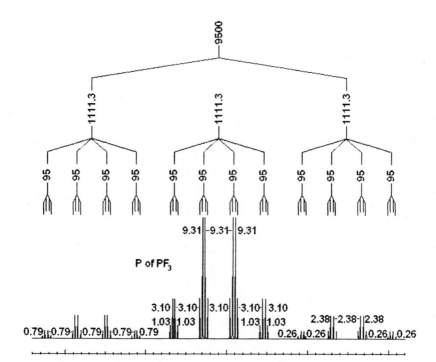

Figure 8. Simulated $^{31}P\{^1H\}$ spectrum of the PF_3 phosphorus in (triphos)Pt(PF$_3$). For clarity, the intensities of the smallest(0.26) and largest (2.38) peaks in the satellites are shown on the right while the intensity of the other peaks (0.79) in the satellites is shown on the left.

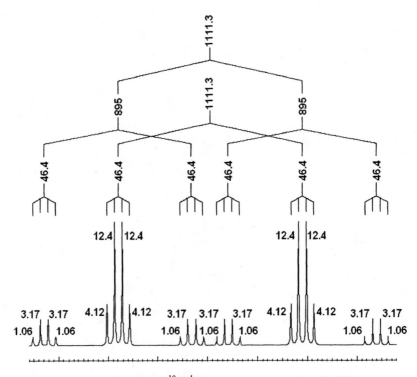

Figure 9. Simulated $^{19}F\{^{1}H\}$ spectrum of (triphos)Pt(PF$_3$).

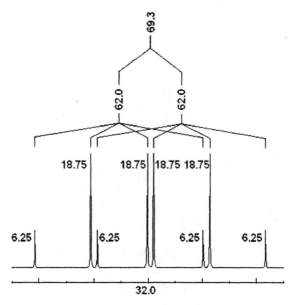

Figure 10. Simulated $^{31}P\{^1H\}$ spectrum of $Rh(CO)_3(PPh_3)CF_3$.

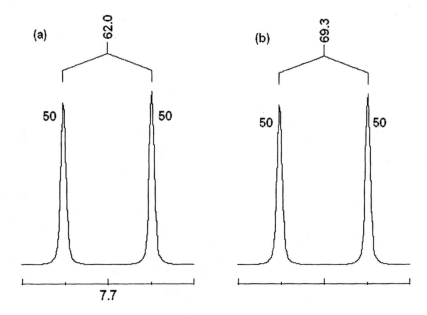

Figure 11. Simulated $^{19}F\{^1H\}$ (a) and $^{103}Rh\{^1H\}$ (b) spectra of Rh(CO)$_3$(PPh$_3$)CF$_3$.

Note: In the original article, the observed coupling constants were not identical when looking at two different nuclei.(2) For example, in the ^{19}F spectrum, the $^3J_{F-P}$ is reported as 61.8 Hz, while in the $^{31}P\{^1H\}$ spectrum the $^3J_{F-P}$ was reported as 62.0 Hz. A single value was chosen for these problems to avoid any potential confusion.

0.79), sixteen medium peaks (integration of 2.34) and eight large peaks (integration of 7.03) .

The $^{13}C\{^1H\}$ is an overlapping doublet of doublet of quartets (Figures 13a). The signal is split into a 1:1 doublet by ^{103}Rh. Each of these peaks is split into a second 1:1 doublet by ^{31}P. Finally, the each peak is split into a 1:3:3:1 quartet which gives eight peaks with a relative integration of 3.12 and eight peaks with a relative integration of 9.38. The $^{19}F\{^1H\}$ is a doublet of quartets of doublets. The initial doublet is due to coupling to phosphorus, the quartet is due to coupling to the three equivalent ^{13}C of the labeled carbonyls and the remaining doublet is due to coupling to ^{103}Rh.

While the ^{103}Rh spectrum was not obtained, the coupling constants that are provided allow for simulation of the spectrum (Figure 14). (2) The signal should be an overlapping quartet of doublets of quartets. In the simulation,

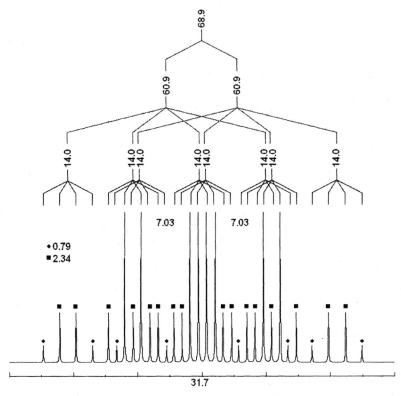

Figure 12. Simulated $^{31}P\{^1H\}$ spectrum of $Rh(^{13}CO)_3(PPh_3)CF_3$.

none of the peaks directly overlap. There are eight peaks with a relative integration of 0.78, sixteen peaks with a relative integration of 2.34 and eight peaks with a relative integration of 7.03.

4. The phosphorus and deuterium atoms are equivalent in this structure, so there is one signal for the phosphorus. It is coupled to deuterium which has a spin = 1. Therefore, the phosphorus signal will be split into a pentet based on $2nI+1$. If the pentet were due to coupling to four equivalent spin $\frac{1}{2}$ nuclei, the intensity of the peaks based on Pascal's triangle would be 1:4:6:4:1. However, this would not be the pattern of the intensities seen due to coupling to deuterium. As previously discussed, the Pascal's triangle for spin 1 nuclei is based on powers of three. For coupling to two deuterium atoms, the sum of the row must be 3^2. This gives a pattern of relative intensities of 1:2:3:2:1. This pattern is also seen in the ^{195}Pt satellites giving the relative intensities shown (Figure 15).

272

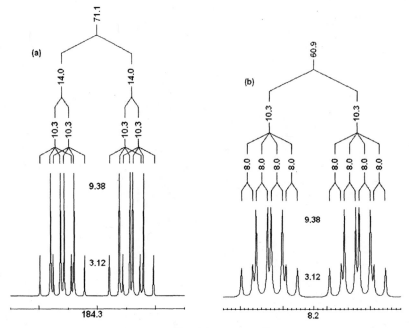

Figure 13. Simulated $^{13}C\{^1H\}$ (a) and ^{19}F (b) spectra of
Rh($^{13}CO)_3(PPh_3)CF_3$.

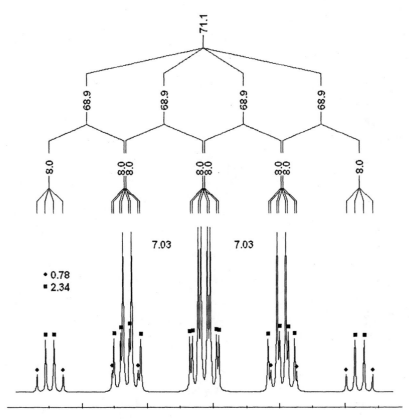

Figure 14. Simulated $^{103}Rh\{^1H\}$ spectrum of $Rh(^{13}CO)_3(PPh_3)CF_3$.

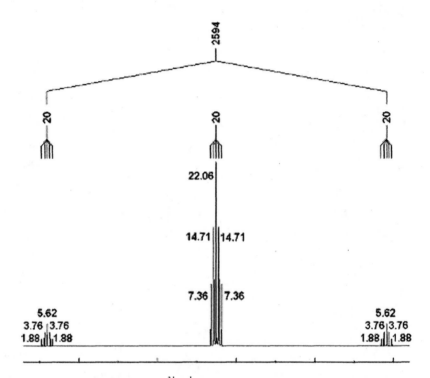

Figure 14.   Simulated $^{31}P\{^1H\}$ spectrum of trans-Pt(PMe₃)₂(D)₂

Note: The actual $^{31}P\{^1H\}$ spectrum of trans-Pt(PMe₃)₂(D)₂ was not reported. (3) The values used for the coupling constant in this example are actually the $^2J_{P-H}$ and $^1J_{P-Pt}$ found for the analogous protium complex, trans-Pt(PMe₃)₂(H)₂. (5) Although the actual coupling constants are anticipated to be different, the actual experimental values are not needed for the purposes of this problem.

Acknowledgement

The authors deeply thank Thao 'Liz' Nguyen for catching a critical mistake a few hours before this appendix was sent to the publisher.

Literature Cited

1. Chatt J.; Mason, R.; Meek, D.W. *J. Am. Chem. Soc.* **1975**, *97*, 3826.
2. Vincente, J.; Gil-Rubio, J.; Guerrero-Leal, J.; Bautista, D. *Organometallics* **2004**, *23*, 4871.
3. Packett, D.L.; Trogler, W.C. *Inorg. Chem.* **1988**, *27*, 1768.
4. All spectra were simulated using WinDNMR. Reich, H.J. WinDNMR: Dynamic NMR Spectra for Windows *J. Chem. Educ. Software* **3D2**.
5. Packett, D.L.; Jensen, C.M.; Cowan, R.L.; Strouse, C.E.; Trogler, W.C. *Inorg. Chem.* **1985**, *24*, 3578.

Chapter 19

Using ^{95}Mo NMR Spectroscopy in the Synthesis and Characterization of Mo(CO)$_{6-n}$(CNR)$_n$ Complexes

Martin Minelli

Department of Chemistry, Grinnell College, 1116 8th Avenue, Grinnell, IA 50112

In the *Advanced Inorganic Chemistry* laboratory students are introduced to different synthetic methods. To become familiar with inert atmosphere synthesis, the students prepare Mo(CO)$_{6-n}$(CNR)$_n$ complexes. ^{95}Mo NMR spectroscopy is used to monitor the reactions and to characterize the products. The characterization also includes ^{14}N NMR, ^{13}C NMR and infrared spectroscopy. An electrochemistry component can be added as well. The experiment illustrates concepts such as ligand π-acceptor properties, the influence of the ligands on Δ$_O$, and the *trans*-effect. The students gain NMR experience working with quadrupolar nuclei at low resonance frequencies.

Upper level undergraduate inorganic chemistry laboratory courses are designed to familiarize students with different synthetic and characterization methods involving inorganic and organometallic complexes. The synthesis of molybdenum carbonyl – isocyanide complexes, $Mo(CO)_{6-n}(CNR)_n$ (equation 1), can be used as an example to familiarize students with inert atmosphere syntheses using Schlenk techniques and a dry box. For the syntheses, a modification of the procedure published by Albers et al. (1) is used. The compounds are air- and moisture-sensitive but can withstand short (less than one minute) exposure to air. This makes them ideal candidates for an introduction to inert atmosphere techniques.

$$Mo(CO)_6 + n\ CNR\ \xrightarrow{PdO} Mo(CO)_{6-n}(CNR)_n + n\ CO \quad \text{(equation 1)}$$

Isocyanides are isoelectronic to carbonyls, which allows the full range of $Mo(CO)_{6-n}(CNR)_n$ to be prepared. Isocyanides with aliphatic R-groups can replace four of the six carbonyl ligands in $Mo(CO)_6$ (1, 2). However, if 2,6-dimethylphenylisocyanide (Figure 1) is employed, all six carbonyl ligands can be replaced. As the substitution proceeds, only cis and fac isomers are observed, illustrating that a carbonyl ligand trans to another carbonyl ligand is replaced faster than a carbonyl ligand trans to isocyanide (trans effect). The mono- to tri-substituted complexes form after about 15 minutes of refluxing in toluene, while the higher substituted complexes need at least an hour or more. Many isocyanides are liquids, but 2,6-dimethylphenylisocyanide is a solid under ambient conditions and easier to handle. Therefore we use it in the laboratory as the ligand of choice. Students work in groups of two and each group synthesizes one of the mono- to tri-substituted complexes and one of the tetra- to hexa-susbtituted complexes.

^{95}Mo NMR spectroscopy is used to monitor the progress of the reactions and to characterize the final product. The students have been introduced to multinuclear NMR spectroscopy in the first week of the course and this experiment gives them some experience in working with quadrupolar nuclei.

Molybdenum has two naturally occurring NMR-active isotopes, ^{97}Mo and ^{95}Mo. Both isotopes have quadrupolar nuclei with a nuclear spin of 5/2. Because of its higher natural abundance (15.72% vs. 9.46%) and lower quadrupole

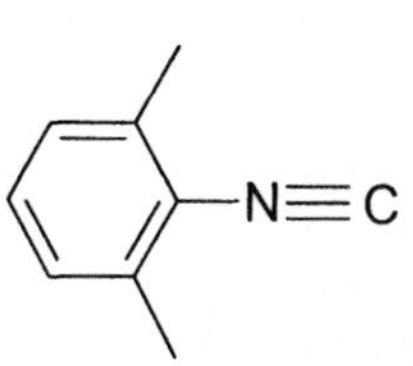

Figure 1. 2,6-dimethylphenylisocyanide

moment (0.12 vs. 1.1 $(Q/10^{-28}m^2)$), ^{95}Mo is the nucleus of choice for molybdenum NMR spectroscopy (*3*). ^{95}Mo NMR spectroscopy provides a direct probe of the metal center.

The known chemical shift range for molybdenum is about 8000 ppm (*4, 5*). Although ^{95}Mo NMR spectroscopy provides a direct probe of the metal center, the chemical shifts of molybdenum complexes with different oxidation states of the metal center often appear in similar regions of the spectrum. However, within a series of related complexes, molybdenum NMR spectroscopy provides a sensitive tool that allows the detection of slight changes within the coordination sphere. For example, when the oxygens in $[MoO_4]^{2-}$ are replaced by sulfurs, a deshielding of the molybdenum nucleus by about 500 ppm accompanies each sequential replacement. (*6*). Substitution of an oxygen by a sulfur in the *N*-salicylidene-2-aminophenol (sapH$_2$) ligand (Figure 2) in MoO$_2$(sap)(DMF) causes a deshielding of the Mo nucleus by about 200 ppm (*7*).

Figure 2. X=O: N-salicylidene-2-aminophenol; X=S: N-salicylidene-2-aminothiophenol

In Mo(VI) imido complexes, changing the position of alkyl substituents on the aryl ring of the imido ligand in Mo(NAr)Cl$_2$(dtc)$_2$ (dtc= diethyldithiocarbamate) complexes results in chemical shifts from -174 ppm to -254 ppm (*8*). A correlation between the ^{95}Mo NMR chemical shift and the angle of the imido linkage could also be shown (Figure 3).

While ^{95}Mo NMR spectroscopy is sensitive to changes in the environment of the molybdenum, the detection of a signal can be difficult. Quadrupolar nuclei (nuclear spin >1/2) have the ability to relax along an electric field gradient as is shown in equation 2 below for the line width, $\Delta v_{1/2}$,of a quadrupolar nucleus (*9*):

$$\Delta v_{1/2} = (\pi T_{2q})^{-1} = (3\pi/10)(2I=3/I^2(2I-1))(e^2 q_{zz}Q/h)^2(1+\eta^2/3)\tau_c \text{ (equation2)}$$

where Q is the nuclear quadrupole moment, q the local electric field gradient and η the asymmetry parameter for q ($\eta= (q_{yy}-q_{xx})/q_{zz}$). This means that molybdenum can relax very fast in complexes with a strong electric field gradient, resulting in

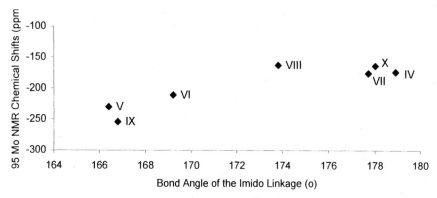

Figure 3. Angles of the imido linkages vs. ^{95}Mo NMR chemical shifts for Mo(NAr)Cl$_2$(dtc)$_2$ complexes (NAr: IV=2,4,6- trimethylphenyl; V= 2,3-dimethylphenyl; VI= 2,4-dimethylphenyl; VII=2,6-dimethylphenyl; VIII= 2,6-diisopropylphenyl; IX= phenyl; X= 2-aminophenyl) (Reproduced with permission from reference 8. Copyright 2002 Elsevier.)

broad line widths. Bulky molecules have long correlation times, τ_c, that will also increase the line width. The choice of the solvent is important; the more viscous the solvent, the longer τ_c. The low resonance frequency of molybdenum (6.516 MHz for ^{95}Mo and 6.653MHz for ^{97}Mo vs. ^{1}H TMS at 100MHz, 3) causes probe ringing. To avoid the effects of probe ringing (rolling baselines), longer pre-acquistion delays are used (Figure 4). These cause the loss of a large part of the signal for a fast relaxing nucleus.

CO and CNR are close in the ligand π-acceptor series. Mo(CO)$_6$ and the isocyanide compounds Mo(CO)$_{6-n}$(CNR)$_n$ are molecules with either no or with only a small electric field gradient. This means that their line widths are narrow (around 10Hz) and they can be easily detected by ^{95}Mo NMR spectroscopy. The ^{95}Mo chemical shifts of the Mo(CO)$_{6-n}$(CNR)$_n$ complexes are listed in Table 1.

The Mo(CO)$_{6-n}$(CNR)$_n$ complexes are synthesized by reacting Mo(CO)$_6$ with stoichiometric amounts of isocyanide in toluene with PdO as catalyst. The students use ^{95}Mo NMR spectroscopy to verify that the reaction is completed. To do this, they cool the reaction mixture so that a 0.5 mL aliquot can be withdrawn with a gas-tight syringe through the side-arm of the Schlenk flask and transferred into a deaereated 5mm NMR tube through a septum. A ^{95}Mo NMR spectrum of the reaction solution can be obtained within a few minutes. The spectra are measured without lock. If only small amounts of the desired product and less substituted complexes are present, the reaction mixture can be refluxed for a longer period of time. Otherwise the reaction mixture can be worked up. Because quadrupolar nuclei can relax faster than dipolar nulcei, it can be pointed out how fast the acquisition proceeds. The importance of the pre-acquistion delay

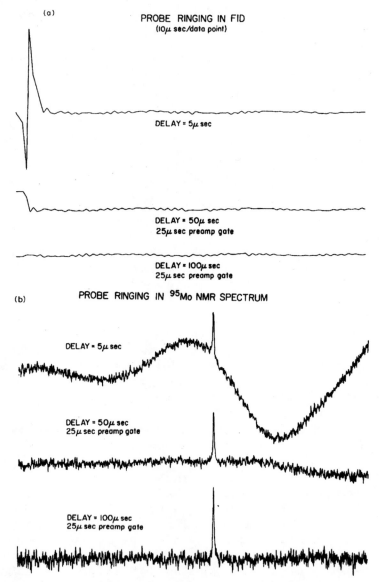

Figure 4. Probe ringing from a modified Bruker WM-250 spectrometer,
MoO$_2$(ONC$_5$H$_{10}$)$_2$ in CH$_2$Cl$_2$ (4); a) FID; b) transformed spectra.
(Reproduced from reference 4. Copyright 1985 American Chemical Society.)

Table 1. ^{95}Mo and ^{14}N NMR Chemical Shifts (ppm) of the Mo(CO)$_{6-n}$(CNR)$_n$ Complexes*

Complex	δ ^{95}Mo	δ ^{14}N
Mo(CO)$_6$	-1856	-
Mo(CO)$_5$(CNR)	-1848	-200 (50)
cis-Mo(CO)$_4$(CNR)$_2$	-1825	-202 (80)
fac-Mo(CO)$_3$(CNR)$_3$	-1786	-202 (100)
cis-Mo(CO)$_2$(CNR)$_4$	-1715	-196 (200)
Mo(CO)(CNR)$_5$	-1626	-203 (330)
Mo(CNR)$_6$	-1525	-197 (350)
CNR	-	-206 (10)

*CNR=2,6-dimethylphenylisocyanide; ^{95}Mo NMR chemical shifts relative to 2M Na$_2$MoO$_4$ in D$_2$O, basic; ^{14}N NMR chemical shifts relative to nitromethane, neat; line widths in Hz in parentheses.
SOURCE: Data from reference 2. Copyright 1989 American Chemical Society.

can also be demonstrated by collecting data with different delays. Shorter pre-acquistion delays will result in a wide range of points at the beginning of the FID. When the data are transformed, a rolling baseline is observed. Since the students synthesize two complexes that need different reaction times, there is enough time for these demonstrations. By the time the second preparation is ready, the students can measure the NMR spectrum on their own.

Figure 5a shows an infrared spectrum of a mixture of the tetra- and penta-substituted complexes. Figure 5b shows the ^{95}Mo NMR spectrum of the same mixture. While infrared spectroscopy is generally a good tool to identify carbonyl complexes, the presence of CO and CN bands in the same region does not allow a diagnostic identification of the components in a mixture.

Once the presence of the product is confirmed, the students remove the solvent from the reaction mixture under vacuum and recrystallize the solid. They characterize the recrystallized products using ^{95}Mo, ^{14}N, ^{13}C NMR spectroscopy and infrared spectroscopy. Like ^{95}Mo, ^{14}N is a quadrupolar nucleus that has a low resonance frequency (7.226 MHz vs. ^{1}H TMS at 100MHz, 3), but it has a high natural abundance (>99%) and the nitrogen from the isocyanide ligand can be detected easily. While there are no significant chemical shifts changes (Table 1), the ^{14}N NMR signals demonstrate how the increasing size of the molecules with increased substitution can increase the correlation time, τ_c, and the line width.

When the groups have obtained NMR data for the whole series of Mo(CO)$_{6-n}$(CNR)$_n$ complexes, they can exchange their data and look at the trends within the series. As the carbonyl ligands in Mo(CO)$_{6-n}$(CNR)$_n$ complexes

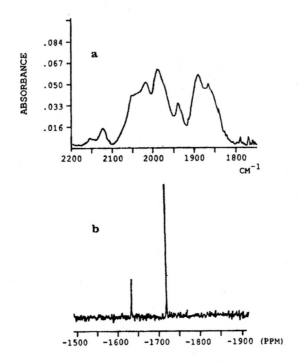

Figure 5. A mixture of Mo(CO)₂(CNR)₄ and Mo(CO)(CNR)₅ in CH₂Cl₂
(R=2,6-dimethylphenyl): a) IR spectrum; b) ⁹⁵Mo NMR spectrum
(Reproduced from reference 2. Copyright 1989 American Chemical Society.).

Table 2. ^{13}C NMR Chemical Shifts (ppm) of the $Mo(CO)_{6-n}(CNR)_n$ Complexes*

Complex	OC-Mo-CO	RNC-Mo-CO	RNC-Mo-CO	RNC-Mo-CNR
$Mo(CO)_6$	201.6			
$Mo(CO)_5(CNR)$	203.6	206.5	165.7	
cis-$Mo(CO)_4(CNR)_2$	206.4	209.5	169.8	
fac-$Mo(CO)_3(CNR)_3$		212.2	173.6	
cis-$Mo(CO)_2(CNR)_4$		215.0	177.1	181.0
$Mo(CO)(CNR)_5$		217.1	180.5	184.8
$Mo(CNR)_6$				185.0

*CNR=2,6-dimethylphenylisocyanide; CDCl₃ was used as an internal reference.
SOURCE: Data from reference 2. Copyright 1989 American Chemical Society.

are replaced by isocyanide ligands, the molybdenum center becomes increasingly more deshielded. Since the π-acceptor capabilities of isocyanides are smaller than those of carbonyls, the color of the complexes changes from white via yellow to orange and red. The chemical shifts of the isocyanide nitrogen change very little (Table 1). The carbonyl carbons and the isocyanide carbon bound to the molybdenum become more deshielded with each substitution (Table 2) just like the molybdenum nucleus.

The laboratory sequence provides an opportunity for the students to become familiar with inert atmosphere synthesis and with the use of multinuclear NMR spectroscopy to monitor a synthesis, to characterize the product and to gain experience with the NMR properties of quadrupolar nuclei. The laboratory can be expanded to include cyclic voltammetry as aryl isocyanides have a good ability to help stabilize the metal center for oxidation states greater than zero (10). The species [Mo(CO)$_4$(CNR)$_2$] are good alkylation catalysts and an investigation to apply these products could be added as well (11).

References

1.   Albers, M.O,; Singleton, E. Coville, N.J. J. Chem. Educ. 1986, 63, 444.
2.   Minelli, M.; Maley, W.J. Inorg. Chem. 1989, 28, 2954.
3.   Brevard, C.; Granger, P. Handbook of High Resolution Multinuclear NMR; Wiley-Interscience: New York, 1981.
4.   Minelli, M.; Enemark, J.H.; Brownlee, R.T.C.; O'Connor, M.J.; Wedd, A.G. Coord. Chem. Rev. 1985, 68, 169.
5.   Malito, J. Molybdenum-95 NMR Spectroscopy, Annual Reports in NMR Spectroscopy 1997, 33, 151.
6.   Lutz, O.; Nolle, A.; Kroneck, P. Z. Naturforsch., Teil A 1977, 32, 505.
7.   Christensen, K.A.; Miller, P.E.; Minelli, M.; Rockway, T.W.; Enemark, J.H. Inorg. Chim. Acta 1981, 56, L27.
8.   Minelli, M.; Hoang, M.L.; Kraus, M.; Kucera, G.; Loertscher, J.; Reynolds, M.; Timm, N.; Chiang, M.Y.; Powell, D. Inorg. Chem. 2002, 41, 5954.
9.   a) Abragam, A. The Principles of Nuclear Magnetism; Oxford University Press, London, 1961, 314. b) Harris, R.K. in NMR and the Periodic Table; Harris, R.K., Mann, B.E., Eds.; Academic Press: London, 1978; 17.
10. Lyons, L.J.; Pitz, S.L.; Boyd, D.C. Inorg. Chem. 1995, 34, 316.
11. Trost, B.M.; Merlic, C.A. J. Amer. Chem. Soc. 1990, 112, 9590.

Supplementary Material

The students work on this experiment during three afternoons. Several tasks need to be carried out prior to the laboratory period (see below). Molybdenum hexacarbonyl and the isocyanide ligand are poisonous and must be handled with care. We store and handle these reagents in a drybox. All manipulations outside the dry box are done in well-ventilated hoods.

The NMR spectra are currently measured on a Bruker 400 MHz DRX spectrometer using a 5mm TBI probehead. Previously, we used a Bruker 300 MHz AC spectrometer with a 10mm broadband probehead. For ^{95}Mo NMR, a 2M Na_2MoO_4 solution basic in D_2O and for ^{14}N NMR, nitromethane, neat, are used as standards. The molybdate standard can be prepared by adding 0.30mL of 50%NaOH and 2.4g of $Na_2MoO_4 \cdot 2H_2O$ to a 5mL volumetric flask and filling it with D_2O. Using the standards, the instrument is tuned and matched for ^{95}Mo and ^{14}N. Since molybdenum has a large chemical shift range that cannot be covered with one collection window, two files, one for the tuning with Na_2MoO_4 and one for the isocyanide complexes are used. A solution of $Mo(CO)_6$ in $CDCl_3$ can also be used for tuning. The chemical shift of the molybdate standard is 0 ppm, the chemical shift of $Mo(CO)_6$ in $CDCl_3$ is -1856 ppm. One collection window is sufficient for the ^{14}N NMR experiment. For ^{13}C NMR, the solvent is used as internal standard .

Schedule

Before the laboratory 1: A Schlenk line with two reflux condensers, two stirrers and two heating mantles should be set up in a hood before the experiment begins. Distilled solvents (toluene dried over Na, methylene chloride dried over P_2O_5 and hexane dried over Na) should be available. The glassware (two Schlenk filters and four 100mL Schlenk flasks with stir bars, two 5mm NMR tubes) needs to be dried in an oven, preferably on the day before the experiment. Around noon on the day of the laboratory, two of the Schlenk flasks with stir bars, two NMR tubes, two 14/20 rubber septa for the Schlenk flasks, two small rubber septa for the NMR tubes and three spatulas should be placed into the antechamber of the dry box. This way, the students can take the equipment into the drybox when the laboratory begins at 1:15pm. If each Schlenk flask is placed into a beaker for support, it is easier to add the reactants. The antechamber is then evacuated for at least 30 minutes. For faster entry into the drybox, three cycles of completely evacuating the antechamber and refilling it with argon can be used.

Laboratory 1 (about four hours): After setting up the Schlenk line (hooking up the trap and cooling it with liquid nitrogen; turning on the argon flow), the glassware is taken from the antechamber into the dry box. Each of the 100-mL Schlenk flasks is charged with $Mo(CO)_6$ (0.52g, 2mmol), the desired number of moles of ligand (it is advisable to add an extra equivalent of ligand for the preparation of the penta- and hexa-substituted compounds) and a spatula tip of PdO. PdO works best of all the catalysts described in reference 1. Adding all the reagents at the same time is different from the original procedure, but it works well and is more convenient than adding the ligand later on the Schlenk line. The Schlenk flasks and the NMR tubes should be closed with the septa (just sticking the septum into the NMR tube and pulling the sides over outside of the dry box is fine).

Outside of the dry box, the Schlenk flasks are hooked up to the Schlenk line via the condenser and 40mL of toluene is added to the flasks through the side arm. The entire set-up is then evacuated and refilled with argon between three times and five times. Now the reaction mixture can be heated. The mono- to tri-substituted complexes should be refluxed for about 15 minutes, the tetra-substituted complex for 30 minutes, the penta- and hexa-substituted complexes for about 90 minutes. After the projected reaction time, the flask is cooled to ambient temperature by taking off the heating device (oil bath or heating mantle). To check if the desired product has been formed, 0.5mL of the solution are withdrawn with a gas-tight syringe and added to the NMR tube. A short needle is briefly inserted into the septum on the NMR tube after the addition to release excess pressure. Then the ^{95}Mo NMR spectrum is measured. If the reaction is not yet completed (only observed for the tetra- to hexa-susbtituted complexes), the heating can resume for for an additional twenty minutes. Otherwise, the solution is filtered through a Schlenk filter. A medium or fine frit should be used, otherwise the PdO, the only solid present, will pass through the filter. The solvent is then evaporated under vacuum and the crude product redissolved in a minimum amount of CH_2Cl_2. Hexane is added until the solution becomes slightly cloudy. The flask is then stored under argon in a freezer. A precipitate generally forms by the next day.

After laboratory 1: The sample in the freezer should be checked for precipitate. If no precipitate forms, more hexane can be added or some of the solvent can be removed.

Laboratory 2 (about three hours): The product should be filtered and dried on the Schlenk line. After about ca. 30-60 minutes of drying under vacuum, the product in the evacuated filter can be taken into the dry box together with a labeled storage vial, a NMR tube and an ampule of $CDCl_3$. The product is weighed in the dry box and stored in the vial. The yields typically range between 40 and 60%. Since the product is light-senstive, the container should be covered or stored away from light. An NMR sample of each compound, as concentrated as possible in $CDCl_3$, should be prepared and the ^{95}Mo and ^{14}N NMR spectra should be measured immediately followed by ^{13}C NMR. To detect all of the quarternary carbons of the 2,6-dimethyphenylisocyanide ligand, a collection overnight may be necessary. Solutions are typically not stable for long periods of time without special purification of the NMR solvent. A green color indicates decomposition.

Laboratory 3: To measure the infrared spectrum, a ziploc bag is taken into the dry box. The KBr pellet is made and placed into the ziploc bag to take it to the infrared spectrometer. The rest of the period can be used to work up the data.

When all students have their data, the worksheet on the following page can be used for the evaluation.

Worksheet
Mo(CO)$_{6-n}$(CNR)$_n$ Complexes

1) ^{95}Mo and ^{14}N NMR Chemical Shifts (ppm) of the Mo(CO)$_{6-n}$(CNR)$_n$ Complexes

Complex	δ ^{95}Mo	δ ^{14}N
Mo(CO)$_6$		-
Mo(CO)$_5$(CNR)		
cis-Mo(CO)$_4$(CNR)$_2$		
fac-Mo(CO)$_3$(CNR)$_3$		
cis-Mo(CO)$_2$(CNR)$_4$		
Mo(CO)(CNR)$_5$		
Mo(CNR)$_6$		
CNR	-	-206 (10)

What are the trends? Add the line widths for the ^{14}N peaks in Hz.

2) ^{13}C NMR Chemical Shifts (ppm) of the Mo(CO)$_{6-n}$(CNR)$_n$ Complexes

Complex	OC-Mo-CO	RNC-Mo-CO	RNC-Mo-CO	RNC-Mo-CNR
Mo(CO)$_6$	201.6			
Mo(CO)$_5$(CNR)				
cis-Mo(CO)$_4$(CNR)$_2$ ·				
fac-Mo(CO)$_3$(CNR)$_3$				
cis-Mo(CO)$_2$(CNR)$_4$				
Mo(CO)(CNR)$_5$				
Mo(CNR)$_6$				

What are the trends?

3) Why are only cis and fac isomers formed?

4) What are the trends for the colors of the complexes? Where is CNR in comparison to CO in the spectrochemical series?

5) Discuss the analytical value of the IR data in comparison to the ^{95}Mo NMR data.

Chapter 20

^{19}F NMR Spectroscopy as a Characterization Tool for Substituted Ferrocene

E. J. Hawrelak

Department of Chemistry, Bloomsburg University of Pennsylvania, Bloomsburg, PA 17815

The reaction of sodium cyclopentadienide (NaCp) and iron(II) bromide (FeBr$_2$) in refluxing THF for 1 h afforded ferrocene (Cp$_2$Fe). The Cp ligands of ferrocene were lithiated via the reaction between ferrocene and n-butyllithium. The dilithioferrocene product ([C$_5$H$_4$Li]$_2$Fe) was reacted with the fluorinated phenyl compounds, octafluorotoluene (CF$_3$C$_6$F$_5$) and hexafluorobenzene (C$_6$F$_6$) to afford two fluoro-susbstituted ferrocene products, 1,1'-octaflourophenylferrocene ([{CF$_3$C$_6$F$_5$}C$_5$H$_4$]$_2$Fe) and 1,1'-pentaflurorphenylferrocene ([{C$_6$F$_5$}C$_5$H$_4$]$_2$Fe. Both ^{1}H and ^{19}F NMR were used to characterize the compounds.

Introduction

This experiment introduces a class of compounds known as **metallocenes**. Metallocenes are compounds formed from two cyclopentadienyl ligands and a metal ion. The cyclopentadienyl (**Cp**) ligand is a monoanionic ligand with the formula C$_5$H$_5^-$. A common starting material for the preparation of metallocenes is sodium cyclopentadienide, NaC$_5$H$_5$, produced via the reaction of cyclopentadiene, C$_5$H$_6$ with sodium hydride, NaH (eq 1). This simple

$$NaH \ + \ C_5H_6 \longrightarrow NaC_5H_5 \qquad (1)$$

"sandwich" structural motif (**A**, Figure 1) is known for many transition metals.

Among these, the best characterized are M = V, Cr, Fe, Co, Ni, Ru, and Os. Others have been observed or proposed as reactive intermediates.

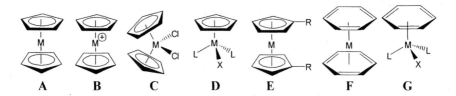

Figure 1. Various Structures of "Metallocenes"

Variations on the metallocene structure are shown in **B** – **G**. The metallocenium ions (**B**) are simply charged metallocene species; examples include M = Fe and M = Co. "Bent" metallocenes (**C**) have been extensively studied with metals of groups 3 – 6, lanthanide elements, thorium, and uranium. Bent metallocenes have applications as catalysts. "Half" metallocenes (**D**) are a huge category that includes any complex with just one cyclopentadienyl ligand. Two especially well studied examples are $CpMn(CO)_3$ and $CpFe(CO)_2X$, where X = Cl, Br, or CH_3. Metallocenes may be ring-substituted (**E**), which changes their physical properties and chemical reactivity. Because of the isoelectronic relationship between Cp and benzene, many of the same structural motifs are found in the compounds formed when benzene bonds to transition metals (**F, G**).

Bonding in Metallocenes

The Cp ligand may bond to metal ions in a variety of ways, η^1 (monohapto), η^3 (trihapto), or η^5 (pentahapto). The most common boding mode for Cp is η^5, for which several different resonance structures can be drawn for the bonding of an η^5-Cp ligand to a transition metal ion (Figure 2). The actual bonding picture of a metallocenes is complex, requiring an analysis of various metal-ligand interactions. The molecular orbital (MO) diagram of ferrocene, which best describes the bonding in a Cp complex, will be used as an example. Consider the three types of MO's of a Cp ligand (20 MO's are possible), zero nodes, one node, and two nodes (Figure 3). Generating all 20 MO's for the Cp ligand is an excellent exercise for students. The mixing of these Cp MO's with the s, p, and d orbitals of iron generate the appropriate molecular orbital diagram for the representative metallocenes, ferrocene. One such interaction is between the d_{zy} orbital of iron and the appropriate group Cp orbital, one of the 1 nodal surfaces (Figure 4). The complete molecular orbital diagram, representing the bonding in

Figure 2. Resonance Structures of Cp Bonding

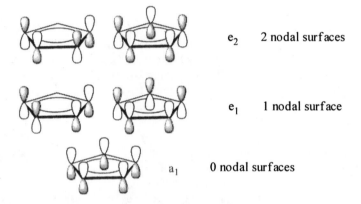

e_2 2 nodal surfaces

e_1 1 nodal surface

a_1 0 nodal surfaces

Figure 3. Molecular Orbital Types for Cp Ligands

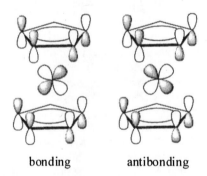

bonding   antibonding

Figure 4. Bonding and Antibonding Orbital Overlap within Ferrocene

ferrocene is shown in Figure 5 (appendix). Generating the MO diagram for Ferrocene is another good exercise for students. More details of the bonding picture can be found in any good inorganic chemistry textbook.

Ferrocene

The synthesis and structural characterization of ferrocene (1,2) was one of a relatively small number of truly surprising discoveries that led to the emergence of organometallic chemistry as a separate discipline with its own guiding principles, empirical knowledge base, practitioners, and applications. The idea of ligand face-bonding was not easily accepted by leading scientists of the day, partly because the initial "proof" of the structure rested heavily on a relatively new technique for the organic chemists of that time -- infrared spectroscopy. Within a few years the molecular structure had been unambiguously determined using X-ray diffraction, and numerous congeners had been prepared, so that even the most skeptical were well satisfied.

The synthesis of ferrocene has been accomplished so many different ways that it would be difficult to know which is best. However, one of the most general and reliable methods of preparing many different metallocene complexes is a simple substitution process, in which halide ligands are replaced by Cp anions. This is the method we will follow in the synthesis of ferrocene (eq 2). In both $FeBr_2$ and Cp_2Fe, iron is considered to have an oxidation state of (+2) and the two ligands are each (–1).

$$2\ NaC_5H_5 + FeBr_2 \longrightarrow (\eta^5\text{-}C_5H_5)_2Fe + 2\ NaBr \qquad (2)$$

Ferrocene is a volatile, air-stable, orange solid, easily purified by sublimation, recrystallization from hexanes, or chromatography on alumina. The best-known reactions of ferrocene are Friedel-Crafts acetylation (eq 3) using aluminum chloride as the catalyst. In fact, the Friedel-Crafts "aromatic" substitution reaction originally led chemists to propose the name "ferrozene" in analogy to benzene, even though most other typical aromatic substitution conditions (bromination, nitration, sulfonation) lead instead to oxidation. The best-controlled of these oxidations uses sulfuric acid (eq 4).

$$Cp_2Fe + CH_3COCl \longrightarrow (\eta^5\text{-}CH_3COC_5H_4)(\eta^5\text{-}C_5H_5)Fe \qquad (3)$$
$$Cp_2Fe + H_2SO_4\ (conc) \longrightarrow [Cp_2Fe]_2SO_4 + H_2 \qquad (4)$$

The oxidation of ferrocene (eq 5) is electrochemically reversible, and its potential can be adjusted by attaching substituents to the Cp ligands. Also, synthetic methods for attaching ferrocene to other organic molecules are well developed. These two facts have enabled organic chemists to prepare substances that have interesting, adjustable, and possibly useful electrochemical properties. This idea probably accounts for most of the ongoing research in ferrocene chemistry today.

$$(\eta^5\text{-}C_5H_5)_2Fe \; + \; e^- \xrightarrow{\hspace{3cm}} (\eta^5\text{-}C_5H_5)_2Fe^+ \quad E^o = 0 \; mV \qquad (5)$$

Experimental Objective

In this sequence, ferrocene (**A** where M = Fe) and two substituted ferrocenes in which each Cp ligand bears one pentafluorophenyl or a perfluorotoluene substituent (**E**, M = Fe, R = C_6F_5 or $CF_3C_6F_4$), will be prepared. These three complexes will be compared using NMR spectroscopy and can also be compared by electrochemistry if desired. This experiment will also introduce ^{19}F NMR. The magnetic properties of ^{19}F are similar to ^1H, resulting in well resolved and easily integrated NMR spectra Table 1.

Table 1. Comparison of Magnetic Properties of ^1H and ^{19}F

Isotope	Natural Abundance	Relative Sensitivity	Frequency 250 MHz	400 MHz
^1H	99.985	1.0	250.00	400.00
^{19}F	100.00	0.83	235.19	376.31

Discussion

While the physical and chemical properties of $\eta^5\text{-}C_5H_5$ metal complexes with varying electron-donating Cp ring substituents are well known (3), complexes with Cp ring substituents of electron-withdrawing character (4) are underrepresented in the literature. The isolation of the Cp ligand as a stable cyclopentadienyl anion, and the stability of the Cp ring substituents toward electrophilic/oxophilic transition metal fragments are two challenges that plague this class of ligands. The pentafluorophenyl (5) and perfluoro-4-tolyl (6) substituted Cp ligands are not hindered by these issues and are straightforward to prepare.

Ferrocene, 1,1'-bis(pentafluorophenyl)ferrocene, and 1,1'-bis(perfluoro-4-tolyl)ferrocene are each prepared via a ligand-substitution procedure (7) (eq 2) with the appropriate ligand sodium salt. Each compound can be isolated as an air-stable reddish-orange solid. While ferrocene remains stable in solution, the pentafluorophenyl- and perlfuoro-4-tolyl-susbstituted ferrocenes will slowly decomposed in solution.

The ^1H NMR spectrum of ferrocene (Figure 6) shows a singlet at 4.16 ppm corresponding the ten equivalent hydrogen atoms on the two Cp rings. The Cp rings can easily "spin" above and below the iron center (barrier to rotation ~1 kcal/mol). The symmetry of ferrocene is lowered by the addition of either

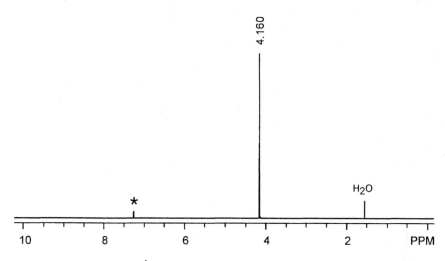

Figure 6. ¹H NMR Spectrum of (C₅H₅)₂Fe in CDCl₃

the pentafluorophenyl or perfluoro-4-tolyl substituent on the Cp ring. The respective ¹H NMR spectra demonstrate this fact, showing two peaks for the two sets of equivalent hydrogen atom on the Cp rings (Figure 7 and 8). While the ¹H NMR spectra provides evidence of a mono-substituted Cp ring, they provide no evidence as to the identity of the substituents. The question that needs to be answered is "did the fluorinated groups remain intact during the synthesis of the ferrocenes?"

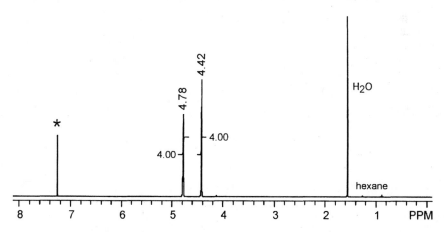

Figure 7. ¹H NMR Spectrum of [(C₆F₅H₄]₂Fe in CDCl₃

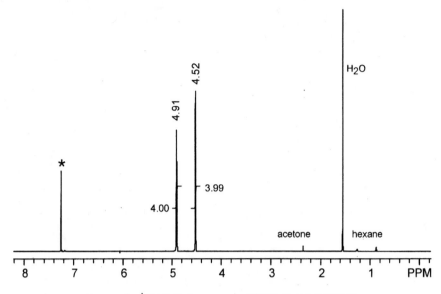

Figure 8. ^1H *NMR Spectrum of* $[(CF_3C_6F_4)C_5H_4]_2Fe$

Analysis of 1,1'-bis(pentafluorophenyl)ferrocene, and 1,1'-bis(perfluoro-4-tolyl)ferrocene using ^{19}F NMR will address the unanswered question. Similar to ^1H NMR and contrary to ^{13}C NMR, ^{19}F NMR peaks can be integrated to determine the number of fluorine atoms corresponding to a particular peak. Also, ^{19}F NMR peaks demonstrate similar splitting patterns as ^1H NMR peaks, following the n+1 rule. This is true when the spectra are proton-decoupled and the only splitting is the result of neighboring fluorine atoms. Each of the spectra reported here are proton-decoupled.

The ^{19}F NMR spectrum of 1,1'-bis(pentafluorophenyl)ferrocene shows three peaks corresponding to the three fluorine environments of the pentafluorophenyl moiety (Figure 9). The ortho-fluorine atoms can be assigned to the doublet at – 140.8 ppm due to the integration of 4 (2 fluorine atoms per ring) and splitting by the meta-fluorine. The multiplet at -159.3 ppm with an integration of 4 fluorine atoms corresponds to four meta fluorine atoms of the two rings. The final peak, a triplet with an integration value of 2 represents the para-fluorine atoms for each ring.

The ^{19} NMR spectrum of 1,1'-bis(perfluoro-4-tolyl)ferrocene also contains three peaks (Figure 10). The ortho-fluorine atoms are assigned to the multiplet at -139.5 ppm and the meta-fluorine atoms to the mutliplet at -143.1 ppm due to the integration values of 4 and the trend that ortho-fluorine atoms are further downfield than meta-fluorine atoms. The trifluoromethyl moiety is represented by the peak at -57.4 ppm based on the integration value of 6.

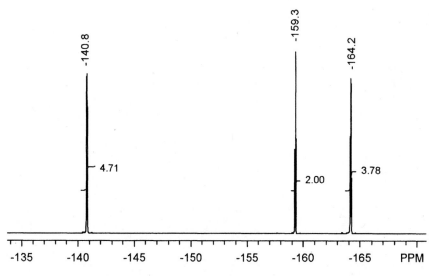

Figure 9. ^{19}F NMR Spectrum of $[(C_6F_5)C_5H_4]_2Fe$ in $CDCl_3$

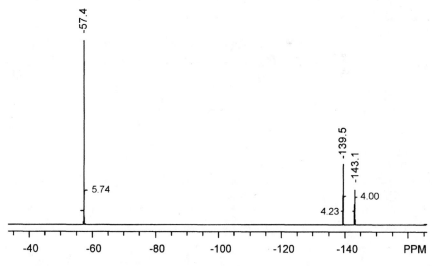

Figure 10. ^{19}F NMR Spectrum of $[(CF_3C_6F_4)C_5H_4]_2Fe$ in $CDCl_3$

296

References:

1. Kealy, T.J.; Pauson, P.L. *Nature*, **1951**, *168*, 1039.
2. Wilkinson, G.; Rosenblum, M.; Whiting, M.C.; Woodward, R.B. *J. Am. Chem. Soc.*, **1952**, *74*, 2125.
3. (a) Möhring, P.C.; Covill, N.J. *Coord. Chem. Rev.* **2006**, *250*, 18. (b) Zachmanoglou, C.E.; Docrat, A., Bridgewater, B.M. *J. Am. Chem. Soc.* **2002**, *124*, 9525. (c) Meier, E. J. M.; Koźmiński, W.; Linden, A.; Lusternberger, P.; von Philipsborn, W. *Organometallics*, **1996**, *15*, 2469. (d) Möhring, P.C.; Coville, N. J. *J. Organometal. Chem.* **1994**, *479*, 1. (e) Maitlis, P. M. *Acc. Chem. Res.* **1978**, *11*, 301. (f) King, R.B. *Coord. Chem. Rev.* **1976**, *20*, 155.
4. (a) Baschky, M. C.; Sowa, J. R., Jr.; Gassman, P. G.; Kass, S. R. *J. Chem. Soc. Perkin Trans 2* **1996**, 213. (b) Herberich, G. E.; Fischer, A. *Organometallics* **1996**, *15*, 58. (c) Schut, D. M.; Weakley, T. J. R.; Tyler, D. R. *New J. Chem.* **1996**, *20*, 113. (d) Hughes, R. P.; Trujillo, H. A. *Organometallics*, **1996**, *15*, 286. (e) Oberoff, M.; Duda, J. K.; Mohr, R.; Erker, G.; Fröhlich, R.; Grehl, M. *Organometallics* **1996**, *15*, 4005. (f) Barthel-Rosa, L. P.; Catalano, V. J.; maitra, K.; Nelson, J. H. *Organometallics*, **1996**, *15*, 3924. (g) Hauptman, E.; Waymouth, R. M.; Ziller, J. W. *J. Am. Chem. Soc.* **1995**, *117*, 11586. (h) Miquel-Garcia, J. A.; Adams, H.; Bailey, N. A.; Mailtis, P. M. *J. Chem. Soc. Dalton Trans.* **1994**, *116*, 385. (i) Lee, I.-M.; Gauthier, W. J.; Ball, J. M.; Iyengar, B.; Collins, S. *Organometallics*, **1992**, *11*, 2115. (j) Wei, C.; Aigbirhio, F.; Adams, H.; Bailey, N. A.; Hempstead, P. D.; Maitlis, P. M. *J. Chem. Soc., Chem. Commun.* **1991**, 883. (k) Lichtenberger, D. L.; Renshaw, S. K.; Basolo, F.; Cheong, M. *Organometallics*, **1991**, *10*, 148. (l) Piccolrovazzi, N.; Pino, P.; Consiglio, G.; Sironi, A.; Moret, M. *Organometallics*, **1990**, *9*, 3908. (m) Burk, M. J.; Arduengo, A. J., III; Calabrese, J. C.; Harlow, R. L. *J. Am. Chem. Soc.* **1989**, *111*, 8938. (n) Finch, W. C.; Anslyn, E. V.; Grubbs, R. H. *J. Am. Chem. Soc.* **1988**, *110*, 2406. (o) Bernheim. M.; Boche, G. *Angew. Chem. Int. Ed. Engl.* **1980**, *19*, 1010.
5. (a) Deck, P.A. *Coord. Chem. Rev.* **2006**, in press. (b) Deck, P.A.; Jackson, W.F.; Fronczek, F.R. *Organometallics*, **1996**, *15*, 5287.
6. (a) Deck, P.A.; McCauley, B.D.; Slebodnick, C. *J. Organomet. Chem.* **2006**, *691*, 1973.
7. (a) Gubin, S. P. *Dokl. Akad. Nauk SSSR Ser. Khim.* **1972**, *205*, 346. (b) *Organomet. Synth.* **1965**, *1*, 64. King, R. B., ed. New York: Academic Press. (c) Newmeyanov, A. N. *Dokl. Akad. Nauk SSSR Ser. Khim.* **1964**, *154*, 646. (d) Fischer, E. O.; Fellmann, W. *J. Organomet. Chem.* **1963**, *1*, 191.
8. (a) Deck, P.A.; Lane, M.J.; Montgomery, J.L.; Slebondnick, C.; Fronczek, F.R. *Organometallics*, **2000**, *19*, 1013. (b) Guillaneux, D.; Kagen, H.B.; *J.*

Org. Chem. **1995**, *60*, 2502. (c) Riant, O.; Argouarch, G.; Guillaneux, D.; Samuel, O.; Kagan, H.B. *J. Org. Chem.* **1998**, *63*, 3511. (d) Dietz, S.D.; Bell, W.L.; Cook, R.L. *J. Organomet. Chem.* **1997**, *546*, 67. (e) Rausch, M.D.; Moser, G.A.; Meade, C.F. *J. Organomet. Chem.* **1973**, *51*, 1. (f) Slocum. D.W.; Englemann, T.R.; Ernst, C.; Jennings, C.A.; Jones, W.; Koonsvitsky, B.; Lewis, J.; Shenkin, P.J. *Chem. Educ.* **1969**, *46*, 145.

Appendix:

Experimental Section

Preparation of Ferrocene, $(\eta^5\text{-}C_5H_5)_2Fe$

A dry 100-mL Schlenk flask is fitted with a stir bar, sealed with a rubber septum, and flushed with nitrogen. The flask is evacuated and transferred to the glove box. Inside the glove box, the flask is charged with 5.0 mmol of $FeBr_2$. Outside the box, the flask is connected to a nitrogen line, and about 50 mL of THF is admitted using a canulla and "two-needle" technique. Then, 11.0 mmol of NaCp is added as a 2.0 M solution in THF (5.5 mL) using a syringe and "two-needle" technique. A condenser is fitted, and the solution is stirred under reflux, under N_2 for about 1 h and then cooled. The solvent is then removed simply by connecting the Schlenk flask to the rotary evaporator (only the reactants are air sensitive -- the product can survive exposure for short periods). Apply the vacuum cautiously to avoid bumping. After evaporating all of the THF, add about 10 mL of hexane to the flask and evaporate again. This is a useful trick for removing the last traces of THF. About 50 mL of dichloromethane is then added to the flask to dissolve the product. The dark residue is then filtered through a short (3-5 cm) column of alumina. You should use another 50 mL portion of dichloromethane to rinse out the flask and make sure all the product is eluted. The eluent is evaporated in a (pre-weighed) round-bottom flask to obtain the ferrocene product. Sublime a small sample (perhaps 20-50 mg -- enough for NMR) and characterize the compound by ^1H NMR in $CDCl_3$ or C_6D_6.

Preparation of 1,1'-dilithioferrocene, $1,1'\text{-}(\eta^5\text{-}C_5H_4Li)_2Fe$[8]

A dry 50 mL Schlenk flask is fitted with a stir bar, sealed with a rubber septum, and flushed with nitrogen. The flask is charged with 3.0 mmol of $(C_5H_5)_2Fe$, 900 μL of freshly distilled tetramethylethylenediamine, TMEDA, and approximately 15 mL of pentane. The solution is cooled in an ice bath and

6.6 mmol of 1.6 M n-BuLi in hexanes is syringed in dropwise. The ice bath is removed and the reaction mixture is stirred for 18 h at ambient temperature. When the stirring is stopped, the orange solid settled to the bottom of the flask, and the solvent is carefully decanted off under a nitrogen purge. The solid is washed with pentane (2 x 20 mL) to remove excess TMEDA and n-BuLi. The dilithioferrocene product can be isolated and stored under an inert atmosphere or it can be used immediately for the preparation of 1,1'-[(C_6F_5)C_5H_4]$_2$Fe or 1,1'-[($CF_3C_6F_4$)C_5H_4]$_2$Fe.

Preparation of 1,1'-Bis(perfluoro-4-tolyl)ferrocene, 1,1'-[η^5-($CF_3C_6F_4$)C_5H_4]$_2$Fe[6]

A dry 100-mL Schlenk flask is fitted with a stir bar, sealed with a rubber septum, and flushed with nitrogen. The flask is charged with 21.0 mmol of perfluorotoluene, $CF_3C_6F_5$ and approximately 20 mL of THF. The solution is cooled in an ice bath and 3.0 mmol of 1,1'-[η^5-C_5H_4Li)$_2$Fe as a slurry in approximately 40 mL of pentane is added using a canulla and "two-needle" technique. The ice bath is removed and the reaction is stirred for 18 h at ambient temperature. The solvent is then removed simply by connecting the Schlenk flask to the rotary evaporator (only the reactants are air sensitive -- the product can survive exposure for short periods). The dark residue is dissolved in a mixture of hot hexane/toluene and filtered through a short (3-5 cm) column of alumina. The solution is transferred to a pre-weighed round-bottom flask and placed in the freezer until the next lab period. Isolation of the deep reddish-orange solid (68% literature yield, 21 % undergraduate student yield) can be achieved via vacuum filtration. Finally, the substituted ferrocene should be characterized by ^{1}H and ^{19}F NMR in $CDCl_3$ or C_6D_6. The NMR spectrum should be obtained quickly as the product will slowly decompose in solution. ^{1}H NMR ($CDCl_3$) δ 4.93 (m, 4H), 4.52 (t, 4H). ^{19}F NMR ($CDCl_3$) δ -55.6 (t, 6F), -138.6 (m, 4F), -142.3 (m, 4F).

Synthetic Procedure for 1,1'-Bis(pentafluorophenyl)ferrocene, 1,1'-[η^5-(C_6F_5)C_5H_4]$_2$Fe[5]

For the substituted ferrocene, ($C_6F_5C_5H_4$)$_2$Fe, the procedure is essentially the same as for 1,1'-[η^5-($CF_3C_6F_4$)C_5H_4]$_2$Fe, a few details are different. First, 21.0 mmol of hexafluorobenzene, C_6F_6, is used instead of octafluorotoluene. The 1,1'-[(C_6F_5)C_5H_4]$_2$Fe is best purified by recrystallization from hexane. Dissolve your sample in a minimal amount of warm hexane, stopper the flask, label it, and place it in the refrigerator until the next lab period. Isolation of the deep reddish-orange solid (88% literature yield, 25 % undergraduate student yield) can be achieved via vacuum filtration. Finally, the substituted ferrocene should also be characterized by ^{1}H NMR, also in $CDCl_3$ or C_6D_6. The NMR spectrum should be obtained quickly as the product will slowly decompose in solution. ^{1}H NMR ($CDCl_3$) δ 4.78 (tt, 4H), 4.42 (t, 4H). ^{19}F NMR ($CDCl_3$) δ-140.67 (d, 4F), -159.17 (t, 2F), -164.07 (m, 4F).

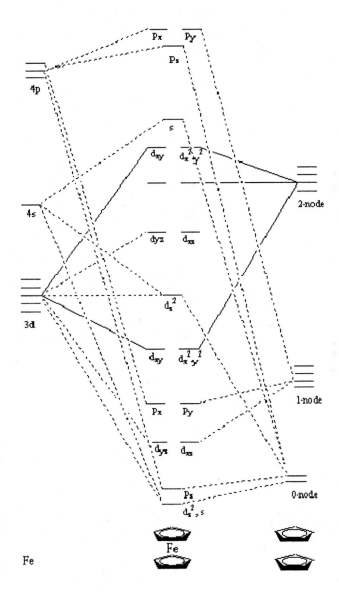

Figure 5. Molecular Orbital Diagram of Ferrocene

Chapter 21

NMR Studies of Polyamide Macrocyclic Tetradentate Ligands and Their Complexes Relevant to Green Chemistry (with Other Inorganic/Organometallic Experiments)

Erich S. Uffelman

Department of Chemistry, Washington and Lee University,
Lexington, VA 24450

Over the past several years, new NMR experiments have been developed in our advanced undergraduate curriculum and our undergraduate research made possible by NSF-sponsored acquisition of a JEOL Eclipse+ 400 MHz multinuclear FT NMR (NSF-ILI) and a subsequent upgrade to an inverse gradient probe (NSF-MRI). NMR work done in class-based and research-based inorganic laboratories at the junior and senior level is summarized.

Context

Over the past six years we have developed several NMR-based experiments across our undergraduate curriculum that include an NMR Smell Module for General Chemistry (*1*), an experiment involving terpene unkowns for organic spectroscopy (*2*), two experiments for an upper level synthetic inorganic course (Chem 252) (*3,4*), and two experiments which link our sophomore level organic course to Chem 252 (*5,6*). This chapter reviews our work geared toward upper level inorganic chemistry classroom and inorganic research experiences.

NOTE: This chapter is adapted with permission from references 1–4. Copyright 2003–2004 W.T. Lippincott.

Undergraduate Inorganic Synthesis Laboratory

Chem 252 is geared for junior and senior chemistry majors headed to graduate school. It has been created as a problem-based learning environment that integrates lab and lecture and builds on foundations laid in lower level chemistry courses. The lab features weekly use of the NMR in tandem with synthetic work. Our Department's contributions at this level involve a novel ring opening metathesis polymerization (ROMP) lab sequence (*6*) and two lab sequences with a unique macrocyclic tetraamide ligand (*3,4*).

ROMP Lab

The Romp lab (Figure 1) uses France's work with Nobel laureate Robert Grubbs's (*9*), and is pedagogically useful in linking Chem 242 (our second semester sophomore organic course) and Chem 252. Students in 242 perform the two step process of (i) running the Diels-Alder reaction of furan with maleic anhydride and (ii) the methanolic acid-catalyzed ring opening reaction of the anhydride to give the diester (*5*). In 252, students take the diester produced by the organic students and perform the ROMP reaction with the commercially available ruthenium carbene (*6*). Students in 252 obtain ^{1}H NMR spectra on the diester, the ruthenium catalyst, and the reaction mixture prior to cleaving the polymer from the catalyst. In addition to reviewing their organic reaction mechanisms and learning ROMP mechanisms, students make the following NMR observations: (i) The line widths of the peaks for the organic monomer are narrower than the line widths for the polymer. (ii) The carbene proton peak for the catalyst comes in a, for them, unexpected region of the NMR spectrum (20.63 ppm). (iii) The carbene peak for the catalyst is different than the carbene peaks for the propagating polymer.

Experiments Illustrating Ligand Principles for Green Oxidation Catalysts

Two experiments (*3,4*) have been developed based on work that led to Terry Collins's Presidential Green Chemistry Challenge Award. These experiments support W&L projects in Green Chemistry (*10*).

The macrocyclic ligands developed by Uffelman in Collins's group (Figure 2), when tetradeprotonated, are powerful sigma electron donors to metal ions. The X groups can be varied to tune the electronic properties, while the R and R' groups can be varied to alter the steric, solubility, and oxidation resistance properties of the complexes. The Y axial ligand is typically a chloride ion or water molecule. The complexes resist oxidative and hydrolytic decomposition *remarkably* well and thus: (i) make it possible to isolate and study unusually high oxidation states of first row transition metals (*11-23*) (ii) permit unusual coordination geometries of the first row transition metals to be studied (*11,12*)

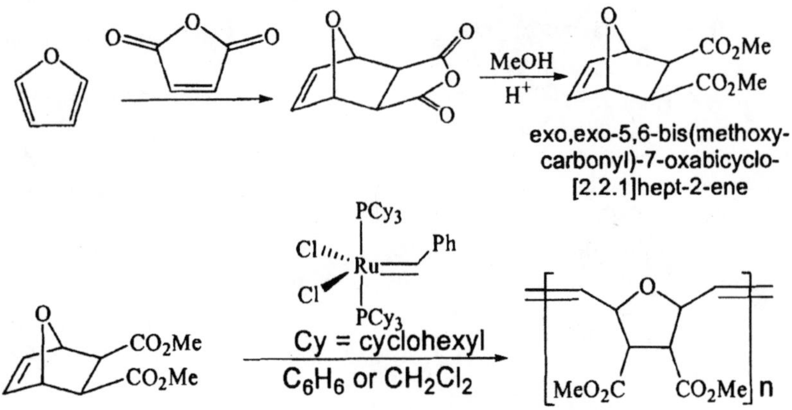

Figure 1. Top: Chem 242 reaction sequence. Bottom: Chem 252 ROMP lab.

(iii) are crucial to new environmentally friendly industrially viable catalytic processes in which the relevant macrocyclic iron complexes activate hydrogen peroxide (21,22) and molecular oxygen (23). These experiments illustrate the remarkable redox and aqueous acid/base stability that make the macrocycles so useful when applied to iron-based Green Oxidations, and the students see classic aspects of square planar d^8 transition metal chemistry via the novel example of Cu(III) (Figure 3), and the classic aspects of low spin octahedral d^6 transition metal chemistry starting from highly unusual square planar intermediate spin Co(III) (Figure 4).

The syntheses of all of the copper (Figure 3) and cobalt (Figure 4) complexes can be accomplished with only 6 mg of $H_4 1$ for each entire sequence, using standard microscale glassware kits and commercially available Schlenck lines as described in the pedagogical literature (3,4); we will focus here on how the NMR aspects of the experiments can be related to many chemical concepts.

Problem-Based Learning Environment

Students are given in-house lab manuals and three papers (19,21,22) on Green Chemistry and high-valent transition metal complexes. The points below are the focus of student-student and student-professor discussions during the labs, and they are also an integral component of the student lab reports.

(1) *Brønsted acid-base chemistry*: Students consider that the pKa of an organic amide is 18-22, the pKa of a protonated tertiary amine is 10-11, and the pKa of diisopropylamine is 36. Students calculate that very little $H_4 1$ is deprotonated by diisopropylethylamine and are asked how it can be efficacious in preparing [Cu(1)]⁻. Students calculate that hydroxide is sufficiently basic to

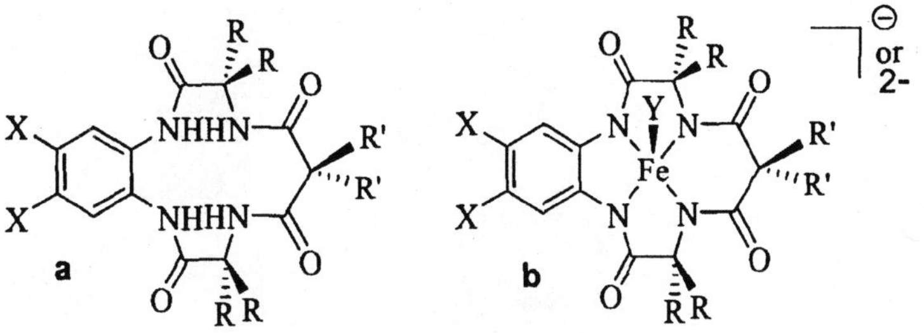

Figure 2. (a) Tetraamide macrocycles protected against oxidative decomposition. (b) Iron complexes that activate hydrogen peroxide for industrially important, environmentally friendly oxidations (21,22).

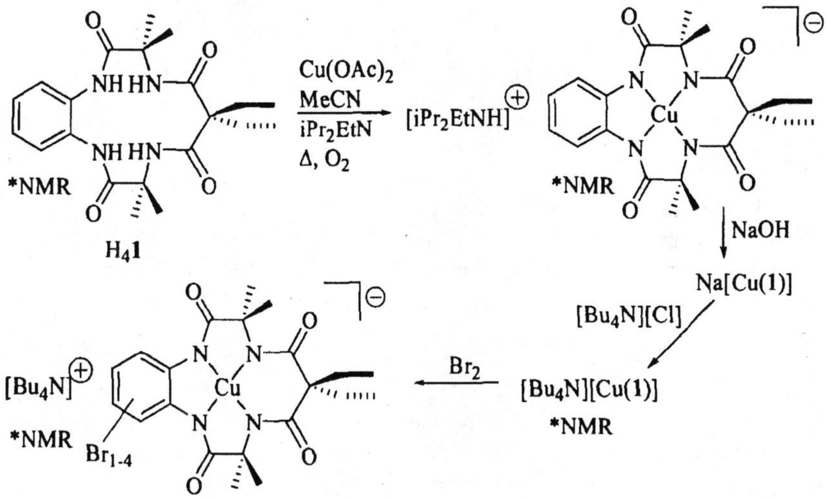

Figure 3. Chemical transformations in the copper experiments.

304

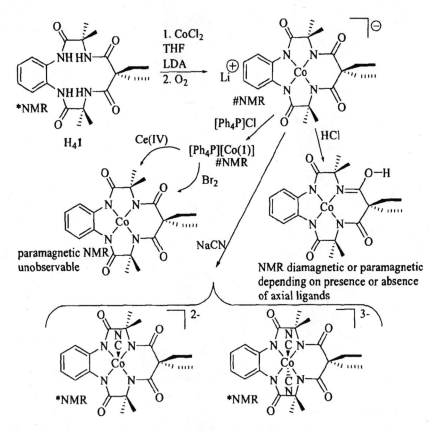

Figure 4. Chemical transformations in the cobalt experiments. Diamagnetic NMR spectra = *NMR. Paramagnetic NMR spectra = #NMR.

deprotonate [Is₂EtNH]⁺. They then decide if LDA can significantly deprotonate H₄1. The students recall the pKa of water and explain why it is excluded from the cobalt insertion reaction. They discuss the protonation of [Co(1)]⁻ with HCl and the deprotonation of H[Co(1)] with NaOH. Students elucidate these proton transfer reactions from the ¹H NMR spectra of the different complexes and H₄1.

(2) *Sterics*: Students are asked if diisopropylethylamine is effective in the copper insertion, why not use NH₃? Students discuss the formation constants for transition metal complexes of ammonia vs. hindered trialkylamines. Students consider in the cobalt pre-lab if NaNH₂ would be a good substitute for LDA in the procedure. They evaluate both attack at the Co²⁺ starting material and attack at the amide carbonyl groups of the macrocycle for NaNH₂ vs. LDA, thus reviewing the organic concept of basicity vs. nucleophilicity.

(3) Redox: ^{1}H NMR spectroscopy is crucial to students following the redox processes in the copper and cobalt experiments. $H_4 1$ and some of the complexes are diamagnetic; some of the complexes are paragmagnetic and NMR observable; some of the complexes are paramagnetic and NMR unobservable. Because $Cu(OAc)_2$ contains Cu(II) and $[Cu(1)]^-$ contains Cu(III), students are asked why a Cu(III) complex is not chosen as a starting material (it is a rare oxidation state for copper), and they are asked what oxidizes the copper during the reaction (O_2). These questions make the students consider "reagents" they might have in a reaction that are not listed in the procedure. The students are also asked to ponder the fate of the superoxide byproduct. Because $CoCl_2$ is Co(II) and $[Co(1)]^-$ is Co(III), students are asked why a Co(III) complex is not chosen as a starting material (it is typically less substitutionally labile, and the Co(III) could oxidize the deprotonated ligand system prior to complexation). Also, given that the color change on exposing yellow-green $[Co(1)]^{2-}$ to air to give dark purple $[Co(1)]^-$ is so dramatic, students readily identify O_2 as the oxidizing agent, but they are asked again to consider the fate of the superoxide byproduct generated. One **key** element of discussion in the lab manual and the lab lecture is the viability or non-viability of a clean oxidation number assignment to the dark blue neutral [Co(1)] species produced by oxidizing purple $[Co(1)]^-$. Students compare X-ray crystal structures of $[Co(1)]^-$ and [Co(1)] provided in the lab material, and they see that the bond distances in the benzene ring of $[Co(1)]^-$ do not vary much, consistent with an unperturbed pi-system, whereas the bond distances in the benzene ring of dark blue [Co(1)] do vary in a manner that is analogous to the bonding pattern of an orthoquinone. [Co(1)] is indefinitely stable in the absence of moisture (moisture does not decompose it, but rather reduces it to H[Co(1)]). Students explain, using the background reading (*19*) the design features of the macrocycle that protect it from decomposition under the different oxidizing conditions present in the copper and cobalt experiments.

(4) *Le Chatelier's Principle*: Connected with the Brønsted acid-base and redox ideas above, the students explain how the irreversible oxidation of the metallated macrocycle from Cu(II) to Cu(III) pulls the reaction to completion. Students can determine the purity of their product by NMR (typically very pure).

(5) *Aqueous transition metal chemistry*: During the Co insertion, if students use excess $CoCl_2$ and LDA they have to filter off black cobalt oxides formed after exposing the reaction mixture to air and moisture. Students must determine what the black solid might be, and how it appeared in the reaction.

(6) *Nonaqueous chemistry*: The Cu insertion does not require anhydrous solvents or drying tubes, but it does not work in aqueous solution. Students consider the fate of Cu(II) dissolved in water or wet organic solvents in the presence of hydroxide (flocculent blue $Cu(OH)_2$ precipitates).

(7) *Solubility product constants*: Students track the different counter ions present with $[Cu(1)]^-$ and $[Co(1)]^-$ using NMR. Students explain the solubility properties of the different salts employed and formed in these experiments based on the "organic" or "inorganic" character of the individual ions.

(8) *Nernst equation*: The cyclic voltammogram of [Ph$_4$P][Co(1)] can be run in CH$_2$Cl$_2$, affording the students experience with this important technique.

(9) *Paramagnetic vs. diamagnetic NMR spectroscopy*: The Cu(III) complexes are all square planar diamagnetic d^8 compounds. However, [Co(1)]⁻ is a square planar d^6 complex with S = 1. Students run the ^{1}H NMR spectrum, but change the following parameters: (a) Sweep width---as a paramagnetic complex, the paramagnetic shift causes the signals to appear between 10 ppm and -50 ppm. (b) Pulse delay---given that T$_1$ is much shorter for the signals in a paramagnetic environment, no pulse delay is needed to prevent signal saturation, thus significantly accelerating the accumulation of scans. (c) Pulse angle--- because T$_1$'s are short, full 90° pulses can be used to increase signal. (d) Number of scans---since paramagnetic line broadening causes noise to interfere more with observation of signal, students are told to run 250 scans on their sample, assuming roughly 2 mg of [Co(1)]⁻ are present. With a 60 ppm sweep width, no pulse delay, and zero filling, excellent spectra are acquired in 3-4 minutes.

(10) *Periodic trends*: Students explain why the d^8 square planar Cu(III) complex brominates on the macrocycle's aromatic ring, rather than oxidatively adding at the metal, as Ir(I) or Pt(II) d^8 square planar complexes might.

(11) *Organic reactivity*: Benzene requires a catalyst to be brominated by Br$_2$, [Bu$_4$N][Cu(1)] brominates on the aromatic ring, and [Ph$_4$P][Co(1)] undergoes electron transfer with Br$_2$. Students deduce the amount of ring bromination of [Bu$_4$N][Cu(1)] using ^{1}H NMR and explain the different reactivities exhibited.

(12) *Spectroscopy*: Students recall the ligand field diagram for d^8 square planar complexes and explain the diamagnetic NMR spectra obtained for the Cu(III) complexes. Most students initially forget they are not preparing a neutral complex and are thus surprised and initially nonplused by the appearance of the protonated trialkylammonium cation signals in the ^{1}H NMR. Also, because students use different amounts of [Bu$_4$N][Cu(1)] and Br$_2$ in their bromination experiment, they obtain different levels of aromatic substitution (one student running this experiment brominated all four available ring positions, causing the aromatic region of her spectrum to disappear). Students thus get different mixtures of compounds with different spectra. This forces students to interpret their own results. Students interpret their paramagnetic ^{1}H NMR of [Co(1)]⁻ by using the integral values. Students discover they have to ignore much taller peaks from diamagnetic impurities in order to correctly interpret the spectrum. Once CN⁻ is added, the students can form the diamagnetic [Co(1)(CN)]$^{2-}$, [Co(1)(CN)$_2$]$^{3-}$, or both. Students explain how the strong field cyanide ligand changes the spin state of the complexes. Because [Co(1)(CN)]$^{2-}$ has C$_s$ symmetry and [Co(1)(CN)$_2$]$^{3-}$ has C$_{2v}$ symmetry, and because students make different mixtures of both complexes, the different spectra obtained again generate useful discussion and foster more independent thinking.

(13) *IR and UV-vis spectroscopies*: Students make many important observations about the different complexes in their different oxidation states by comparing the IR and UV-vis characteristics of the complexes and H$_4$1.

Current Research

Our faculty-led student research at W&L often requires NMR for synthesis (*24,25*) and kinetics (*26*). In addition, collaborations with groups at Stanford and the University of Virginia have generated measurements of zirconocene dynamic processes relevant to polymerization processes using the powerful $T_{1\rho}$ method (*27*). In this concluding section, we focus on our recent research that has started impacting our teaching labs and promises to impact them further soon.

Potential New Green Oxidation Catalysts

The first derivative of $H_3 2$ (Figure 5) was preliminarily reported (*28*) along with preliminary studies with iron (*29*). The synthesis of $H_3 2$ (*30*) is being adapted as a multi-week experiment in our Chem 242 lab (*31*) because it can be performed using common second semester sophomore organic reactions. The iron complexes continue Uffelman's ongoing program in Green Chemistry oxidation catalysis (note the relationship to complexes such as $[Fe(1)(Cl)]^{2-}$ shown in Figure 2). The synthesis of the iron complexes (*30*) is being adapted to the Chem 252 lab. With very labile axial water or axial chloride ligands, iron(III) products of 2^{3-} are either intermediate or high-spin d^5 complexes that give extremely broad paramagnetically shifted peaks, of which only a few are observable. Upon addition of cyanide, the $[Fe(2)(CN)_2]^{2-}$ complex gives classic low-spin d^5 narrow-line paramagnetically shifted ^1H NMR spectra (*32,33*) in which all peaks can be observed, and the R-group alkyl peaks can be assigned by integration values (Figure 6). The new complexes of $H_3 2$ are currently being tested for their catalytic efficacy.

Applications of HMQC and HMBC to Inorganic Complexes

It will become readily apparent to the reader that our latest research results with HMQC (Heteronuclear Multiple-Quantum Coherence) and HMBC (Heteronuclear Multiple-Bond Correlation) spectroscopies (*34,35*) will soon be transferable to our undergraduate courses. HMQC and HMBC both correlate coupled heteronuclear spins and employ detection of the high-sensitivity nuclide. The difference is that the instrument parameters in HMQC detect ^1J heteronuclear couplings, while the parameters in HMBC detect $^{2\ or\ 3}J$ heteronuclear couplings. In classical 2D heteronuclear correlation spectroscopies, the protons were pulsed, the evolution and mixing periods transferred spin information to the ^{13}C or ^{15}N nucleus, and then the ^{13}C or ^{15}N nucleus was detected (Figure 7b). The signal to noise ratio, S/N, is given as $S/N = NAT^{-1}B_0^{3/2}\gamma_{exc}\gamma_{obs}^{3/2}T_2^*(NS)^{1/2}$, where N = molecules in sample volume, A = abundance of spins term, T = temperature, B_0 = static magnetic field, γ_{exc} = magnetogyric ratio of the excited nucleus, γ_{obs} = magnetogyric ratio of the

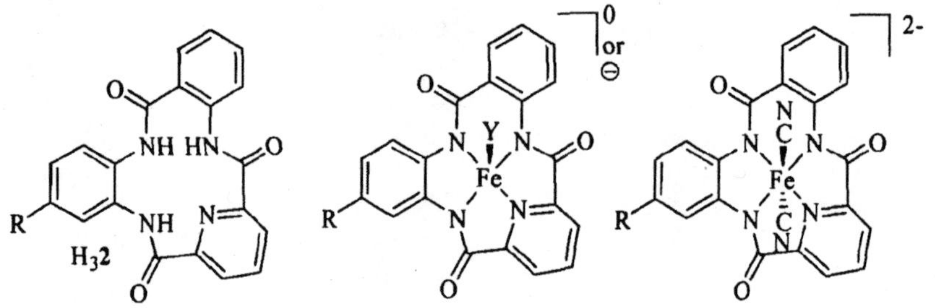

Figure 5. A new macrocyclic triamide ligand resistant to oxidative
decomposition and its iron complexes.

observed nucleus, T_2^* = effective transverse relaxation time, and NS = number of scans. Clearly, exciting and observing the nucleus with the highest magnetogyric ratio will generate significant S/N improvement. In the HMQC or HMBC inverse methods, the protons are pulsed, the evolution period transfers spin information to the ^{13}C or ^{15}N nucleus, but then during the mixing time, that spin information gets transferred back to the protons, and the protons are detected (Figure 7c). Compared to pulsing and detecting the ^{13}C or ^{15}N nucleus, this yields theoretical gains in sensitivity for ^{13}C and ^{15}N by a factor of 32 and 300, respectively (Figure 7a and 7c). Even though the practical gains in sensitivity may be reduced by relaxation and other factors, the inverse methods dramatically increase the sensitivity of heteronuclear correlation spectroscopy, and these gains are multiplied by having more equivalent protons coupled to the heteronucleus.

Two other points are worth briefly mentioning (34,35). First, because these methods detect protons, rather than the heteronucleus, a so-called inverse probe gives the greatest sensitivity. Traditional probes are designed with the X-nucleus receiver coil closest to the sample and the H-nucleus receiver coil furthest from the sample. This is because in conventional spectroscopy on X-nuclei, the X-nucleus was detected directly, and typically had a lower magnetogyric ratio and natural abundance than protons. Hence, the maximum amount of signal was obtained by placing the X-nucleus receiver coil close to the sample, thus sacrificing sensitivity to protons, which are more readily observed. However, since HMQC and HMBC detect protons, it makes more sense to use an inverse probe, in which the proton receiver coils are closer to the sample than the X-nucleus receiver coils. Second, the success of this method depends on gradient pulses. In HMQC and HMBC spectroscopy, it is essential to suppress the hugely dominant ^{1}H-^{12}C or ^{1}H-^{14}N proton signals, so that the ^{1}H-^{13}C or ^{1}H-^{15}N proton signals may be observed, and gradient pulses can achieve attenuation ratios of 1000:1 (35). Phase cycling methods on non-gradient probes are extremely ineffective at this suppression. For non-gradient probes, the so-called "BIRD" (BIlinear Rotation Decoupling) pulse sequence accomplishes pre-saturation of

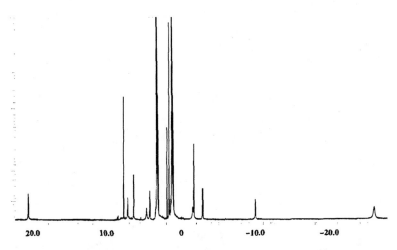

Figure 6. [Et₄N]₂[Fe(CN)₂(2)] (CD₃CN, 400 MHz). Assignments are for the 10 aromatic protons of [2]³⁻ and integrate to 1, unless otherwise noted (in ppm). 20.7, 7.8 (diamagnetic contaminant), 7.3, 6.4, 4.7, 4.3, 3.2 (methylene protons of Et₄N⁺,), 1.95 (CHD₂CN), 1.7 (3H, methyl group of [2]³⁻), 1.3 (methyl protons of Et₄N4⁺), -1.4, -1.7(2H, methylene group of [2]³⁻),-2.9, -9.9, -25.9, -25.9 (the last two peaks are separate in D₂O).

the unwanted ^{1}H-^{12}C or ^{1}H-^{14}N proton signals, beginning with a selective inversion of only those protons. Although BIRD methods sometimes are moderately effective at the needed suppression, they are extremely sensitive to T_1 relaxation values and are not nearly as effective as gradients (*35*). Without ^{1}H-^{12}C or ^{1}H-^{14}N proton signal suppression via gradient pulses, t_1-noise (t_1 here referring to the evolution time in the pulse sequence and not the T_1 relaxation time) can overwhelm the spectrum, as illustrated with examples from our research (Appendix spectra 1 and 2) (*36*). Note the totally unsatisfactory results after 60 h using a conventional probe without gradient suppression versus the outstanding results (Appendix spectra 3 and 4) after 1 h using an inverse probe with gradient suppression on a sample that was less concentrated! We have found in many ^{1}H-^{15}N HMBC experiments that, using 20 mg samples of complexes such as [Bu₄N][Cu(**1**)], we can get excellent resolution of the nitrogens ^{3}J-coupled to the 12 methyl group protons using less than 30 min of instrument time. We are developing ^{1}H-^{15}N HMBC spectroscopy as a means of obtaining ^{15}N chemical shifts in a variety of complexes (Schiff base, cyclam, porphyrin complexes, etc.) to see if the ^{15}N chemical shifts can be correlated with redox potentials, infrared vibrations, Mössbauer isomer shifts, etc. as a means of gauging the electron density at a transition metal center.

310

A final note, the greater proton sensitivity of our inverse probe has, not surprisingly, significantly saved valuable instrument time by giving us better signal to noise in ^1H NMR of our paramagnetic molecules.

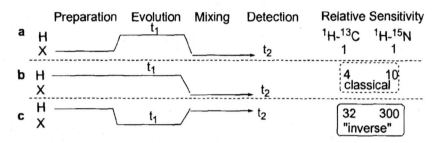

Figure 7. Three methods of obtaining Heteronuclear coupling information (35).

Hazards

Students with pacemakers, metal implants, etc. are kept away from the NMR magnet. Observe standard precautions involving organic solvents and glassware under vacuum. Sodium cyanide is extremely toxic (LD_{50} orally in rats: 15 mg/kg) and reacts with acid to generate highly toxic hydrogen cyanide gas (average fatal dose 50-60 mg); each student is given $\leq$ 1 mg of NaCN. LDA is a very caustic base. Br_2 is a caustic, volatile liquid. See our original *J. Chem. Educ.* supplementary materials and papers for full procedures.

Acknowledgments

ESU thanks the editors and publishers of *J. Chem. Educ.* for permitting adaptation of published material (*3,4*). The National Science Foundation ILI program (Grant No. DUE-9650033) helped fund our JEOL Eclipse+ 400 MHz spectrometer and the MRI program (Grant No. CHE-0319528) funded the inverse gradient probe. ESU thanks the W&L Class of '65 for two Excellence in Teaching Awards that supported development of new NMR experiments across our curriculum. Related work was funded by the Research Corp. (Grant No. CC3870), and the donors of the Petroleum Research Fund, administered by the American Chemical Society (Grant No. 29495-GB3). The Jeffress Foundation (Grant No. J-698) currently supports ESU's projects described in the "Current Research" section. ESU thanks his colleagues and co-authors who appear in the

"References" section. ESU warmly acknowledges ongoing collaborations with Dr. Terrence Collins, Dr. Colin Horwitz, and other members of the Collins group at Carnegie Mellon University. Ashok Krishnaswami of JEOL is gratefully thanked for his extraordinary customer support of our work and instrument. Some of this work has been supported by W&L Glenn Grant, Christian A. Johnson, and R. E. Lee summer stipends.

References

1. Uffelman, E. S.; Cox, E. H.; Goehring, J. B.; Lorig, T. S.; Davis, C. M. *J. Chem. Educ.*, **2003**, *80*, 1368-1371. Our NMR-smell module for General Chemistry extends others' earlier work: see references 7 and 8.
2. Alty, L. T. *J. Chem. Educ.*, **2005**, *82*, 1387-1389.
3. Uffelman, E. S.; Doherty, J. R.; Schulze, C.; Burke, A. L.; Bonnema, K.; Watson, T. T.; Lee, D. W., III *J. Chem. Educ.* **2004**, *81*, 182-185.
4. Uffelman, E. S.; Doherty, J. R.; Schulze, C.; Burke, A. L.; Bonnema, K.; Watson, T. T.; Lee, D. W., III *J. Chem. Educ.* **2004**, *81*, 325-329.
5. France, M. B.; Alty, L. T.; Earl, T. M. *J. Chem. Educ.* **1999**, *76*, 659-660.
6. France, M. B.; Uffelman, E. S. *J. Chem. Educ.* **1999**, *76*, 661-665.
7. Murov, S. L.; Pickering, M. *J. Chem. Educ.*, **1973**, *50*, 74-75.
8. Kegley, S.; Stacey, A. M. "How Do We Detect Odors?" PKAL Conference, Hendrix College; Conway, AR, September, 1995.
9. Schwab, P.; France, M. B.; Ziller, J. W.; Grubbs, R. H. *Angew. Chem., Int. Ed. Engl.* **1995**, *34*, 2039-41.
10. Uffelman, E. S. *J. Chem. Educ.* **2004**, *81*, 172-176.
11. Uffelman, E. S. "Macrocyclic Tetraamido-*N* Ligands that Stabilize High-Valent Complexes of Chromium, Manganese, Iron, Cobalt, Nickel, and Copper," Ph. D. Dissertation, California Institute of Technology, Pasadena, CA, 1991.
12. Collins, T. J.; Uffelman, E. S. *Angew. Chem. Int. Ed. Engl.* **1989**, *28*, 1509-1511.
13. Collins, T. J.; Powell, R. D.; Slebodnick, C.; Uffelman, E. S. *J. Am. Chem. Soc.* **1991**, *113*, 8419-8425.
14. Collins, T. J.; Powell, R. D.; Slebodnick, C.; Uffelman, E. S. *J. Am. Chem. Soc.* **1990**, *112*, 899-901.
15. Collins, T. J.; Kostka, K. L.; Münck, E.; Uffelman, E. S. *J. Am. Chem. Soc.* **1990**, *112*, 5637-5639.
16. Collins, T. J.; Nichols, T. R.; Uffelman, E. S. *J. Am. Chem. Soc.* **1991**, *113*, 4708-4709.
17. Collins, T. J.; Slebodnick, C.; Uffelman, E. S. *Inorg. Chem.* **1990**, *29*, 3432-3436.

312

18. Collins, T. J.; Kostka, K. L.; Uffelman, E. S.; Weinberger, T. *Inorg. Chem.* **1991**, *30*, 4204-4210.
19. Collins, T. J. *Acc. Chem. Res.* **1994**, *27*, 279-285.
20. Bartos, M. J.; Gordon-Wylie, S. W.; Fow, B. G.; Wright, L. J.; Weintraub, S. T.; Kauffmann, K. E.; Münck, E.; Kostka, K. L.; Uffelman, E. S.; Rickard, C. E. F.; Noon, K. R.; Collins, T. J. *Coord. Chem. Rev.* **1998**, *174*, 36 1-390.
21. Gupta, S. S.; Stadler, M.; Noser, C. A.; Ghosh, A.; Steinhoff, B. A.; Lenoir, D.; Horwitz, C. P.; Schramm, K. W.; Collins, T. J. *Science* **2002**, *296*, 326-328.
22. Collins, T. J. *Acc. Chem. Res.* **2002**, *35*, 782-790.
23. Ghosh, A.; Tiago de Oliveira, F.; Yano, T.; Nishioka, T.; Beach, E. S.; Kinoshita, I.; Muenck, E.; Ryabov, A. D.; Horwitz, C. P.; Collins, T. J. *J. Am. Chem. Soc.* **2005**, *127*, 2505-2513.
24. Rorrer, L. C.; Hopkins, S. D.; Connors, M. K.; Lee, D. W., III; Smith, M. V.; Rhodes, H. J.; Uffelman, E. S. *Organic Letters* **1999**, *1*, 1157-1159.
25. France, M. B.; Milojevich, A. K.; Stitt, T. A.; Kim, A. J. *Tetrahedron Lett.* **2003**, *44*, 9287-9290.
26. Holland, M. G.; Griffith, V. E.; France, M. B.; Desjardins, S. G. *J. Polym. Sci., Part A: Polym. Chem.* **2003**, *41*, 2125-2131.
27. Wilmes, G. M.; France, M. B.; Lynch, S. R.; Waymouth, R. M. *Organometallics* **2004**, *23(10)*, 2405-2411.
28. Uffelman, E. S.; Shreves, A. E.; Cartwright, A. R.. Book of Abstracts, 219th ACS National Meeting, San Francisco, CA, March 26-30, 2000 (2000), CHED-714.
29. Uffelman, E. S.; Meyer, M. E.; Coleman, J. T. Abstracts of Papers, 227th ACS National Meeting, Anaheim, CA, March 28-April 1, 2004 (2004), INOR-606.
30. Uffelman, E. S.; Shreves, A. E.; Cartwright, A. R.; Burke, A.; Coleman, J. T.; Reilley, M. J.; Lloyd, J. C.; Meyer, J. J.; Smith, M. V.; Aday, R. M.; Meyer, M. E.; Forstmann, C. L.; McElhaney, A. E. *Organic Letters* to be submitted.
31. Uffelman, E. S.; France, M. B. *J. Chem. Educ.* to be submitted.
32. Bertini, I.; Luchinat, C. In *Physical Methods for Chemists*; Drago, R. S., Ed.; Sanders College Publishing: New York, 1992; Chapter 12.
33. Bertini, I.; Luchinat, C.; Parigi, G. *Solution NMR of Paramagnetic Molecules: Applications to Metallobiomolecules and Models*; Elsevier: New York, 1985; Chapter 5.
34. An outstanding review is: Martin, G. E.; Hadden, C. E. *J. Nat. Prod.* **2000**, *63*, 543-585.
35. An excellent treatment of this material is: Claridge, T. D. W. *High-Resolution NMR Techniques in Organic Chemistry*; Pergamon: New York, 1999; pp 223-251.
36. Uffelman, E. S. to be submitted.

Appendix Spectra and Brief Experimental Procedures for: NMR Studies of Polyamide Macrocyclic Tetradentate Ligands and Their Complexes Relevant to Green Chemistry (with Other Inorganic/Organometallic Experiments)

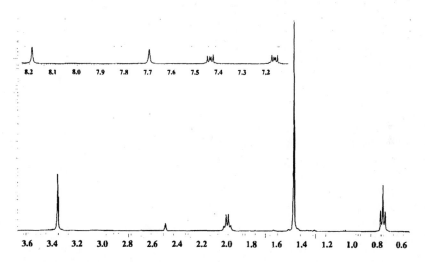

Appendix spectrum 1. ^1H NMR spectrum of H_41 (X = H, R = Me, R' = Et, 47 mg, DMSO-d6, 400 MHz). Assignments (in ppm): 8.19 (s, amide, 2H), 7.70 (s, amide, 2H), 7.43 (m, aromatic, 2H), 7.18 (m, aromatic, 2H), 3.45 (H_2O contamination), 2.45 (DMSO-d6 residual protons), 2.02 (q, ethyl group methylene protons, 4H), 1.57 (s, methyl group protons, 12H), 0.88 (t, ethyl group methyl protons, 6H).

314

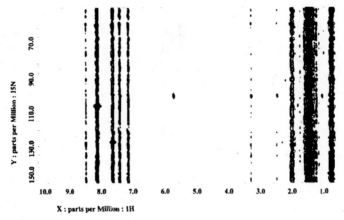

Appendix spectrum 2. ^1H-^{15}N *HMBC NMR of* H_41*(61 mg, DMSO-d6, 400 MHz); 60 h of acquisition time,* **NO** *gradient pulses. The t_1 noise (vertical bands of signals) renders the spectrum uninterpretable. The signals slightly emerging at 106.3 and 126.7 ppm* (^{15}N) *and 7.70 and 8.19 ppm* (^1H) *are not ^3J couplings sought in this HMBC spectrum, but ^1J couplings to the amide protons. Because, complexes of* 1^{4-} *have no amide protons, detecting the ^3J couplings is crucial.*

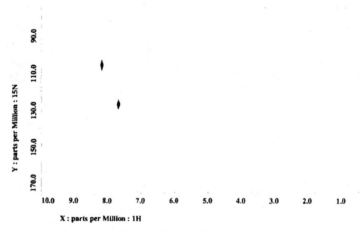

Appendix spectrum 3. ^1H-^{15}N *HMQC NMR of* H_41 *(X = H, R = Me, R' = Et, 47 mg, DMSO-d6, 400 MHz) after 1 h of acquisition time* with *gradient pulses. The settings are for ^1J couplings to the amide protons, and these peaks emerge completely cleanly at 106.3 and 126.7 ppm* (^{15}N) *and 7.70 and 8.19 ppm* (^1H).

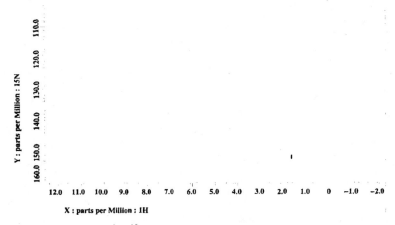

*Appendix spectrum 4. $^1H\text{-}^{15}N$ HMBC NMR spectrum of [Bu$_4$N][Cu(1)] (X = H, R = Me, R' = Et, **25 mg**, DMSO-d6, 400 MHz) after 1 h of acquisition time with gradient pulses. The settings are for ^3J couplings to the amide protons, and the peak representing the coupling of the aliphatic amide nitrogens to the 12 methyl protons emerges completely cleanly at 152 ppm (^{15}N) and 1.57 ppm (^1H).*

Appendix Experimental

This appendix experimental briefly summarizes procedures for making Cu(III) and Co(III) complexes of H$_4$1. For full procedures, please see the papers and their accompanying supplementary material (3,4).

Cu: H$_4$1[3,4] (6-7.5 mg) and anhydrous Cu(OAc)$_2$ (5 mg) are weighed into a 3 mL 14/10 conical vial. A conical mini-stir bar is added, followed by MeCN (0.6 mL) and ethyldiisopropylamine (4 drops). A 14/10 straight tube is attached and the reaction mixture is heated at gentle reflux (50 min) with stirring, while the blue suspension changes to a brown solution. TLC (silica gel/acetone) reveals a brown spot near the solvent front and copper byproducts at the origin. The solution is evaporated to dryness under vacuum and the residue is dissolved in acetone (0.6 mL) and filtered through a 1.5 cm silica gel pad in a glass pipet. Acetone (1-2 mL) is used to flush the brown product through the column. The acetone solution is evaporated to dryness and the residue is washed with hexanes (2 x 1 mL). The brown solid is dried briefly under vacuum, dissolved in CDCl$_3$ (0.8 mL), and filtered into an NMR tube. The CDCl$_3$ product solution is returned to a conical vial, diluted with CH$_2$Cl$_2$ (0.5 mL), and treated with aqueous NaOH (1 M, 1 mL). The brown product extracts into the aqueous layer. The aqueous solution is washed with CH$_2$Cl$_2$ (2 x 1 mL). To aqueous Na[Cu(1)] add a few drops of concentrated aqueous tetrabutylammonium chloride ([Bu$_4$N]Cl). Brown [Bu$_4$N][Cu(1)] is centrifuged, the supernatant is carefully discarded, and the solid is washed and centrifuged twice more with water (2 x 1 mL). Dried [Bu$_4$N][Cu(1)] is dissolved in CDCl$_3$, and the ^{1}H NMR spectrum is obtained. To the CDCl$_3$ solution is added a tiny drop of Br$_2$/CDCl$_3$ solution. If precipitation occurs, dilute aqueous NaOH (2 drops) is added and then shaken until the precipitate goes back into solution. The $CDCl_3$ solution is dried over $MgSO_4$ (if it was treated with aqueous base) and filtered into an NMR tube for a ^1H NMR spectrum.

Co: $H_41^{3,4}$ (6-7.5 mg) and a magnetic stir vane are added to a 3 mL 14/10 conical vial. Students load their vials into the glove box and each adds anhydrous $CoCl_2$ (a few mgs). The vials are removed from the glove box and connected to a N_2 gas manifold by needle and hose. Students take turns syringing dry THF (0.6 mL) into their vials. The solutions are stirred (1 min), and then LDA solution (1.5 M, 0.15 mL) (*Caution! Caustic!*) is syringed into each reaction; the students wear disposable gloves for this step. A yellow-green precipitate forms as the reaction stirs (10 min). The septum cap is removed from the vial, exposing the solution to air, and a dark purple color develops (1 min). After stirring (5 min), TLC (silica gel/acetone) reveals a purple spot moving near the solvent front and cobalt byproducts stuck at the origin. The reaction solution/suspension is filtered through a 1.5 cm silica gel pad in a glass pipet. Acetone (1-2 mL) is used to elute the purple Li[Co(1)] product. The purple acetone/THF solution is evaporated to dryness under vacuum. The purple residue is washed with CH_2Cl_2 (2 x 1 mL) and dissolved in acetone. The solution/suspension is filtered through another 1.5 cm silica gel pad in a glass pipet, using acetone (1-2 mL) to elute Li[Co(1)]. The acetone solution is evaporated to dryness and the purple residue is dissolved in D_2O to give a brown solution. This D_2O solution is filtered into an NMR tube and an NMR spectrum is obtained, making sure that the sweepwidth, pulse delay time, and number of accumulations are modified appropriately for a paramagnetic sample. The D_2O solution of Li[Co(1)] is split into two portions. One portion is left in the NMR tube. The instructor supplies each student with a vial that contains 1 mg of NaCN (*Caution! Toxic! Students are never given access to the NaCN supply!*) which the students dissolve using three or four drops of D_2O, wearing disposable gloves. This $NaCN/D_2O$ solution is added to the Li[Co(1)] solution; only 1-2 drops is sufficient to cause the solution to transform from brown to colorless. Students each take a ^1H NMR of their now diamagnetic solutions.

The other portion of Li[Co(1)] in D_2O is treated with a minimum of gaseous HCl (*Caution!*) generated by adding H_2SO_4 (1-2 mL) to NaCl (75-200 mg) in a micro flask equipped with a gas outlet hose. The HCl gas evolved is added to the D_2O solution of Li[Co(1)], stopping when H[Co(1)] precipitates. Aqueous NaOH is added to dissolve the precipitate and restore the original color. Concentrated aqueous [Ph₄P]Cl is added. The precipitate is washed with H_2O (2 x 0.5 mL). The [Ph₄P][Co(1)] is dissolved in CH_2Cl_2, the purple solution is dried over $MgSO_4$ or Na_2SO_4, and the solution is filtered into two small screw cap vials.

To the first CH_2Cl_2 portion of [Ph₄P][Co(1)] is added a whiff of Br_2 vapor (*Caution!*). Several color changes occur rapidly in succession. If the vapor has been added carefully, the dark intense blue color of square planar neutral Co(1) is transiently observed. To the second portion of [Ph₄P][Co(1)] is added solid $(NH_4)_4[Ce(SO_4)_4]$. Agitation of this solution transiently generates the same dark intense blue color of square planar neutral Co(1).

Permanent Magnet Fourier Transform–NMR

See also Chapters 5 and 7.

Chapter 22

Annotated Bibliography: Permanent Magnet Fourier Transform–NMR Applications in the Laboratory

Amy Abe[1] and Frank B. Contratto[2]

[1]Department of Chemistry, Lake Forest College, 555 North Sheridan Road, Lake Forest, IL 60045
[2]Anasazi Instruments, Inc., 4101 Cashard Avenue, Indianapolis, IN 46203

In this chapter, we present an annotated bibliography of permanent magnet FT-NMR applications in the laboratory. Topics include ^{1}H and ^{13}C spectra of organic molecules, multinuclear spectra of inorganic molecules, spectra of biochemical molecules, spectra of consumer products, MRI, and NMR pedagogy.

For students and educators, NMR spectroscopy is one of many available *tools* to learn chemistry. In a hands-on approach, students see the *unique relationship* between a molecule's *structure* and, in this case, its spectral *properties* (number and intensity of resonances, chemical shift, coupling constants, relaxation.) This article summarizes recent laboratory experiments designed for modern, permanent magnet FT-NMR spectrometers, typically at 60 MHz for ^{1}H or 15 MHz for ^{13}C. The majority of information is presented as abstracted papers and presentations and arranged by molecular entities: organic molecules, inorganic molecules, molecules of biochemical interest and consumer products. In addition we include discussions of educational MRI imaging experiments and introductory NMR concepts for undergraduates and high school students. First, we address the issue of who the users are for this class of low field 60 or 90 MHz permanent magnet, FT-NMR spectrometers.

The customer base for Anasazi Instruments, Inc.,[1] a supplier of spectrometers for the educational market, shows some 300 low field 60 or 90 MHz permanent magnet, FT-NMR spectrometers in use in undergraduate chemistry departments and science high schools, Figure 1. Also commercial, permanent magnet spectrometers developed for monitoring industrial processes[2] or physiology[3] can find pedagogical applications in undergraduate education.[4]

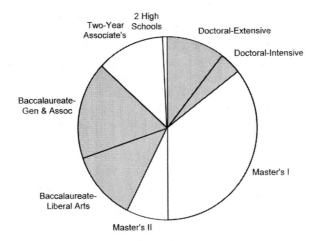

Figure 1. Distribution of Anasazi Instruments, Inc. FT-NMR Spectrometers in Academic Institutions by Carnegie classification.

Chemistry departments that upgrade from ^{1}H continuous wave NMR (CW NMR) spectrometers realize gains in ease of use, throughput and resolution. There is a seamless transition for previously developed laboratory exercises and the larger body of laboratory exercises in the chemical education literature that dates back to the 1960's introduction of Varian's ^{1}H 60 MHz A-60. Institutions that shift undergraduate teaching laboratory experiments from high field superconducting magnet NMR systems recover valuable spectrometer time. Loss of chemical shift dispersion in moving from high field ^{1}H spectra to a FT-NMR spectrometer operating at 1.4 Tesla can often be addressed by including proton decoupled ^{13}C, DEPT, COSY, HETCOR or NOESY spectra.

^{1}H and ^{13}C Spectra of Organic Molecules

NMR Spectroscopy in the Introductory Laboratory[5]

L. J. Kateley, Department of Chemistry, Lake Forest College, Lake Forest, Illinois 60045
Presented at the American Chemical Society 30th Great Lakes Regional Meeting, Chicago IL, May 1997
Nuclei: ^{1}H at 60 MHz[5]
Target audience: general chemistry laboratory

Kateley describes a 2-week introduction to proton NMR. The initial laboratory session introduces chemical shifts and integrals with compounds that present only singlets at 60 MHz. Students use Aldrich Chemical Catalog to find condensed structural formulas for a list of 10 compounds and identify the number of unique hydrogen environments in each molecule. They then predict intensity and chemical shift of the protons resonances. Students acquire spectra of two or three "unknowns" and identify them from the tables they have prepared. In a parallel format, the second laboratory session introduces four common 1st order spin-spin splitting patterns with 10 additional compounds.

Structure and Nuclear Magnetic Resonance. An Experiment for the General Chemistry Laboratory[6]
Rosa M. Dávila and R. K. Widener, College of Southern Idaho, Twin Falls, ID 83301
J. Chem. Educ. **2002,** *79(8)* 997–999
Nuclei: ^1H at 60 MHz, ^{13}C at 15 MHz
Target audience: general chemistry laboratory

During a 2-laboratory exercise, students learn about and interpret simple ^1H and ^{13}C spectra. In the 1st session, NMR features (numbers of peaks, chemical shifts, ^1H multiplets) of C_3H_7O isomers are predicted and used to assign actual spectra. The spectral characteristics of four classes of functional groups are introduced: alcohols, aldehydes, ketones and carboxylic acids. "Unknowns" with methyl and ethyl moieties, are identified in the 2nd session.

A General Chemistry Laboratory Theme: Spectroscopic Analysis of Aspirin[7]
Houston Byrd and Stephen E. O'Donnell, Department of Biology, Chemistry & Mathematics, University of Montevallo, Montevallo, AL 35115
J. Chem. Educ. **2003** *80(2)* 174–176
Nuclei: ^1H at 60 MHz
Target audience: general chemistry laboratory

During a semester of aspirin themed laboratories, students prepare and characterize the structure and purity of their aspirin. In a 2-week exercise, students use IR and ^1H NMR to verify the identity of the salicylic acid reactant and the aspirin they prepared.

The Systematic Identification of Organic Compounds, **8th Edition[8]**
Ralph L. Shriner, Christine K. F. Hermann, Terence C. Morrill, David Y. Curtin, Reynold C. Fuson; Wiley & Sons: New York; 2003
Nuclei: ^1H at 60 MHz, ^{13}C at 15 MHz
Target audience: organic chemistry students

In Chapter 6 of this textbook, Hermann includes several dozen 1.4 Tesla ^1H, ^{13}C, DEPT, COSY and HETCOR spectra to illustrate NMR techniques for molecular structure elucidation. Interpretation of DEPT, COSY and HETCOR are illustrated with spectra of ethyl *p*-aminobenzoate (benzocaine.)

Student-Determined Values for the Calculation of Chemical Shifts of Methylene Protons in Different Chemical Environments[9]

Gary W. Breton, Department of Chemistry, Berry College, Mount Berry, GA 30149-5016

J. Chem. Educ. **2000** *77(1)* 81–83

Nuclei: [1]H at 60 MHz

Target audience: organic laboratory

In an introductory exercise to gain experience with obtaining and interpreting ^{1}H spectra, students examine substituent effects on chemical shifts for a series of 12 *n*-propyl compounds. The substituents exhibit a wide range of electronegativity and anisotropy effects.

Acid-Catalyzed Isomerization of Carvone to Carvacrol[10]

Richard A. Kjonaas and Shawn P. Mattingly, Department of Chemistry, Indiana State University, Terre Haute, IN 47809

J. Chem. Educ. **2005** *82(12),* 1813–1814

Nuclei: [13]C at 15 MHz

Target audience: organic chemistry laboratory

^{13}C NMR spectra are used to verify the changes in molecular structure in the microscale isomerization of spearmint oil (*R*-carvone) to carvacrol. When IR spectra are combined with the NMR spectra, the identity of the product becomes apparent and the authors suggest using the isomerization as a "discovery experiment."

Free Radical Bromination of 2-Methylbutane and Analysis by ^{1}H NMR Spectroscopy[11]

Gary W. Breton, Department of Chemistry, Berry College, Mount Berry, GA 30149

Chem. Educator **2005**, *10(4)*, 298–299

Nuclei: [1]H at 60 MHz

Target audience: organic chemistry laboratory

In a microscale experiment, students photolyze a mixture of 2-methylbutane and Br_2. After predicting the spectral features of the four possible bromination products, ^{1}H NMR is used to identify the product.

Investigating a Reaction of N–Methyltriazolinedione[12]

Gary W. Breton, Department of Chemistry, Berry College, Mount Berry, GA 30149–5016

Chem. Educator **1999**, *4(4)*, 134–136

Nuclei: [1]H at 60 MHz, ^{13}C at 15 MHz

Target audience: organic, advanced organic laboratory

Students are presented with three mechanistic pathways for the addition of 2,3-dimethyl-1,3-butadiene to N-methyl-1,2,4-triazoline-3,5-dione. They isolate the reaction product and use spectroscopic data to choose among the ene, [2+2] and [2+4] bicycloaddition products.

322

Molecular Modeling and Nuclear Overhauser Enhancement Spectroscopy (NOESY): Tools for Studying the Regioselective Bromination of 3-Bromoanisole[13]
Brad Andersh, Department of Chemistry, Bradley University, Peoria, IL 61625, *Chem. Educator* 2000, *5(1)*, 20–23

Nuclei: ^{1}H at 60 MHz, NOESY at 300 MHz
Target audience: organic laboratory

The reaction of 3-bromoanisole with N-bromosuccinimide yields three isomeric dibromoanisoles with one present at >90%. Students use SPARTAN to calculate relative energies of the three sigma-complex intermediates and find that the 3,4-dibromoanisole should predominate under kinetic controlled conditions. NMR is used to confirm the identity of the major product: the 2,3-isomer is eliminated based on splitting patterns of the aromatic protons. Students use NOESY to differentiate between the 3,4- and 2,5-dibromoanisoles.

The Heck Reaction: A Microscale Synthesis Using a Palladium Catalyst[14]
William B. Martin and Laura J. Kateley, Department of Chemistry, Lake Forest College, Lake Forest, Illinois 60045
J. Chem. Educ. 2000, *77(6)*, 757–759

Nuclei: ^{1}H at 60 MHz or 400 MHz
Target audience: organic laboratory

Acrylic acid is coupled with *o*- or *m*-bromoiodobenze in a palladium catalyzed Heck reaction. In principle, the reaction can occur between either end of the alkene and either aromatic halogen. Isolated yields of the major product are in excess of 60% and students are asked to identify the product. The regioselectivity of alkene addition and the ring substitution pattern can be determined from the high field NMR spectrum or with a combination of 60 MHz and IR spectra.

Microscale Synthesis of a Diphenylisoxazoline by a 1,3-Dipolar Cycloaddition[15]
William B. Martin, Laura J. Kateley, Dawn C. Wiser and Catherine A. Brummond, Department of Chemistry, Lake Forest College, Lake Forest, IL 60045
J. Chem. Educ. 2002, *79(2)*, 225–227

Nuclei: ^{1}H at 60 MHz or 400 MHz
Target audience: organic laboratory

The [4+2] cycloaddition of benzonitrile oxide and styrene can produce two products depending on the relative orientation of the aromatic ring of each reactant. The regioselectivity predictions of a SPARTAN comparison of frontier molecular orbital interactions for each reaction pathway are verified by interpretation of the NMR spectrum. In 3,4-diphenylisoxazoline there are two diastereotopic protons with chemical shifts indicative of neighboring oxygen and in 3,5-diphenylisoxazoline there is one proton.

Multi-Week Synthesis of a Sunscreen Agent for the Organic Chemistry Laboratory[16]

Gary W. Breton and Melanie K. Belk, Department of Chemistry, Berry College, Mount Berry, GA 30149

Chem. Educator **2004**, *9(1)*, 27–29

Nuclei: [1]H at 60 MHz

Target audience: organic chemistry laboratory

Octyl 4-methoxycinnamate is prepared in a 3-step procedure over 3 laboratory sessions. Starting with anisyl alcohol, students perform an oxidation, a Wittig and a transesterification in microscale procedures. ^{1}H NMR is used to characterize reactants, intermediates and the product.

The Mosher Method of Determining Enantiomeric Ratios: A Microscale NMR Experiment for Organic Chemistry[17]

Steve Lee, School of Science and Mathematics, Roosevelt University, Chicago and Schaumburg, IL

Chem. Educator **2004**, *9(6)*, 359–363

Nuclei: [1]H at 60 MHz, ^{19}F at 56 MHz

Target audience: organic chemistry laboratory

Lee has adapted Mosher's method for determining enatiomeric purity to the undergraduate laboratory. Students react mixtures of enatiomeric chiral amines or alcohols with Mosher's fluorine containing, enantiopure chiral acid chloride. The resulting mixture of diastereomers is quantified by ^{1}H or ^{19}F NMR. A racemic mixture of the amine or alcohol is also run to establish that the microscale reaction conditions reach completion. Lee has identified three chiral alcohols and an amine that yield diastereomers with resolvable resonances in ^{1}H or ^{19}F spectra at 1.4 Tesla.

Titration of Organolithium Reagents: Handling Air and Moisture Sensitive Compounds in an Organic Experiment[18]

Steve Lee, School of Science and Mathematics, Roosevelt University, Chicago and Schaumburg, IL

Chem. Educator **2005**, *10(5)*, 357–358

Nuclei: [1]H at 60 MHz

Target audience: organic chemistry laboratory

In a 3-week laboratory sequence, students prepare and characterize N-pivolaloyl-*o*-benzylamine or N-pivolaloyl-*o*-toluidine. The secondary amides are identified by ^{1}H and ^{13}C NMR where the amide proton resonance is assigned by H/D exchange. During the 3rd week, students titrate air-sensitive organolithium reagents using their amides as end-point indicators.

Instrumental Proficiency Program for Undergraduates[19]

Duane E. Weisshaar, Gary W. Earl, Milton P. Hanson, Arlen E. Viste, R. Roy Kintner and Jetty L. Duffy-Matzner, Department of Chemistry, Augustana

College, Sioux Falls, SD 57197
J. Chem. Educ. **2005,** *82(6),* 898–900
Nuclei: ^1H at 60 MHz, ^{13}C at 15 MHz
Target audience: chemistry majors

Weisshaar and colleagues developed 1-credit courses that train students in the critical evaluation of data from each of six techniques: GC/MS, HPLC, FTIR, NMR, Raman, AA and UV/vis. NMR course topics include sample preparation, FID shimming, relaxation measurements and techniques for identifying organic molecules.

Incorporation of FT-NMR throughout the Chemistry Curriculum[20]
D. Scott Davis and Dale E. Moore *J. Chem. Educ.* **1999,** *76(12)* 1617–1618
A Research-Based Sophomore Organic Chemistry Laboratory[21]
D. Scott Davis, Robert J. Hargrove and Jeffrey D. Hugdahl, *J. Chem. Educ.*
1999, *76(8)* 1127–1130
Department of Chemistry, Mercer University, Macon, GA 31207

Nuclei: ^1H at 60 MHz and a VT, multinuclear 7 T spectrometer
Target audience: chemistry departments

Davis and colleagues chronicle their success in integrating a permanent magnet FT-NMR and a higher field superconducting magnet FT-NMR into structured and project based laboratory courses by combining a series of innovative and literature experiments. As examples, in the 1st year course, students examine the effect of halogen electronegativity on alkane chemical shifts with the permanent magnet spectrometer. Midway through the sophomore organic course, students use the ^1H 60 MHz with ^{13}C, HETCOR and NOSEY on the higher field spectrometer to characterize products in a multi-step reaction sequence projects. At the advanced level, students measure activation energies in variable temperature kinetics experiments.

Implementation and Use Of Selective Shaped Pulses on Lower-End NMR Spectrometers[22]
Daniel Lim and Guillermo Moyna, Department of Chemistry & Biochemistry, University of the Sciences in Philadelphia, Philadelphia, PA 19104-4495
Journal of Undergraduate Chemistry Research **2004,** *3(3),* 96–102.

Nuclei: ^1H at 90 MHz
Target audience: instrumental analysis, spectroscopy laboratory

Moyna chronicles the development of techniques to generate *shaped pulse* pulse sequences that excite NMR transitions over a narrow 5 Hz window on an Anasazi Instruments, Inc. spectrometer. In a demonstration of the *shaped pulse* pulse sequence, one line of an 18 Hz AX doublet is irradiated. The response of the two lines in the second doublet allows one to determine the sign of the coupling constant.

Multinuclear NMR Spectra of Inorganic Molecules

Multinuclear FT-NMR Using the Anasazi EFT-600 Instrument[23]
S. Moeckly, G.W. Earl, A. Viste, Department of Chemistry, Augustana College
Sioux Falls, SD
Proceedings of the South Dakota Academy of Science, **1999**, *78*, 63–69
Nuclei: ^1H (60 MHz), ^{19}F (56 MHz), ^{31}P (24.3 MHz), ^{11}B (19.2 MHz)
Target audience: undergraduate laboratory

Students are exposed to 1st order splitting from nuclei other than ^1H, including the $I=3/2$ ^{11}B. Compounds were chosen to have two or more coupled NMR active nuclei. Spectra are obtained and analyzed for each nucleus (e.g. $NaPF_6$ where $^1J_{PF} = 714 - 717$ Hz in spectra of the ^{19}F doublet and the ^{31}P septet.)

Infrared Spectroscopy Determination of Lead Binding to Ethylenediamine-tetraacetic Acid[24]
Simona Dragan and Alanah Fitch
Loyola University Chicago, Department of Chemistry, Chicago, IL 60626
J. Chem. Educ. **1998**, *75(8)*, 1018–1021
Nuclei: ^1H at 60 MHz
Target audience: instrumental laboratory

In a series of spectroscopic laboratory experiments, Pb^{+2} is examined with UV-vis, IR and NMR. The focus of this paper is quantitation of PbEDTA by FT-IR. A 60 MHz NMR laboratory experiment is referenced in which the isotopic abundance of ^{207}Pb is measured from the $^3J_{PbH}$ satellites of the proton resonances in PbEDTA.

PbEDTA complexes at 60 and 400 MHz[25]
Presented at Anasazi Instruments, Inc. co-sponsored *FT-NMR Workshops*
Nuclei: ^1H at 60 MHz and 400 MHz
Target audience: inorganic, instrumental analysis laboratory

Interpreting the ^1H spectra of the 1:1 complex of PbEDTA allows students to review a number of NMR spectral assignment concepts. ^{207}Pb has a natural abundance of ~22% and the NMR active, spin ½ nucleus exhibits ~20 Hz three- and four-bond coupling constants to ^1H. The presence of ^{207}Pb satellites of both methylene resonances in ^1H PbEDTA spectra provides evidence that the ligand has 6 coordination sites for Pb^{+2}. The higher field spectra are complicated by the chemical shift difference of the diastereotopic methylene hydrogens and allow students to assign limits to ligand exchange rates. Chemical shift anisotropy effects on NMR relaxation can also be observed.

Spectra of Biochemical Molecules

Di- and tri-peptide "sequencing"[26]
Presented at Anasazi Instruments, Inc. co-sponsored *FT-NMR Workshops*,

Nuclei: ^{1}H at 60 MHz

Target audience: biochemistry, instrumental analysis laboratory

The N-terminal and C-terminal amino acid residues in a small peptide can be identified in 60 MHz spectra. The chemical shift of the α-hydrogen is pH dependent, moving to a more shielded, upfield value when the adjacent protonated amine or carboxylic acid moiety loses a proton. By comparing spectra of dipeptides at pH 3, 7 and 10 students can differentiate the N- and C-terminal residues. A set of COSY spectra will allow sequencing of selected tripeptides.

NMR at Work...
- Titration monitoring of ATP by ^{31}P NMR[27]
Frank B. Contratto, Anasazi Instruments, Inc., 4101 Cashard Avenue, Indianapolis, IN, 46203

Nuclei: ^{31}P at 24 MHz

Target audience: biochemistry laboratory

In this applications note, Contratto describes a titration experiment where ionization is monitored by changes in the chemical shift of ^{31}P resonances. The NMR resonances of three linear phosphates in ATP are distinguishable and can be assigned to the α, β αvδ γ positions from couplings patterns and by decoupling the ribose protons from the a-phosphate. Aqueous solutions (not D_2O) and a conventional pH meter are used to assign the pKa ~ 6 to the γ-phosphate.

Spectra of Consumer Products

Some principles of NMR spectroscopy and their novel application[4]
Shelton Bank, Department of Chemistry, State University of New York at Albany, Albany, NY 12222
Concepts in Magnetic Resonance **1997**, *9(2)* 83–93

Nuclei: ^{1}H at 60 MHz

Target audience: general

Bank is making two points by obtaining spectra of neat samples of beer, hot dogs and other comestibles. He is showing students that these food products can be viewed as simple chemical systems by estimating the ethanol content in beer and identifying the fat, water and protein in hot dogs. And he is demonstrating that messy, real world samples lend themselves to pedagogical exercises. Several of the scenarios Bank describes involve solids and spectrometers designed for CP-MAS or for short spin echo or T_2 experiments.

NMR at Work...
- Determination of the Ethanol Content of Selected Consumer Products by Proton NMR[28]
- Determination of Sodium Content of Selected Consumer Products by Sodium NMR[29]
- Analysis of Edible Oils by ^{1}H and ^{13}C NMR[30]
Frank B. Contratto, Anasazi Instruments, Inc., 4101 Cashard Avenue, Indianapolis, IN, 46203

Nuclei: ^{1}H at 60 or 400 MHz, ^{13}C at 15 or 100 MHz, ^{23}Na at 15.9 MHz
Target audience: non-majors, general chemistry, food or allied health sciences laboratory

These application notes contain data and detailed instructions to construct quantitative or qualitative laboratory exercises for students. Sample preparation consists of adding the neat food or beverage to an NMR tube. Quantitation involves either comparison of integral ratios of oil or ethanol resonances or the preparation of external calibration curves for the sodium.

Ethanol: Ethanol content over a range of concentrations was measured in products such as soy sauces (3-5%), vanilla flavorings (15-35%), cooking wines (12-18%), mouth rinses (20%) and liquor (30-40%). The integrals for the methyl resonance of C_2H_5OH are compared to the water or exchangeable hydrogen resonance. Standards were prepared from 0.1 to 90% ethanol in water to establish the validity of the method.

Edible Oils: The 60 MHz ^{1}H spectra of cooking oils (e.g. corn, canola, olive), specialty seed oils (e.g. walnut, avocado) and triglyceride standards (C18:1, C18:2, C18:3) exhibit well resolved peaks for the unsaturated olefinic protons. The iodine value, a food industry measure of unsaturation, can be calculated from the spectra. The presence of trans fats in partially hydrogenated fats and oils is detected in the ^{13}C spectra.

Sodium: The sodium content of condiments was measured by integrating the ^{23}Na NMR resonance of various soy sauces and Worchester sauces. The sodium concentrations measured by external calibration with NaCl solutions agreed well with stated amounts on the product's nutritional label.

Grocery stores: Sources of NMR samples for non-majors![31]
Frank Contratto, Anasazi Instruments, Inc., 4101 Cashard Avenue, Indianapolis, IN, 46203 and Amy Abe, Department of Chemistry, Lake Forest College, Lake Forest, IL 60045.
Presented at the 230th National Meeting of the American Chemical Society, Washington, DC, August 2005, Paper CHED 127.

Examples are presented to highlight the relationship among components of food, their chemical composition and their detectability in the NMR spectrum. Unsaturation across a series of cooking oils is discussed. The levels of fat and water are examined in foods with a high caloric content from fats (mayonnaise, salad dressing, hot dogs, potato chips.) The type of fluoride is identified in ^{19}F spectra of toothpastes and quantitation of [23]Na levels in condiments is demonstrated.

MRI (magnetic resonance imaging)

The MRI image is a 3-dimensional map of protons in a semi-solid material where properties of the proton resonances are exploited to maximize *contrast* among regions of unique composition and morphology. This differs from the 1D proton NMR spectrum where signal intensity is, in general, representative of the *quantity* of nuclei present throughout the liquid sample.

Many medical MRI imaging systems operate at or below 1.5 Tesla. There is interest in using low field, permanent magnet FT-NMR spectrometers for teaching or demonstrating MRI imaging concepts to physics and chemistry majors and medical technology students. Conradi and co-workers[32] describe an undergraduate experiment at 60 MHz that develops the connection between T_1 relaxation concepts and an inversion recovery weighted MRI image. Students create a 1-dimensional image or slice of a phantom consisting of side-by-side capillaries of water in a 5 mm NMR tube. One of the capillary tubes has trace amounts of $CuSO_4$ to shorten relaxation times. Related undergraduate experiments,[33,34] that provide 3D images from reconstruction of radial gradient projections, can also be implemented on permanent magnet FT-NMR spectrometers.

Anasazi Instruments, Inc.[35] is developing an imaging accessory for their permanent magnet FT-NMR spectrometers that provides precise control of gradients and maximum entropy image reconstruction software. Figure 2 is a 1.4 Tesla image of uncooked bacon where regions of fat and meat (hydrated protein) are distinguished. With such an accessory, students can explore the effects of changes in the NMR parameters (T_1 vs. T_2; tau variation; contrast agents) on image contrast and quality.

For additional background on relaxation, physical chemistry texts emphasize T_1 relaxation in their discussions of NMR spectroscopy. Introductory experiments on transverse (T_1) and longitudinal (T_2) relaxation can be found in the literature.[36, 37, 38]

NMR Pedagogy

In this section, we collect a variety of approaches for introducing NMR concepts and modern FT-NMR techniques to students. We start with techniques for novice chemists or non-majors and progress through advanced NMR techniques. While the cited literature may describe experiments developed on

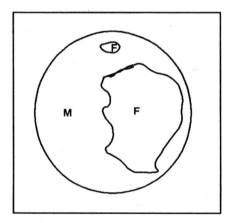

Figure 2. Uncooked bacon. (top) Slice from a 1.4 Tesla MRI image of bacon. (bottom) Line drawing indicating regions of white fat, F, and reddish meat, M, or hydrated protein. The ^1H image was acquired of a 4.2 mm disc of uncooked bacon positioned inside and perpendicular to the axis of a 5 mm NMR tube. (Images from Anasazi Instruments, Inc.)

either superconducting or permanent magnet spectrometers, we chose examples that are suitable for both types of magnets.

There is increasing interest in introducing pre-college students to NMR as part of high school organic chemistry programs or in conjunction with academic or corporate outreach programs. We see three types of curricular situations in which high school students are introduced to NMR: (a) schools such as the Illinois Math and Science Academy[39] and the Oklahoma School of Science & Mathematics,[40] that have on-site FT-NMR spectrometers and facilities to conduct organic chemistry laboratories, (b) partnerships or consortia where the students receive content instruction from their high school chemistry teacher and conduct hands-on laboratories at a local college or university NMR facility and (c) high schools where students receive content instruction only.

Outreach programs with high school students take many forms including: multi-week summer enrichment programs,[41] recruiting science majors and science day programs. Introducing NMR concepts to novices usually involves providing a framework to visualize or discuss three dimensional molecular shapes and structure. In two examples, Wiser and Cody[42] and independently Uffelman and colleagues[43] use models (molecular graphics or space filling) for a small set of molecules that allow high school students to identify symmetry features and the number of unique carbon environments. The specific molecular features are then "verified" against the student acquired ^{13}C NMR spectra. Wiser and Cody have students examine similarities and differences among over-the-counter "pain killers." Uffelman and colleagues compare sensitivities and the types of information gleaned from two molecular detectors for enantiomers of carvone: ^{13}C NMR and the human nose.

Chapman has advocated for using students' intuitive understanding of symmetry to examine the structure of organic molecules.[44] Early in the sophomore organic lecture, he and Russell[45] uses the "one carbon type, one peak" aspect of ^{13}C NMR and DEPT as supporting experimental evidence for learning about the structure of isomeric alkanes. Similarly, Reeves and Chaney[46] developed an introductory ^{13}C NMR laboratory exercise with acyclic hexanes and heptanes.

We observed that as students become comfortable using spin-spin coupling, from ^{1}H NMR, to discern numbers of vicinal hydrogens they often ask how information from $^1J_{CH}$ couplings can be encoded to give inverted and upright peaks in ^{13}C DEPT spectra. During Anasazi Instruments, Inc. co-sponsored *FT-NMR Workshops*, we engage the attendees in a *gedanken* experiment to show that multiplet patterns for doublets, triplets and quartets can be collapsed into the phase relationships seen in DEPT 45, 90 and 135 spectra. We use mechanical pointers to represent rotating frame magnetization vectors and mimic the motions of vectors in the APT (attached proton test) pulse sequence as described in several introductory texts.[47, 48, 49]

Esters can be used as illustrative compounds for practice in interpretation of COSY spectra. The two isolated spin systems from the parent acid and from the parent alcohol are usually separable at higher fields or suitable molecules can be chosen for use in low field permanent magnets. Branz and colleagues[50] developed a laboratory exercise based on esters derived from C_{10} aromatic carboxylic acids and butyl alcohols. We have found that multiplets separated by 0.1 ppm will give rise to resolvable cross peaks in a 60 MHz COSY spectrum and that esters with ethyl and n-propyl moieties can be used in this type of laboratory exercise. Other molecules for demonstrating COSY peak assignments at low field are $trans$-cinnamaldehyde and ibuprofen.

Terpenes are often employed as organic "unknowns" for institutions with access to superconducting magnet spectrometers. The assignment of bridgehead and of exo and $endo$ protons in bicyclic terpenes can prove challenging from just the COSY spectrum. Mills[51] provides examples of bicyclic[3.1.1]terpenes (verbenone, α-pinene, myrtenol, myrtenal) amenable to assignment from COSY and NOESY spectra. Mosher and Roark[52] encourage students to use COSY and long range COSY to identify bicyclic[2.2.1]terpenes (camphor, ketopinic acid, borneol, isoborneol.) Holder and colleagues[53] provide a list of a dozen, commercially available terpenes suitable for identification by a combination of DEPT, COSY, COSY-45 and HETCOR.

References

1. Anasazi Instruments, Inc., 4101 Cashard Avenue, Suite 103, Indianapolis, IN 46203 URL: http://www.aiinmr.com/
2. For example, Process NMR Associates, LLC, 87A Sand Pit Road, Danbury, CT 06810 URL: http://www.process-nmr.com/index.html
3. Anderson, Marvin H.; Schleich, Thomas W.; John, Boban K.; Shoolery, James N. U.S. Patent 6,163,154, 2000.
4. Bank, S. *Concepts Magn. Reson.* **1997**, *9*, 83-93.
5. Kateley, L. J. presented at the American Chemical Society 30th Great Lakes Regional Meeting, Chicago IL, May 1997. The student handout is available upon request to FContratto@aol.com.
6. Dávila, R. M.; Widener, R. K. *J. Chem. Educ.* **2002**, *79*, 997-999; *J. Chem. Educ.* [Online] **2002**, *79*, 997-999, supplement
7. Houston Byrd, H.; O'Donnell, S. E. *J. Chem. Educ.* **2003**, *80*, 174-176; *J. Chem. Educ.* [Online] **2003**, *80*, 174-176, supplement
8. Shriner, R. L.; Christine K. F. Hermann, C. K. F.; Terence C. Morrill, T. C.; David Y. Curtin, D. Y.; *The Systematic Identification of Organic Compounds*, 8th Edition; Wiley & Sons: New York; 2003
9. Breton, G. W. *J. Chem. Educ.* **2000**, *77*, 81-83; *J. Chem. Educ.* [Online] **2000**, *77*, 81-83, supplement

10. Kjonaas, R. A.; Mattingly, S. P. *J. Chem. Educ.* **2005**, *82*, 1813-1814; *J. Chem. Educ.* [Online] **2005**, *82*, 1813-1814, supplement
11. Breton, G. W. *Chem. Educator* **2005**, *10*, 298-299; *Chem. Educator* [Online] **2005**, *10*, 298-299, supporting materials
12. Breton, G. W. *Chem. Educator* **1999**, *4*, 134-136; *Chem. Educator* [Online] **1999**, *4*, 134-136, supporting materials
13. Andersh, B. *Chem. Educator* **2000**, *5*, 20-23; *Chem. Educator* [Online] **2000**, *5*, 20-23, supporting materials
14. Martin, W. B.; Kateley, L. J *J. Chem. Educ.* **2000**, *77*, 757-759
15. Martin, W. B.; Kateley, L. J; Wiser, D. C.; Brummond, C. A. *J. Chem. Educ.* **2002**, *79*, 225-227; *J. Chem. Educ.* [Online] **2002**, *79*, 225-227, supplement
16. Breton, G. W.; Belk, M. K. *Chem. Educator* **2004**, *9*, 27-29; *Chem. Educator* [Online] **2004**, *9*, 27-29, supporting materials
17. Lee, S. *Chem. Educator* **2004**, *9*, 359–363; *Chem. Educator* [Online] **2004**, *9*, 359-363, supporting materials
18. Lee, S. *Chem. Educator* **2005**, *10*, 357–358; *Chem. Educator* [Online] **2005**, *10*, 357-358, supporting materials
19. Weisshaar, D. E.; Earl, G. W.; Hanson, M. P.; Viste, A. E.; Kintner, R. R.; Duffy-Matzner, J. L. *J. Chem. Educ.* **2005**, *82*, 898-900; *J. Chem. Educ.* [Online] **2005**, *82*, 898-900, supplement
20. Davis, D. S.; Moore, D. E. *J. Chem. Educ.* **1999**, *76*, 1617-1618
21. Davis, D. S.; Hargrove, R. J.; Hugdahl, J. D.; *J. Chem. Educ.* **1999**, *76*, 1127-1130
22. Lim, D.; Moyna, G. *J. Undergraduate Chemistry Research* **2004**, *3*, 96-102.
23. Moeckly, S.; Earl, G. W.; Viste, A. *Proceedings of the South Dakota Academy of Science*, **1999**, *78*, 63-69
24. Dragan, S.; Fitch, A. *J. Chem. Educ.* **1998**, *75*, 1018-1021
25. Abe, A.; Contratto F. B. presented at Anasazi Instruments, Inc. co-sponsored *FT-NMR Workshops,* 2000-2005 *PbEDTA complexes at 60 and 400 MHz*; http://www.lakeforest.edu/academics/programs/chem/nmr.asp
26. Abe, A.; Contratto F. B. presented at Anasazi Instruments, Inc. co-sponsored *FT-NMR Workshops,* 2000-2005 *Di- and tri-peptide "sequencing"*; http://ww.lakeforest.edu/academics/programs/chem/nmr.asp
27. Contratto F. B., Anasazi Instruments, Inc., 4101 Cashard Avenue, Indianapolis, IN, *NMR at Work... - Titration monitoring of ATP by ^{31}P NMR*; http://www.lakeforest.edu/academics/programs/chem/nmr.asp
28. Contratto F. B., Anasazi Instruments, Inc., 4101 Cashard Avenue, Indianapolis, IN, *NMR at Work... - Determination of the Ethanol Content of Selected Consumer Products by Proton NMR*; http://www.lakeforest.edu/academics/programs/chem/nmr.asp
29. Contratto F. B., Anasazi Instruments, Inc., 4101 Cashard Avenue, Indianapolis, IN, *NMR at Work... - Determination of Sodium Content of Selected Consumer Products by Sodium NMR*; http://www.lakeforest.edu/academics/programs/chem/nmr.asp

333

30. Contratto F. B., Anasazi Instruments, Inc., 4101 Cashard Avenue, Indianapolis, IN, *NMR at Work...* - *Analysis of Edible Oils by ^1H and ^{13}C NMR*; http://www.lakeforest.edu/academics/programs/chem/nmr.asp

31. Contratto, F. B.; Abe, A. presented at the 230th National Meeting of the American Chemical Society, Washington, DC, August 2005, Paper CHED 127; http://www.lakeforest.edu/academics/programs/chem/nmr.asp

32. Cornell, D. A.; Clewett, C. F. M.; Conradi, M. S. *Concepts Magn. Reson.* **2000**, *12*, 257-268.

33. Hull, L. A. *J. Chem. Educ.* **1990**, *67*, 782-783.

34. Quist, Per-Ola *J. Chem. Educ.* **1996**, *73*, 751-752.

35. This article was written in March 2006. Current information about the imaging accessory can be found at www.aiinmr.com.

36. Williams, K. R.; King, R. W. NMR Relaxation Times. In *Physical Chemistry: Developing a Dynamic Curriculum;* Schwenz, R. W.; Moore, R. J., Eds.; American Chemical Society: Washington, DC, 1993; Chapter 21.

37. Lorigan, G. A.; Minto, R. E.; Zhang, W. *J. Chem. Educ.* **2001**, *78*, 956-958; *J. Chem. Educ.* [Online] **2001**, *78*, 956-958, supplement.

38. Gasyna, Z. L.; Jurkiewicz, A. *J. Chem. Educ.* **2004**, *81*, 1038-1039; *J. Chem. Educ.* [Online] **2004**, *81*, 1038-1039, supplement.

39. Mary van Verst and Joel Ray described student's experiences with NMR at Illinois Math and Science Academy at a 2003 Aii sponsored *FT-NMR applications* workshop at Illinois Wesleyan University. The workshop handout is available upon request to FContratto@aol.com.

40. http://www.ossm.edu/chem/

41. Richard K. Shoemaker initiated a successful program in the 1990's at University of Nebraska. An example of the introductory NMR handout is at http://www.wollernet.com/writing.html

42. Wiser, D. C.; Cody, J. A. presented at the 225th National Meeting of the American Chemical Society, New Orleans, LA, March 2003; Paper CHED 625.

43. Uffelman, E. S.; Cox, E. H.; Goehring, J. B.; Lorig, T. S.; Davis, C. M. *J. Chem. Educator* **2003**, *80*, 1368-1371; *Chem. Educator* [Online] **2003**, *80*, 1368-1371, supporting materials

44. Chapman, O. L. *J. Applied Developmental Psychology,* **2000**, *21*, 97-108, 593. For an earlier version of the paper, see: http://coke.physics.ucla.edu/laptag /Standards/Chapman_NAS_Cog_Sci.pdf

45. Chapman, O. L.; Russell, A. A. *J. Chem. Educ.* **1992**, *69*, 779-782.

46. Reeves, P. C.; Chaney, C. P. *J. Chem. Educ.* **1998**, *75*, 1006-1007.

47. Pavia, Donald L.; Lampman, Gary M.; Kriz, George S.; *Introduction to Spectroscopy;* 3rd edition; Harcourt College Publishers: Fort Worth, TX, 2001 pp 536-540

48. Kemp, William; *NMR in Chemistry: A Multinuclear Introduction*; Macmillan: New York, 1988 pp 141-145

49. Farrar, Thomas C.; *Introduction to Pulse NMR Spectroscopy;* Farragut Press: Madison, WI, 1989 p. 176

50. Branz, Stephen E.; Miele, Robert G.; Okuda, Roy K.; Straus, Daniel A. *J. Chem. Educator* **1995**, *72*, 659-661.
51. Mills, N. S. *J. Chem. Educ.* **1996**, *73*, 1190-1192
52. Holder, G. N.; Farrar, D. G.; Gooden, D. M. *Chem. Educator* **1999**, *4*, 173-176; *Chem. Educator* [Online] **1999**, *4*, 173-176, supplementary material
53. Roark, J. L.; Mosher, M. D. *Chem. Educator* **1998**, *3(4)*, 1-12

Chapter 23

Showcasing 2D NMR Spectroscopy in an Undergraduate Setting: Implementation of HOMO-2D *J*-Resolved Experiments on Permanent Magnet NMR Systems

Brian K. Niece[1] and Guillermo Moyna[2]

[1]Department of Natural Sciences, Assumption College, 500 Salisbury Street, Worcester, MA 01609–1296
[2]Department of Chemistry and Biochemistry, University of the Sciences in Philadelphia, 600 South 43rd Street, Philadelphia, PA 19104–4495

Rigorous descriptions of two-dimensional (2D) NMR spectroscopy are oftentimes absent from the chemistry curriculum. This is due in part to the lack of suitable instrumentation required to present these techniques in a laboratory setting. Furthermore, the complex mathematical tools needed to explain the theory behind many of these experiments makes their presentation in undergraduate chemistry courses impractical. In this chapter we show that techniques which can be understood using classical vector models, such as 2D *J*-resolved (2DJ) experiments, can be used to discuss the main aspects of 2D NMR spectroscopy in detail at the undergraduate level. We also outline the implementation of these experiments in the Anasazi Instruments Eft-series of permanent magnet NMR spectrometers, present typical results obtained for a number of samples, and discuss general guidelines to follow in the selection of additional examples.

NMR spectroscopy is arguably the instrumental technique with the widest utility across all branches of chemistry. Its uses range from the estimation of moisture content in foods and the elucidation of the covalent structure of small molecules, to the determination of the three-dimensional fold of proteins and their complexes. Consequently, students going on to academic and industrial careers in chemistry are likely to be exposed to a variety of NMR techniques, and it is thus critical that their applications, as well as the concepts governing them, are discussed throughout the undergraduate curriculum. Indeed, this view is shared by the ACS Committee on Professional Training, whose guidelines now state that students must "use and understand" NMR spectroscopy (*1*).

The vast majority of the NMR experiments discussed in the chemistry curriculum involve one-dimensional (1D) techniques. On the other hand, and not counting the examples presented elsewhere in this volume, there are only a few well-documented experiments aimed at introducing two-dimensional (2D) NMR spectroscopy to undergraduate students (*2*). Furthermore, these laboratories concentrate primarily on interpreting the results obtained from the 2D spectra and, with counted exceptions, avoid in-depth discussions of the pulse sequences employed to acquire them (*3*). Taking into account that most of the recent advances in NMR spectroscopy entail multi-dimensional, multi-pulse experiments, it would be desirable to cover both experimental as well as theoretical aspects of 2D NMR in better detail at the undergraduate level. Several factors preclude this type of endeavor. First, the instruments available to undergraduate students are usually dated and either lack the capability to perform 2D experiments or require a considerable amount of time to carry them out, therefore limiting hands-on exposure to these techniques in a laboratory setting. Even in cases where access to adequate instrumentation and time are not an issue, the experiments chosen to introduce 2D methods to chemistry majors are normally COSY, HETCOR, and other correlation-type techniques. While they are ideal for illustrating the power of 2D NMR spectroscopy, understanding the theoretical foundations of these experiments requires a considerable knowledge of quantum mechanics. As a result, the actual pulse sequences used to record these 2D spectra are usually treated as "black-boxes."

In this chapter, we show that homonuclear 2D *J*-resolved (HOMO-2DJ) techniques can be used successfully to present the most important concepts of 2D NMR spectroscopy in detail to an undergraduate audience. While these methods lack the appeal of modern correlation experiments, they share basic building blocks with all multi-dimensional techniques and their inner-workings can be fully understood through the use of classical vector models which are appropriate for beginning students. In addition, we outline the implementation of the HOMO-2DJ pulse sequence on the Anasazi Instruments Eft-series of permanent magnet NMR spectrometers, instruments which have gained enormous popularity among instructors in colleges and universities across North

America in the past decade. In the final section of the chapter we present typical results obtained for a number of samples and experimental settings, and discuss general guidelines that should be followed in the selection of further examples to use in the demonstration of different applications of the technique.

Classical Description of 2DJ Spectroscopy with Vector Models

2DJ NMR experiments were among the first multi-dimensional techniques employed routinely by chemists (*4*). As shown in more detail below, the method allows for the separation of chemical shift (δ) and scalar coupling (J) information in two independent dimensions of the spectrum. Therefore, these experiments can be used to resolve overlapped spin systems, and as a result they were quite popular with those working in structure elucidation problems until the advent of more powerful 1D and 2D correlation-type techniques.

2DJ methods are based on the Hahn spin-echo experiment (*5*). Analysis of this simple 1D pulse sequence, presented below, is critical to understand these techniques, and also serves to establish some of the notation used throughout this chapter:[†]

$$90_{+y} - t_d/2 - 180_{+y} - t_d/2 - Acquire_{-x}$$

90+y and *180+y* represent 90° ($\pi/2$) and 180° (π) pulses applied along the <+*y*>-axis of the rotating frame, t_d is a time delay, and *Acquire_x* indicates the acquisition of the NMR signal, or free-induction decay (FID), on the <−*x*>-axis. For a system of isolated spins (i.e., a singlet), the 90° pulse aligns equilibrium magnetization originally along the <+*z*>-axis, or M_0, with the <+*x*>-axis (**M**, Figure 1). During the first half of t_d, **M** precesses on the <*x,y*> plane under the effects of its chemical shift or Larmor frequency, Ω, expressed in Hz. The angle of **M** with respect to its original position after this delay, θ, will depend on Ω and t_d, and its value in radians can be computed as $\pi \cdot \Omega \cdot t_d$. The 180° pulse then rotates **M** around the <+*y*>-axis until it rests exactly on the opposite side of the <*x,y*> plane. In other words, while the <*y*> component of **M** remains unchanged, its <*x*> component has the same magnitude but changes sign. Following the inversion pulse, **M** precesses on the <*x,y*> plane under the effects of Ω in the

---

[†] A detailed description of the NMR phenomenon using the classical vector model and its associated notation is beyond the scope of this chapter, and we direct the reader to several authoritative volumes that present this topic in detail (*6-8*). These background concepts should also be discussed with students if the techniques outlined here are to be incorporated into lectures or laboratories.

338

same direction and for the same time, and therefore θ angle, as it did during the first $t_d/2$ delay. It can be clearly seen that irrespective of the length of t_d or Ω frequency, **M** will be aligned exactly along the $<-x>$-axis right before the acquisition of the NMR signal. After Fourier transformation (FT) of the resulting FID, this technique will therefore yield peaks of purely absorptive phase regardless of their chemical shift. Repeating the experiment for different values of t_d gives signals that are only affected by spin-spin, or transverse, relaxation (T_2). Consequently, one of the most common applications of spin-echoes is the determination of accurate T_2 times for systems of isolated spins in inhomogeneous magnetic fields (6-8).

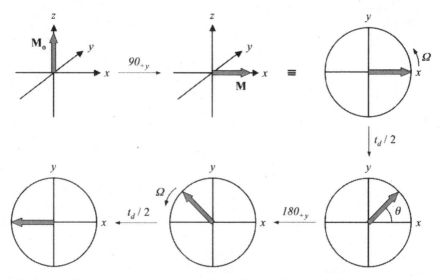

Figure 1. Classical vector analysis of the effects of the spin-echo sequence on a system of isolated spins with a Larmor frequency of Ω Hz. Only the $<x,y>$ plane of the coordinate system is considered after the 90° pulse.

Markedly different results are obtained when the Hahn spin-echo is applied to a homonuclear *J*-coupled spin system. The analysis in this case will be carried out for a weakly coupled, or first order, two-spin system (i.e., a doublet). In addition, and since we showed above that the spin-echo sequence is insensitive to the effects of chemical shift, an on-resonance signal will be employed in this example (i.e., $\Omega = 0$). It is important to point out that these assumptions only make the overall analysis easier to comprehend, but the same conclusions will be reached if signals of other multiplicities that are either on-or off-resonance are considered. As was the case for the system of isolated spins, the 90° pulse along the <+y>-axis aligns **Mo** with the <+x>-axis. Given that the system is on-resonance, the precession during the first half of the t_d delay will be governed solely by scalar coupling, and the components of **M** corresponding to the two signals of different energies in the doublet, $\mathbf{M}_\alpha$ and $\mathbf{M}_\beta$, will rotate on the <x,y> plane in opposite directions at rates of $+J/2$ and $-J/2$ Hz (Figure 2). It should be stressed that the α and β labels refer to the energy levels of the coupling partner of the spin being observed. The θ angle separating the two **M** vectors after the first delay is in this case proportional to t_d and J, and its magnitude in radians is equal to $\pi \cdot J \cdot t_d$. The 180° inversion pulse once again rotates the $\mathbf{M}_\alpha$ and $\mathbf{M}_\beta$ vectors to the opposite side of the <x,y> plane. However, this pulse also has the effect of inverting the populations, and thus energy levels, of the coupled spin. Therefore, the α and β labels of the two **M** vectors are also inverted, and so is their direction of gyration. As a result, the slowest moving component of **M** becomes the fastest and vice-versa, and the two will continue to dephase instead of refocusing during the second half of t_d.

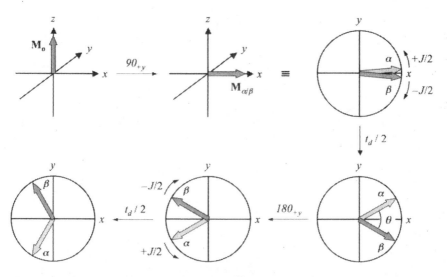

Figure 2. Classical vector analysis of the effects of the spin-echo sequence on an on-resonance J-coupled doublet. Only the <x,y> plane of the coordinate system is considered after the 90° pulse.

At the end of the second delay, the θ angle between $\mathbf{M}_\alpha$ and $\mathbf{M}_\beta$ will be equal to $2\pi \cdot J \cdot t_d$. In contrast to what was observed for isolated spins, the signals for the two vectors on the <−x>-axis will now depend not only on T_2 relaxation, but also on J and t_d. Indeed, if the experiment is repeated for different

t_d delays, the initial signal intensities for $\mathbf{M}_\alpha$ and $\mathbf{M}_\beta$ will vary periodically with J and t_d as a function of $\cos(\pi \cdot J \cdot t_d)$. This phenomenon is known as J-modulation and is better depicted in Figure 3, where experimental results of the spin-echo sequence at selected values of t_d are presented for a doublet and a triplet. Inspection of the data obtained for the doublet reveals that the intensities of the signals at $\Omega + J/2$ and $\Omega - J/2$ are at a maximum at $t_d \approx 0$, cross a null at $t_d = 1/2J$, reach a minimum at $t_d = 1/J$, go through a second null at $t_d = 3/2J$, and so on (Figure 3a). For the triplet, the intensity of the center line at Ω will only be affected by transverse relaxation, and those for the signals at $\Omega + J$ and $\Omega - J$ will be maximum, null, minimum, and null at t_d values of ~ 0, $1/4J$, $1/2J$, and $3/4J$, respectively (Figure 3b). The analyses for systems of other multiplicities are analogous, and we encourage the reader to carry them out (8).

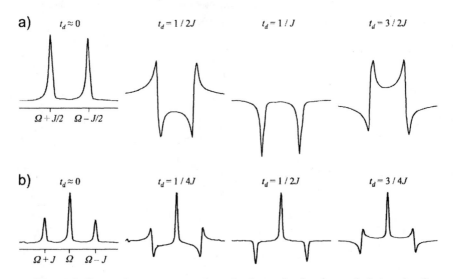

*Figure 3. Spin-echo spectra at selected values of t_d for the methyl signals of isopropanol (doublet, J = 6.2 Hz, **a**) and ethyl acetate (triplet, J = 7.2 Hz, **b**) in CDCl3 solution. Recorded on an Anasazi Eft-90 NMR spectrometer.*

The way in which these J-modulation effects can be exploited in the generation of a 2D spectrum can be conveniently explained if results for a large number of spin-echo experiments, collected by progressively changing t_d in small increments, are presented as a stack plot (Figure 4a). We can once again corroborate that in the case of the triplet shown, the lines at $\Omega + J$ and $\Omega - J$ go through periodic cycles and that the one at Ω is only affected by T_2 relaxation. If the stack plot is now sliced with vertical planes running along $\Omega + J$, Ω, and $\Omega -$

J, the variation in the intensities of peaks at these frequencies as a function of t_d is obtained (Figure 4b). The modulation of the signals at $\Omega + J$ and $\Omega - J$ is obvious in these plots. Further inspection shows that, barring relaxation effects, their intensities change as a function of $\cos(2\pi \cdot J \cdot t_d)$. Another salient aspect of these plots is their resemblance to FIDs acquired along a virtual time domain t_d.

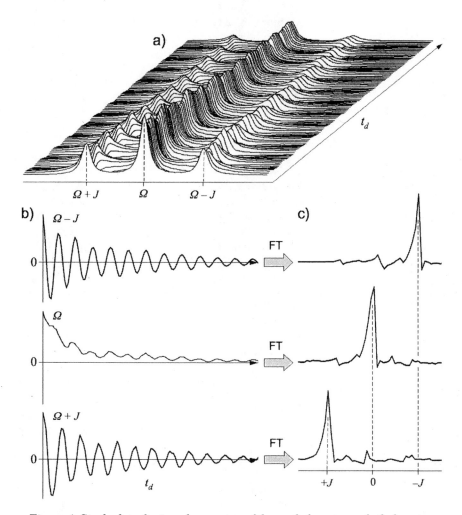

*Figure 4. Stack plot of spin-echo spectra of the methyl protons of ethyl acetate collected by incrementally changing t_d (**a**). J-modulation as a function of t_d for signals at $\Omega - J$, Ω, and $\Omega + J$ (**b**), and their corresponding spectra after FT (**c**). Recorded on an Anasazi Eft-60 NMR spectrometer.*

As a matter of fact, if the J-modulated data are treated as FIDs and used as input of a FT in the time domain t_d, the resulting frequency spectra will have peaks at $+J$, 0, and $-J$ Hz, respectively (Figure 4c).

If the same process is repeated for a sufficient number of vertical slices of the stack plot and the data in the frequency domain are processed in magnitude mode (*vide infra*), a 2D spectrum with cross-peaks at $[+J, \Omega + J]$, $[0, \Omega]$, and $[-J, \Omega - J]$ is obtained (Figure 5a). Notice that as is customarily done in 2D NMR spectroscopy, the results are shown as a contour plot for clarity. The stepwise process outlined above is precisely what is done to record an HOMO-2DJ spectrum. However, and as described in more detail in the following section, the experiment is fully automated so that after selection of appropriate acquisition and processing parameters little user intervention is required. At this point it is also convenient to adopt the standard notation used in 2D experiments to describe this 2D variant of the spin-echo pulse sequence:

$$90_{+y} - t_1/2 - 180_{+y} - t_1/2 - Acquire\ (t_2)$$

The varying t_d delay is replaced by the t_1 delay, and is now referred to as the incremental delay of the 2D experiment. FT along the virtual time domain generated by t_1 will give rise to the indirect, or f_1, dimension of the 2D spectrum. In addition, the acquisition time in which the actual NMR signals for different t_1 delays are collected is now referred to as t_2. FT along t_2 yields the direct, or f_2, dimension of the 2D spectrum. It should also be pointed out that the basic building blocks of every multidimensional NMR experiment are present in the 2DJ experiment, including a preparation step (the $90°$ pulse), an evolution period (the combined $t_1/2$ delays), a mixing event (the $180°$ inversion pulse), and an acquisition period (the acquisition of the NMR signal, or FID, in t_2).

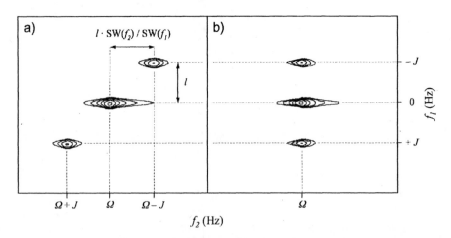

*Figure 5. HOMO-2DJ spectrum of the methyl protons of ethyl acetate obtained as described in the text before (**a**) and after the tilting operation (**b**). Recorded on an Anasazi Eft-60 NMR spectrometer.*

We now return our attention to the HOMO-2DJ spectrum shown in Figure 5a. A quick inspection of the data reveals that Ω and J information for the spin system is present on the f_2 dimension. This is consistent with the fact that during the t_2 acquisition time, the system was allowed to evolve under the effects of both chemical shift and scalar coupling. On the other hand, the f_1 dimension is centered exactly at 0 Hz and holds only J-coupling information. This can be easily understood if one recalls that the data in the t_1 time domain consists of J-modulated intensities originated by spin-echoes. As described earlier, the sequence refocuses chemical shift effects in t_1, and thus Ω information is lost from the f_1 dimension. Although the spectrum is not symmetric, scalar coupling data are present in both frequency domains, and the cross-peaks for every spin system are slanted at a $45°$ angle. Having J-couplings in both dimensions is redundant, and they can be removed from f_2 using a simple processing routine known as a tilting operation. Since the 2D spectrum is stored in the computer as a matrix, the algorithm shifts data points in rows along f_2 so as to align all cross-peaks corresponding to a spin on a single column along f_1 at its chemical shift in the direct dimension. In general, tilting involves shifting all data points in the 2DJ spectral matrix by $l \cdot SW(f_2) / SW(f_1)$ Hz in the f_2 dimension, where l is their distance in Hz from the center of the f_1 axis and $SW(f_2)$ and $SW(f_1)$ are the spectral widths of the direct and indirect dimensions, respectively (Figure 5a). Data points are shifted left or right along f_2 rows depending on their position in the spectrum with respect to the center of the f_1 dimension (i.e., the sign of l). Figure 5b shows the cross-peaks for the triplet after the tilting operation, and additional examples are presented below.

Some features of HOMO-2DJ spectra deserve particular attention. First, the cross-peaks arising from signals of any given multiplet appear at unique positions on the f_2 dimension before the tilting operation. Consequently, slices in t_1 and f_1, before and after the second FT respectively, contain a single frequency. By showing the correspondence between the J-modulated intensity in t_1 and the resulting frequency signal in $f1$ the origin of the indirect dimension can be easily demonstrated. Second, J and Ω information will appear separated on the two dimensions after tilting, and thus multiplets for each spin system will be completely isolated on different columns of f_1 with their chemical shifts well resolved along f_2. As stated earlier, the ability to separate these two parameters is particularly useful if the 1D ^1H spectrum shows overlapped signals, a scenario that is commonly encountered when using permanent magnet NMR spectrometers operating at lower fields. In addition, if all the rows in the tilted HOMO-2DJ spectral matrix are projected onto the f_2 axis, a 1D spectrum in which signals for every spin system appear as singlets is obtained. Interestingly, this projection is equivalent to a ^1H-decoupled 1D ^1H spectrum, an experiment that cannot be recorded by means of any other pulse sequence or decoupling scheme. Finally, it should be noted that the spectral width of the indirect dimension only needs to accommodate the widest multiplet in the system, and in most cases a SW(f_1) of 50 Hz is sufficient. Since the number of t_1 increments in HOMO-2DJ spectra is typically 64 (see below), the digital resolution in f_1 will be 50 / 64, or 0.78, Hz per point. This can be improved further if the $f1$ data are zero-filled, and hence coupling constants for most first-order spin systems can be measured with reasonable accuracy in this dimension.

Practical Implementation of 2DJ Pulse Sequences

The pulse program required to implement the HOMO-2DJ experiment on Anasazi Eft-series NMR spectrometers is available as an appendix. It will operate properly on instruments running WinPNMR version 040505 or later (9). Data processing requires the use of the NUTS NMR software package (10). A complete set of files, including the pulse program, parameter file, and processing macro and software are available online (11).

Before using the HOMO-2DJ pulse program, appropriate parameters should be set and saved in the file 'homo2dj.ini.' These include the number of scans, data size, and spectral width (NS, SI, and SW, respectively). NS should be a multiple of four to use a complete phase cycle, and often just four scans are adequate for reasonably concentrated samples. SI should be set to give sufficient digital resolution in f_2, and a value of 2048 has been found to give acceptable spectra. The value of SW will depend on the ^{1}H chemical shift dispersion of the sample. The remaining parameters are set by the pulse program.

When run, the program first prompts for a relaxation delay (line 35). The default value of 2 s has been found to give acceptable results, although degassed and sealed samples may benefit from longer delays if the spectra display appreciable ridges along the f_1 dimension (i.e., "t_1 noise"). The 90° pulse is then read from the spectrometer files, and the 180° pulse length is calculated (lines 43-47). Next, the user is asked for the value of the f_1 spectral width, SW(f_1), which must be sufficiently wide to accommodate the widest J-coupled multiplet in the spectrum (line 56). The reciprocal of SW(f_1) in microseconds is taken as the t_1 increment to apply between slices (line 60). Assuming J-coupling constants of ~ 7 Hz, the default SW(f_1) of 75 Hz is somewhat larger than required for most organic molecules. However, this larger value leads to a smaller t_1 increment and thus a shorter effective t_1 acquisition time. Consequently, fewer data points are recorded in the indirect dimension following complete decay of the NMR signal, and as a result spectra with less noise are obtained in f_1 after FT. The program then prompts the user for the number of t_1 slices to be acquired (line 70). The default value of 64 is adequate for many samples. Due to their longer relaxation times, 128 experiments in t_1 may be helpful with degassed and sealed samples. After an experiment filename is provided (line 84), the program defines the phase cycle (lines 86-91), loads the selected parameters into the pulse programmer (lines 107-119), and loops to collect the desired number of t_1 slices (lines 133-150). The incremental delay is updated within the loop on lines 139-141.

After acquisition, the data can be processed with the macro available online (*11*), which uses standard processing parameters found in the literature (*12*). By default, the data in t_2 are first multiplied by a sine window function, but gaussian functions can be used to enhance sensitivity (*vide infra*). The data are then zero-filled to 2048 points and Fourier-transformed. After transposing, a sine squared window function is applied, followed by zero-filling to 256 points and FT in t_1. The macro then computes a magnitude spectrum, calls an external utility to tilt and symmetrize the data, and presents the result as an intensity plot.

While not presented here, the heteronuclear counterpart of the HOMO-2DJ technique, HETERO-2DJ (*6-8*), can also be implemented successfully in the Anasazi Eft-series of NMR spectrometers. The pulse program, parameter file, and processing macro and software needed to perform this experiment, as well as results obtained for a variety of samples, are available online (*11*).

Typical Results and Suggested Experiments

Since the primary utility of the HOMO-2DJ experiment is the illustration of 2D experiments, it is best to introduce students to the technique with a relatively straightforward sample such as ethyl acetate. Its complete HOMO-2DJ spectrum, collected in approximately 15 minutes on a neat sample containing a few drops of TMS using the default parameters discussed above, is presented in Figure 6. This spectrum and all those presented below were collected on an Anasazi Eft-60 NMR spectrometer, and the projection on f_2 is a high resolution 1D ^{1}H spectrum of the same sample. As mentioned earlier, the multiplets from the projection appear rotated on the f_1 dimension of the 2DJ spectrum.

Instructors wishing to demonstrate the generation of a 2D spectrum in the classroom can collect the data and manually perform the initial processing steps through the first FT. This will allow them to display the stack plot as presented in Figure 4a. In class, they can then select appropriate t_2 slices as those shown in Figure 4b. Completing the second FT will show students how the FIDs in t_1 generate the expected signals. Note that prior to transformation in t_2, the data in Figure 4a were processed with an exponential window function with a line broadening factor of 0.8 Hz rather than the default sine function used to process the HOMO-2DJ spectrum shown in Figure 6. While exponential apodization leads to broadened peaks in the final 2DJ spectrum, it yields spectra in f_2 that lack the ripples characteristic of sine multiplication and can be easily phased. These data will look familiar to students accustomed to 1D spectroscopy, thus helping them to make the connection between the stack plot and the 2D spectrum. In a laboratory course, students could acquire, process, and inspect t_2 slices of this sample themselves.

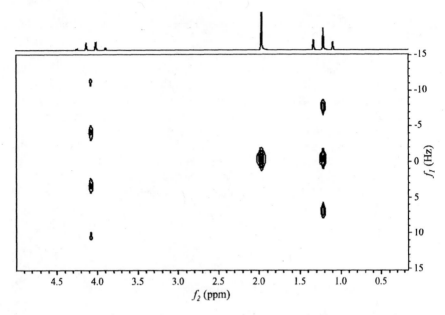

Figure 6. HOMO-2DJ spectrum of neat ethyl acetate.

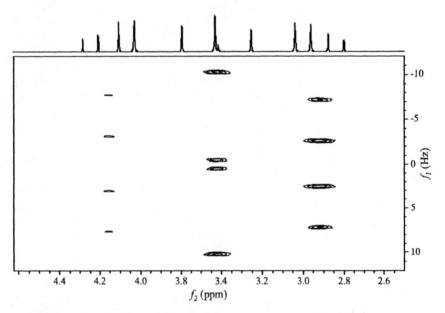

Figure 7. HOMO-2DJ spectrum of 20% 2,3-dibromopropanoic acid in C_6D_6.

The HOMO-2DJ spectrum of 2,3-dibromopropanoic acid (2,3-DBPA) presents students with a practical application of the technique (Figure 7). It was acquired on a commercial degassed and sealed sample of 20% 2,3-DBPA in C_6D_6. A total of 128 slices in the f_1 dimension were collected in about 30 minutes. In this case, SW(f_1) was set to 40 Hz to improve resolution in the indirect dimension, and a sensitivity-enhancement gaussian window function with a line broadening of 2.5 Hz was used to process the data in t_2. At first glance, an undergraduate student might expect the 1D ^{1}H spectrum of 2,3-DBPA to contain a doublet and a triplet arising from the CH_2 and CH protons, respectively. They may be surprised to find that the actual spectrum contains twelve peaks in the aliphatic region. The 2DJ spectrum reveals that these arise from three separate chemical shifts, one corresponding to the CH proton and two to the diastereotopic CH_2 protons.

The HOMO-2DJ spectrum of triethyl phosphite (TEP), shown in Figure 8, illustrates another interesting application of the technique. It was acquired using a solution of 25% TEP and 6% TMS in C_6D_6. The relaxation delay was 5 s, and all other parameters were set to default values. The total acquisition time was

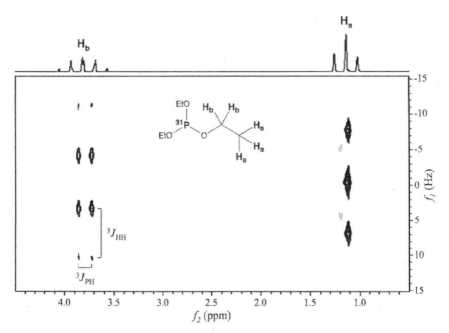

Figure 8. HOMO-2DJ spectrum of 25% triethyl phosphite in C_6D_6. Cross-peaks in grey correspond to symmetrization artifacts.

about 30 minutes. The triplet at 1.12 ppm arises from the methyl protons (H_a), which are uncoupled to the ^{31}P nucleus. The multiplet at 3.79 ppm corresponds to the methylene protons (H_b), which are coupled to both the methyl protons ($^3J_{HH}$ = 7.1 Hz) and the ^{31}P nucleus ($^3J_{PH}$ = 8.0 Hz), and is actually a quartet of doublets. The 2DJ spectrum rotates ^1H-^1H couplings into the f_1 axis and leaves the ^{31}P-^1H coupling on the f_2 axis, thereby simplifying spectral analysis and allowing both coupling constants to be measured in a straightforward manner.

Many other samples can be used to showcase the HOMO-2DJ experiment. In general, molecules bearing no more than three magnetically non-equivalent coupled spins should be employed, particularly if the goal is to analyze the inner-workings of the pulse sequence in detail. Samples in which uncoupled spin systems are overlapped in the 1D ^1H spectrum can be used to demonstrate the utility of the technique in structural elucidation problems. Alkyl cinnamates, many of which are commercially available, show partial overlap between the aromatic and olefinic signals at low fields and are ideal for this purpose. Finally, molecules with strongly coupled spin systems should be avoided, as second order artifacts complicate the analysis of the multiplets on the f_1 dimension.

Conclusions

To summarize, the HOMO-2DJ experiment is ideal for introducing concepts of 2D NMR spectroscopy at the undergraduate level for several reasons. First and foremost, the method can be described solely in terms of the classical vector model, making it accessible to undergraduate students lacking the mathematical and quantum mechanical tools necessary to understand the common correlation-type techniques. Based on our experience, the theoretical aspects of the HOMO-2DJ experiment can be adequately covered in one lecture period in courses focusing on spectroscopic structure elucidation methods. Provided that additional lectures are devoted to present pertinent background NMR material, the technique could also be discussed successfully in other junior-and senior-level chemistry courses, such as Physical Chemistry and Instrumental Analysis. In addition, we have found that by selecting appropriate samples and conditions several HOMO-2DJ spectra can be recorded, processed, and analyzed within one laboratory period, giving students the opportunity to gain hands-on experience on the practical aspects of 2D NMR spectroscopy. Finally, the experiment can be easily implemented on a wide range of NMR spectrometers, and in particular on permanent magnet instruments such as the Anasazi Eft-60 and Eft-90 systems available in many undergraduate programs.

References

1. American Chemical Society, Committee on Professional Training. *Undergraduate Professional Education in Chemistry: Guidelines and Evaluation Procedures*; Washington, DC, 2003; p10.
2. (a) Augé, J.; Lubin-Germain, N. *J. Chem. Ed.* **1998**, *75*, 1285-1287; (b) Dwyer, T. J.; Norman, J. E.; Jasien, P. G. *J. Chem. Ed.* **1998**, *75*, 16351640; (c) LeFevre, J. W. *J. Chem. Ed.* **2000**, *77*, 361-363; (d) Bose, R. N.; Al-Ajlouni, A. M.; Volckova, E. *J. Chem. Ed.* **2001**, *78*, 83-87; (e) Seaton, P. J.; Williamson, R. T.; Mitra, A.; Assarpour, A. *J. Chem. Ed.* **2002**, 79, 106-110; (f) Mak, K. K. W. *J. Chem. Ed.* **2004**, *81*, 1636-1640; (g) Alonso, D. E.; Warren, S. E. *J. Chem. Ed.* **2005**, *82*, 1385-1386; (h) Alty, L. T. *J. Chem. Ed.* **2005**, *82*, 1387-1389.
3. Williams, K. R.; King, R. W. *J. Chem. Ed.* **1990**, *67*, A125-A137.
4. Aue, W. P.; Karhan, J.; Ernst, R. R. *J. Chem. Phys.* **1976**, *64*, 4226-4227.
5. Hahn, E. L. *Phys. Rev.* **1950**, *80*, 580-594.
6. Sanders, J. K. M.; Hunter, B. K. *Modern NMR Spectroscopy: A Guide for Chemists*, 2nd ed.; Oxford University Press: New York, 1993.
7. Günther, H. *NMR Spectroscopy: Basic Principles, Concepts, and Applications in Chemisty*, 2nd ed.; Wiley & Sons: Chichester, UK, 1995.
8. Claridge, T. D. W. *High-Resolution NMR Techniques in Organic Chemistry*; Tetrahedron Organic Chemistry Series, Vol. 19; Pergamon Press: Oxford, UK, 1999.
9. *EFT Spectrometer Software Manuals*; Anasazi Instruments, Inc.: Indianapolis, IN, 2003.
10. *NUTS NMR Data Processing Software*; Acorn NMR, Inc.: Livermore, CA, 2004.
11. The pulse programs, parameter files, and processing macros and software can be downloaded from http://tonga.usip.edu/gmoyna/aii2dj.
12. Berger, S.; Braun, S. *200 and More Basic NMR Experiments: A Practical Course*, 3rd ed.; Wiley-VCH: Weinheim, 2004; Chapter 10, pp 367-369.

Chapter 24

Fourier Transform NMR in the Chemistry Curriculum

An Integrated Approach Using a Permanent Magnet FT-NMR in Conjunction with High Field NMR Data Files and Computational Chemistry

Michael J. Collins and Ronald T. Amel

Department of Chemistry, Viterbo University, La Crosse, WI 54601

This chapter describes the pedagogy of NMR concepts across the undergraduate curriculum using a 1.4T fixed magnet FT-NMR. Experiments range from use of simple C-13, H-1, and DEPT spectra in 2nd year courses to 2D, B-11, F-19, and P-31 spectra in 3rd and 4th year courses. Additionally, molecular modeling has been integrated into several experiments. This chapter also describes our NMR website with which students can compare high field (300 MHz) and low field (60 MHz) spectra of a number of compounds with difficult-to-resolve signals. Examples of student data are included and assessment of outcomes is discussed.

Introduction

This chapter will describe the pedagogy used at Viterbo University to introduce NMR concepts across the undergraduate curriculum, especially those pedagogical changes made possible by implementing low cost, modern FT-NMR methods. The chapter will also describe some specific experiments in organic and inorganic courses. Finally, the chapter will discuss the methodology and outcomes of assessing improvements in the understanding of NMR among our undergraduate majors.

According to the American Chemical Society's Committee on Professional Training, NMR is one of the instruments listed in the first category of "essential instruments," since it is such a powerful analytical tool. NMR finds applications in virtually all areas of chemistry. (1) It is commonly used for structure determination of compounds, conformational analysis, following the progress of reactions, and following reaction kinetics. (2) Graduate schools and industry demand students who have experience with state-of-the-art Fourier transform (FT) NMR techniques and interpretation of spectra.

At the time we received funding from the NSF-CCLI program, we lacked the physical infrastructure and budget to support a high field, superconducting NMR. We chose instead to upgrade our 20 year old Varian EM-360A CW-NMR, which had an excellent 60 MHz (1.4Tesla) fixed magnet, to the Anasazi Instruments, Inc. (AII), EFT-60 NMR with C-13 and broadband probes. At the time of this writing, there were more than 425 such instruments installed, mostly in the USA. The cost of this upgrade was about half the cost of a used refurbished 300 MHz instrument. At Viterbo, one of our main justifications for acquiring and implementing FT-NMR in our curriculum was the ability to incorporate spin-decoupled C-13 NMR into the organic chemistry curriculum. In fact, our plan from the beginning was to introduce C-13 NMR prior to any discussion of proton NMR because of the simplicity of the C-13 spectrum. The large chemical shifts of C-13 compared to H-1 allow full resolution of each C-13 signal, aliphatic and aromatic, even at the low field of our magnet. The lack of spin-spin splitting in C-13, and the ease of use of the DEPT technique, allows students quite quickly to solve structural problems and gain an appreciation of the power of NMR as a structural tool. We begin to introduce H-1 NMR and the concepts of spin-spin splitting only after C-13 competence is gained.

Our experience over the years in introducing NMR concepts in organic chemistry by the traditional approach of starting with H-1 NMR has been that students are confused by the idea of spin-spin splitting - that the splitting pattern of THIS chemical shift is due to the number of equivalent protons three bonds away at THAT chemical shift. Students find C-13 NMR much easier to understand and interpret, and have told us for some time, even before we had an instrument that would do C-13 and they were just working textbook problems, that C-13 NMR should be introduced first, and only later should we present the more complex H-1 NMR concepts. This approach has improved student understanding of NMR structural analysis.

With our permanent magnet EFT-60 NMR, we now introduce our first term organic students to hands-on experiences with C-13 NMR, including DEPT. Once students have gained experience acquiring, processing, and interpreting C-13 spectra (read below for examples of specific experiments), we introduce proton NMR. There are several advantages to deferring H-1 NMR to later in the first semester. (1) It gives the students a chance to gain experience in drawing, visualizing, and naming organic chemical structures. Students process C-13 data and develop critical thinking skills in analyzing the data and drawing conclusions from it. By combining the C-13 NMR results with molecular modeling, students begin to see relations between computed atomic charge and chemical shifts, and improve their ability to pick out equivalent C and H atoms in a structure. (2) It gives the students an opportunity to gain experience in relating C-13 chemical shifts and DEPT phasing to aliphatic organic structures. Again, combining this with molecular modeling boosts student confidence in their problem solving skills using NMR. (3) It allows the instructor to separate the concept of chemical shifts, which is based upon the laboratory magnetic field interacting differently with each chemically distinct spin nucleus according to electron density about the nucleus, from the concept of spin-spin splitting, which is based upon the differences in local magnetic field due to the orientations of nuclear spin adjacent to the nucleus of interest.

More complex concepts are continued in our upper division courses. Advanced organic students build on their knowledge of H-1, C-13 and DEPT by gaining skills in correlation NMR, specifically H-H COSY and C-H HETCOR experiments. We also use the broadband capability of the EFT-60 to introduce F-19, B-11, and P-31 NMR concepts to our Advanced Inorganic students.

To accomplish these ends we (a) upgraded our fixed magnet, 60 MHz (1.4T) CW-NMR to a FT-NMR; (b) added software for data processing of NMR spectrum files from a variety of sources; and (c) acquired high field FID NMR spectrum files from a local university and from the internet for processing and analyzing on campus.

Activities and Experiments

Organic Chemistry

NMR concepts and experiments are introduced in the first semester of organic chemistry courses. In each semester there are two lab activities with NMR as the primary focus. C-13 NMR with DEPT can effectively illustrate the concept of structural isomers, using the chlorobutanes. Students note 1°, 2°, and 3° carbons in each structure and their proximity to Cl, and use that information to try to predict the number of C-13 peaks and their relative chemical shifts.

Because 1-chlorobutane and 2-chlorobutane both give 4 peaks, the DEPT technique is used to distinguish the two structures. Alternatively, instead of using DEPT, students can identify the isomers by using the number of peaks and the chemical shifts of the chlorine containing carbons combined with the electronic charges on those carbons as shown by molecular modeling. This is more difficult for students than using DEPT but it highlights chemical shift more and brings in molecular modeling using HyperChem (Hypercube, Inc., www.hyper.com). FID files of the compounds are posted on a web-based learning system such as Blackboard™. Students process them off-line using the program NUTS (Acorn NMR, Inc., www.acornnmr.com), which is available on computers throughout the science building and other computer labs on campus. The NUTS program is the data processing program for the EFT-60 NMR, so this activity also reinforces their data processing skills. Hard copies of the DEPT spectra are provided. A data table noting the number of peaks, chemical shifts, and electronic charges is given in Table I. Using simple bond polarities based upon electronegativity differences students predict that the chlorine bearing carbons would be δ^+. However, molecular modeling indicates that these carbons are δ^-, although they are the least δ^- of any of the carbons. Students accept the modeling results because the electron withdrawing effects of the chlorines are still evident. This exercise has been very successful in our course for science majors. When faced with the assignment some students have trouble envisioning how they are going to identify the isomers, but as they obtain spectra they see the distinctions that can be made. However, the exercise was only marginally successful when used in the nursing chemistry course and the organic chemistry course for dietetics majors. Too much class time was required to introduce NMR concepts prior to the activity and on the whole these students did not appreciate the importance of basing the concept of isomerism on experimental data. Consequently the experiment has not been continued in those two classes.

Table I. Chemical Shifts of Cl-bound Carbon in the Chlorobutanes

Compound	Number of Carbon Peaks	Chemical Shift of the Cl-bound carbon / ppm	Charge of the Cl-bound Carbon
1-chloro-2-methylpropane	3	52	−0.070
1-chlorobutane	4	50	−0.068
2-chlorobutane	4	60	−0.040
2-chloro-2-methylpropane	2	72	−0.004

Students working in pairs in the first semester of the organic chemistry course for science majors also identify an unknown compound using C-13 (with DEPT), H-1, FT-IR, and boiling point/melting point data. Students obtain all the FIDs and process them on the instrument as well as off-line. This experiment requires about 15 minutes set up time and one half hour on the instrument for each pair of students.

Chemical shift H-1 concepts are used effectively in the organic chemistry course for dietetics majors to determine the fat and water content in butter, lard, and various margarine and "reduced fat" spreads (1). Because the samples are not free flowing liquids, the signals are broadened and so spin-spin splitting is not observed. Although the methyl and methylene protons overlap, the protons on carbons attached to carbonyls or glycerol oxygens, as well as vinylic protons can be observed; and the water peak is obvious. The dietetics students are very interested to find out the results of their experiments because often the results reinforce or call into question their views regarding nutrition. In assessing the experience, students reported unanimously that they loved it. They also enjoy calculating the cost per gram of fat and per gram of water of the various spreads. A description of this experiment has been submitted for publication and is also available from the authors. The experiment requires five to ten minutes of instrument time for each student.

In the second semester of the organic chemistry course for science majors the students, working individually, identify another unknown compound using C-13 (with DEPT), H-1, FT-IR, and boiling point or melting point data. This second unknown builds on their experience in the first semester. This time they are individually responsible for the identification. This experiment requires 15 minutes set up time and about one half hour on the instrument for each student.

In another second semester experiment, decoupled C-13 NMR, and DEPT are combined with HyperChem molecular modeling to study substituent effects on an aromatic ring. Monosubstituted benzenes with electron withdrawing and releasing groups are studied. Students use PM-3 computations in HyperChem to calculate the charges on the various ring carbons, and proton decoupled C-13 and DEPT to identify the chemical shift of the carbon bearing the substituent. Especially instructive is the combined use of C-13 and DEPT to identify this carbon. Students can unambiguously locate it since C-13 will show all carbons while the carbon bearing the substituent doesn't appear in the DEPT spectrum. Analysis of chemical shifts combined with computed charges provides an illustration of the electron withdrawing and releasing effects on an aromatic ring. It should be noted that this experiment does not look at substituent effects on the carbocation intermedicate of an electrophic aromatic substitution reaction. FID files are posted on Blackboard[TM]. Students process them off-line. Hard copies of the DEPT spectra are provided. The HyperChem results and the C-13 chemical shifts are shown in Table II.

In another non-laboratory activity, students use H-1, C-13, DEPT spectra to identify the three isomeric xylenes. The students are told only that the compounds are isomers. FID files of the various compounds are posted on Blackboard[TM]. Students, working in groups, download the files and process them with NUTS.

In advanced organic chemistry students obtain H-1 and C-13 FIDs for 3-heptanone using our 1.4T instrument. The same data has been obtained using the 300 MHz (7.07 T) instrument at the University of Wisconsin—La Crosse. The students then process the data and compare the results. It should be noted

Table II. Effect of Substituents on C-13 Chemical Shifts and Electronic Charges

Compound	Charge on the Carbon Bearing the Substituent/ppm	Chemical Shift of the Carbon Bearing the Substituent/ppm
Ethylbenzene	-0.078	143
Bromobenzene	-0.109	123
Anisole	+0.089	159
Benzene (any carbon)	-0.102	129

that the chemical shifts in routine C-13 spectra are about 20 times those in routine proton spectra, covering a range of about 240 ppm vs. about 10–12 ppm for protons. Thus, even low field instruments can resolve most C-13 peaks with proton decoupling. In this experiment both high and low field instruments give good C-13 results. However, the low field H-1 has considerable overlap of signals. The students are impressed to see how the signals are resolved using the 300 MHz instrument, although the multiplet near 2.5 ppm still results from overlapping signals. This points out that even at 300 MHz overlapping signals are common (see Figure 1).

Students also are introduced to 2D NMR using processed COSY and HETCOR spectra. In addition, they identify more complex unknown compounds using our instrument to obtain H-1, C-13, DEPT, COSY, and HETCOR spectra. They are asked to identify the compound and indicate which atoms of their proposed structure give the various peaks on the spectra. They also indicate with COSY how protons are coupled and with HETCOR how protons correlate with carbons. This experiment requires about 45 minutes on the instrument per student.

Advanced Inorganic Chemistry

In our Advanced Inorganic Chemistry laboratory, students synthesize metal complexes and organometallic compounds and use various techniques to study their properties and reactions. Students use NMR as a routine tool in synthesis and analysis, complementing FT-IR, magnetic susceptibility, UV-Vis, GC-MS, and kinetics. Advanced use of NMR includes not only C-13, H-1, and correlation techniques, but also extensive work with other nuclei, especially with P-31 NMR. Some experiments that we have done that work well with NMR are summarized below.

The first experiment, which can be done easily in one lab period, is the synthesis of ammonium tetrafluoroborate (*2*). This straightforward synthesis is followed by acquiring FT-IR and FT-NMR (F-19) spectra. The vibrations for NH_4^+ and BF_4^- ions are modeled using HyperChem and used to assign IR bands observed for the compound. The sample can be analyzed with F-19 NMR to give an interesting 1:1:1:1 quartet from the interaction of the equivalent F atoms

356

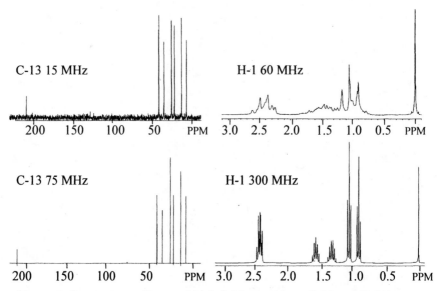

Figure 1. Comparison of low and high field strength on C-13 and H-1 Spectra of 3-Heptanone. Notice the vast improvement in the H-1 spectrum at high field compared to minor change of the C-13 spectrum from low to high field.

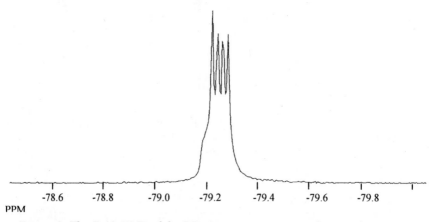

Figure 2. The F-19 NMR of the BF_4^- ion in an aqueous solution of ammonium tetrafluoroborate, showing the four equivalent F nuclei split by the $I = 3/2$ B-11 nucleus, giving rise to a 1:1:1:1 quartet. Chemical shifts are relative to a 0.1M NaF external standard. The B-11 NMR signal shows up as one broad peak on our instrument.

with the four equally populated states of the spin 3/2 B-11 nucleus. See Figure 2. The asymmetry of the peak seems not to be an artifact. It has been consistently present in our samples and may be due to F-19 bonded to B-10 nuclei, which have a natural abundance of 20% (reviewer's comment).

In a second experiment, students synthesize the Schiff base bis-(salicylidene)ethylenediamine (H$_2$salen) and obtain its proton NMR, which is quite a nice spectrum that integrates well. They use this material to synthesize the [Co(II)salen] complex and measure its dioxygen uptake, They also obtain its magnetic susceptibility, FT-IR, and NMR, and compare each with the free ligand. The [Co(II)salen] complex is sparingly soluble in deuterochloroform, and so the signal is weak. Twenty pulses are sufficient to reveal significant changes in the NMR spectrum relative to the free ligand. In addition to strong peaks at +3 and +5 ppm there are weaker peaks at +15, +4, +1.5, +1.2, and —0.8 ppm. The FT-IR mainly reflects the disappearance of the H-bonded OH group (3).

In a third experiment, students synthesize the square planar, diamagnetic (4d^8) Wilkinson's catalyst, [RhCl(PPh$_3$)$_3$], which is important in homogeneous, ambient temperature, low pressure catalysis of olefin hydrogenation. The Advanced Inorganic class had for several years successfully synthesized Wilkinson's catalyst (3) and carried out experiments with it: the oxidative addition of H$_2$ to make the Rh(III) dihydride complex; the catalytic hydrogenation of cyclohexene to cyclohexane (followed by GC-FID); and the preparation of the CO and C$_2$H$_4$ complexes. The capabilities of our FT-NMR have significantly enhanced the student learning experience, allowing us to see the upfield Rh-H hydrogen signals in a H-1 experiment on a 0.1M solution of Wilkinson's catalyst in deuterochloroform. See Figure 3.

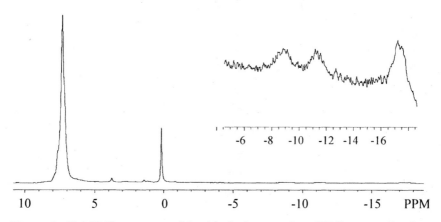

Figure 3. H-1 NMR spectrum of the dihydrido- complex of Wilkinson's Catalyst in CDCl$_3$, Rh(PPh$_3$)$_3$H$_2$Cl. The expanded hydride signals are shown in the inset and is quite similar to the published spectrum. The sample was obtained at ambient temperatures by bubbling H$_2$ gas into an NMR tube containing 0.1M solution of Wilkinson's catalyst, so the spectrum is that of a mixture of the free complex and the hydride complex.

358

We also can see the two different P signals in P-31 NMR in the expected 2:1 ratio for the pair of P atoms trans from each other and the single P atom trans from Cl (4).

A fourth inorganic lab experiment is based on that developed at Washington and Lee University (5). It involves a ring-opening metathesis polymerization (ROMP) experiment that makes use of a 7-oxanorbornene monomer prepared and characterized by the students in our organic chemistry course. It uses Grubbs's catalyst, a well-defined, ruthenium carbene complex that promotes the living polymerization of norbornene derivatives, which can be synthesized or purchased. The single ruthenium carbene proton in Grubbs's catalyst has a characteristic resonance at +21 ppm that can be seen readily on the FT-NMR. The metathesis polymerization reaction progress can be followed by proton NMR spectroscopy as well. See Figure 4.

All of the inorganic experiments into which we have incorporated an NMR component have all appropriate safety and waste disposal information within the original references, which are included here in our reference section. One of the advantages of working with a fixed magnet, low field NMR instrument is that the magnetic field is contained within the housing and so poses less of a hazard to individuals and objects that can be magnetized.

Assessment and Feedback

Student competence in various NMR techniques was assessed through written exams, laboratory reports, written questionnaires, and oral interviews. Assessment of student learning has been a very positive experience for the faculty involved. It has been rewarding in that we have gotten very favorable responses and good learning outcomes. It has also given us insights into the problems students face in learning NMR concepts and applying them to structure problems, and we have used these insights to begin to make changes in our curriculum that will facilitate student learning.

The students enrolled in Advanced Inorganic Chemistry in the Fall of 2000 had not previously used any of the FT-NMR techniques, having been the last class to use our old proton CW-NMR in their organic chemistry course. A pre-test was given to the class on heteroatom NMR concepts, spectral interpretation, spectral prediction, and structure determination.[1] These included spectra of boron fluorine and boron hydride compounds using B-11, H-1, and F-19 spectra;

---

[1] Questions include (a) making qualitative sketches of the F-19 and B-11 NMR spectra of the BF_4^- ion and relating the magnitudes of any splitting in the two spectra to each other; (b) making sketches of the H-1 and P-31 NMR spectra of trimethylphosphine, given P-31 has spin ½, again including spin-spin splitting and relating magnitudes of splitting in the two spectra; and (c) given that the P-31 NMR (without splitting) of $RhCl(PPh)_3$ has two peaks in a 2:1 intensity, pick a structure from a list that is consistent with this (e.g. cis- and trans-square, tetrahedral, trigonal pyramid with three equivalent P).

(a)

Ru=C-H proton signal

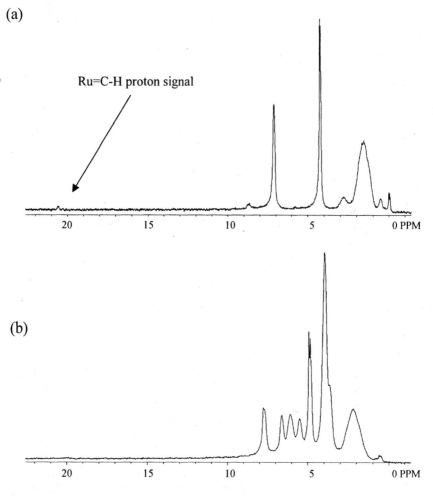

20   15   10   5   0 PPM

(b)

20   15   10   5   0 PPM

Figure 4. (a) Grubbs's catalyst, RuI(=CHPh)(Pcy₃)₂Cl₂, in C₆D₆. Despite the huge signals from the cyclohexyl and phenyl groups, the "carbene" proton at +21ppm can be seen with multiple pulses. This spectrum results from 32 pulses with a 3 sec relaxation delay. (b) Grubbs's catalyst and polymer mixture after 6 hours in the NMR tube at 25°C. The Ru carbene signal is lost over time due to averaging of the environments in polymers of different lengths.

360

and the P-31 spectrum of Wilkinson's catalyst. None of the students got even one question correct. At the end of the term, after short lectures on NMR of other nuclei with spin ½ and spin 3/2, and after doing B-11 and F-19 experiments on ammonium tetrafluoroborate and H-1 and P-31 experiments on Wilkinson's catalyst and substrate complexes, the same test was administered, unannounced, and everyone of the students was able to answer all of the questions correctly. In the pre-test given to the 2002 class, the first to have been introduced to FT-NMR in organic, four of the five students answered two of the three questions correctly, and the fifth answered one of the three correctly. In the post test, all students answered all of the questions correctly. Anecdotal reports from graduates tell us that they were well prepared for graduate level work using FT-NMR concepts and using NMR as a structural tool. These results demonstrate the impact of the new curriculum on our students in preparing them for more advanced work in NMR.

Website

As part of our NSF-funded curricular enhancements, we developed a rudimentary web site that has NMR FID and processed spectrum files of about 50 compounds (www.viterbo.edu/nmr). The files are compatible with the NUTS format. We did this for two reasons, one local and one more global. First, we wanted our students to be able to interpret high field NMR spectra and to be able to see visually the difference between spectra obtained at 60 MHz and 300 MHz. This would reinforce key ideas about the different ways chemical shift and spin-spin splitting respond to increased magnetic fields: (1) that chemical shift in an absolute sense is proportional to the field, and, hence, is unchanged in relative terms (ppm) at higher fields; and (2) that spin-spin splitting is independent of the size of the magnetic field. These two concepts taken together mean that high fields can reduce the overlap of closely spaced signals. This simplifies spectral analysis, especially for complex proton NMR spectra.

Our second reason for posting the web site was to give faculty and students who do not have access to an NMR the ability to download and process C-13 and H-1 NMR from our 60 MHz instrument and from a 300 MHz instrument for direct comparison and for developing problems that include NMR results.

Acknowledgements

- Dr. Ron McKelvey and the University of Wisconsin - La Crosse Chemistry department for obtaining FIDs on their Brucker 300 MHz NMR over many days.

- Anasazi Instruments, Inc. and staff for their assistance throughout in training, curriculum development, regular software upgrades, and helpful suggestions.
- NSF-CCLI for their support in funding the purchase of the EFT-60 upgrade and in training Ron and Mike in the use and implementation of the instrument.
- Viterbo University, particularly Dr. William Medland, President, Dr. Jack Havertape, Academic Vice President, and Dr. Mary Hassinger, Dean of the School of Letters and Sciences, for their support and encouragement in the grant process.
- Dr. Pamela Maykut of the Viterbo University Psychology Department for helping us to develop and implement our assessment plan. Dr. Maykut led all of the student interviews and wrote a summary report of her qualitative analysis of the results.
- Viterbo students of the Organic classes 2000-2004 and Inorganic classes 2000-2004 for cooperating both during the implementation phase when Ron and Mike were both learning as we went and during the assessment period for giving us such useful and constructive feedback.

References

1. Collins, M.J.; Amel, R.T., Unpublished, Viterbo University, La Crosse, WI, 2006.
2. Lange, W. *Inorganic Syntheses;* Fernelius, W.C. Ed.; McGraw Hill: New York, NY, 1946; Vol. II, pp. 23-24.
3. Szafran, Z.; Pike, R. M.; Singh, M. M. *Microscale Inorganic Chemistry*; John Wiley & Sons, Inc.: New York, 1991.
4. P. Meakin, J.P. Jesson, and C.A. Tolman, *J. Amer. Chem. Soc.*, **1972**, *94(9)*, 3240-3241.
5. France, Marcia B.; Uffelman, Erich S. Ring-Opening Metathesis Polymerization with a Well-Defined Ruthenium Carbene Complex: An Experiment for the Undergraduate Inorganic or Polymer Laboratory *J. Chem. Educ.* **1999**, *76*, 661.

Indexes

Author Index

Subject Index

368